Biological Rhythms

Biological Rhythms

Editor
Vinod Kumar

Springer-Verlag

Narosa Publishing House

EDITOR
Vinod Kumar
Professor
Department of Zoology
University of Lucknow
Lucknow-226 007, India

ISBN 3-540-42853-4   Springer-Verlag Berlin Heidelberg New York
ISBN 0-387-42853-4   Springer-Verlag New York Berlin Heidelberg
ISBN 81-7319-385-1   Narosa Publishing House, New Delhi

Printed in India.

Dedicated to

Professor **Maroli K. Chandrashekaran,** *FNA, FNASc, FASc, FTWAS*
the pioneer and champion of Indian Chronobiology,
in recognition of his original and outstanding
contributions to biological rhythm research
and of his lifelong stimulation of the
science of biological rhythms

Preface

Time, in a simple way, is defined as a measure of the interval between two events. The time interval between two successive events of a recurring process is described as 'rhythm', which in the context of a biological process is called the biological rhythm. Biological rhythmicity disallows an organism to perform its physiological functions at the time, which is unfavourable. To ensure that an organism does not end up performing at the wrong time, it must have a "clock "ticking and telling when 'ideal time' is imminent. There can be a debate on the type of the clock (ultradian, daily, infradian or seasonal) or the nature of the clock-system (single unit or multiunit), but there is consensus that the clock is endogenous, resides within the organism and is corrected (fine-tuned) by exogenous environment to remain in synchrony with the 'ideal time' of the day, of the season and of the year. Natural selection ensures that the environmental synchronizer(s) selected by the animal is (are) most reliable; the daily cycle of changes in illumination qualifies as one of the most reliable exogenous synchronizers (*zeitgeber*; *zeit*=time, *geber*=giver).

The study of the biological clock can be traced back to as early as fourth century BC, when Greek philosopher Androsthenes recorded sleep movements of leaves of the tamarind tree. However, modern era of biological rhythms began in 1729 with the demonstration of endogenous rhythmicity in plants by de Mairan, a French astronomer. Nineteenth and twentieth centuries witnessed more focussed research with strong zeal, and last four decades have contributed to the phenomenal growth of the subject. At the end of the millennium, a "biological clock" has been inferred in so many and such diverse tissues and organisms that a summary of the most interesting and important observations in the field of biological rhythm would be a tedious, voluminous and daunting task. In recent years, many new titles have appeared but most of them devote considerably either to the clinical aspects or to a specialised aspect which is of limited interest to a general chronobiologist, especially to the students. The scattered nature of the vast literature on the subject at the organismal level makes it difficult for a student, teacher and scientist to go through them so easily. The purpose of this book is, therefore, to present a timely review on some important aspects of biological rhythms both at the organismal and cellular/molecular levels under one cover. With suitable illustrations and important references, this book is an easy text primarily focusing on the topics, which have largely been ignored in recent publications.

The present book includes topics, which will be of interest alike to students, teachers and scientists. The core of this volume is a systematic consideration of the relevance of the topics in biological rhythm studies at a level intermediate between an elementary textbook and a research review. There are twenty chapters in the book, and they focus on four main themes. The first chapter is an overview of the progress that has been made in the area of biological rhythm research. Chapters 2 to 4 focus on the evolution and basic features of biological rhythms while Chapters 5 to 7 deal with the circadian system at genetic, molecular and cellular levels. Chapters eight to ten detail mechanisms involved in the perception and light-entrainment of the clock in invertebrates and vertebrates. Next four chapters

(Chapters 11 to 14) summarise important features of the biological clock at the level of whole animal covering all vertebrate classes (fish to mammal). Chapters 15 and 16 are on long term (seasonal) rhythms in plants and higher vertebrates. Short term rhythms (ultradian rhythms), the significance of having a clock system in animals living in extreme (arctic) environments, and the diversity of circadian responses to melatonin, the key endocrine element involved in regulation of biological rhythms, have been discussed in Chapters 17 to 19. Finally, a chapter on sensitivity to light of the photoperiodic clock is added which, using vertebrate examples, illustrates the importance of wavelength and intensity of light on circadian and non-circadian functions.

A well-known expert writes each chapter. When presenting information, the text provides consistent thematic coverage and feeling for the methods of investigation. Reference citation within the body of the text adequately reflects the literature as subject is developed. A chapter begins with an abstract that enables a reader to know at the first glance the important points covered in that chapter. The chapter concludes with a full citation of references included in the text, which could be useful for further reading. The book ends with a comprehensive subject index that may be useful for quick searches.

I have undertaken this exercise as a token of my deep respect to Professor Maroli Krishnayya Chandrashekaran, FNA, FNASc, FASc, FTWAS, Bhatnagar Laureate, and the AstraZeneca Research Professor of Life Sciences at the Jawaharlal Nehru Centre for Advanced Scientific Research, Bangalore, India. Professor Chandrashekaran has made many original and outstanding contributions to the science of biological rhythms, and has several firsts to his credit. He is the first recipient of Aschoff's rule and the pioneer and champion of Indian Chronobiology. He gave Indian Chronobiology an international standing. Above all, Professor Chandrashekaran is a good human being. I am privileged to be among those few whom he inspired to take up research in the field of biological rhythms.

The present text is by no means a comprehensive account of the subject on biological rhythms. In a work of this nature, sins of omissions and commissions are inevitable and for which I take full responsibility. I cordially thank all the authors who have agreed to be a part of my endeavor in bringing out this book. I express my gratitude to Professor K.C. Pandey, Professor Sir Brian K. Follett and Professor E. Gwinner for their continuous encouragement in my academic pursuits. I thank my friends, Dr. B.P. Singh, Department of Science and Technology, New Delhi and Dr. Roland Brandstätter, Max-Planck Institute for Ornithology, Andechs, Germany, for useful discussions in planning of the book. My colleague, Sangeeta Rani, has been involved at every stage of the preparation of this book, and I appreciate her very keen interest and continuing support. I have a sincere word of praise for my students, Sudhi Singh, Manju Misra, Shalie Malik and Amit Kumar Trivedi who helped me in various ways in organising the final version of the manuscript. I am thankful to Mr. N.K. Mehra, M/s Narosa Publishing House, New Delhi, for bringing out the book timely and in an excellent format. My parents and all family members have been constantly supporting my academic endeavors all these years, and I am sure they will derive some satisfaction that their supportive efforts allowed me to produce something worthwhile.

Lucknow, India   VINOD KUMAR

Contributors

Domien G. M. Beersma, Zoological Laboratory, University of Groningen, P.O. Box 14, 9750 AA Haren, The Netherlands

George E. Bentley, Department of Psychology, Behavioral Neuroendocrinology Group, Johns Hopkins University, 3400 North Charles Street, Baltimore, Maryland 21218 U.S.A.

Cristiano Bertolucci, Department of Biology, University of Ferrara, Ferrara, Italy

Roland Brandstätter, Department of Biological Rhythms and Behaviour, Max-Planck Research Centre for Ornithology, Von-der-Tann-Straße 7, D-82346 Erling-Andechs, Germany

Gregory M. Cahill, Department of Biology and Biochemistry, University of Houston, Houston, TX, U.S.A.

Andrew N. Coogan, School of Biological Sciences, 3.614 Stopford Building, Oxford Road, University of Manchester, Manchester, UK M13 9PT

Rodolfo Costa, Department of Biology, University of Padova, Italy

David J. Cutler, School of Biological Sciences, 3.614 Stopford Building, Oxford Road, University of Manchester, Manchester, UK M13 9PT

Serge Daan, Zoological Laboratory, University of Groningen, P.O. Box 14, 9750 AA Haren, The Netherlands

Wolfgang Engelmann, Department of Animal Physiology and Department of Botany, University of Tübingen (Germany)

Gerta Fleissner, Zoologisches Institut, J.W. Goethe-Universität, Siesmayerstr. 70, D-60054 Frankfurt a. M., Germany

Günther Fleissner, Zoologisches Institut, J.W. Goethe-Universität, Siesmayerstr. 70, D-60054 Frankfurt a. M., Germany

Augusto Foa, Department of Biology, University of Ferrara, Ferrara, Italy

Russel G. Foster, Department of Integrative & Molecular Neuroscience, Division of Neuroscience & Psychological Medicine, Imperial College School of Medicine, Charing Cross Hospital, Fulham Palace Road, London, W6 8RF, UK

Chiaki Fukuhara, Neuroscience Institute, Morehouse School of Medicine, 720 Westview dr., Atlanta, 30310, Georgia, U.S.A.

Menno P. Gerkema, Zoological Laboratory, Univerisity of Groningen, P.O. Box 14, Nl 9750 AA Haren, Netherlands

Charlotte Helfrich-Förster, Department of Animal Physiology and Department of Botany, University of Tübingen (Germany)

Anders Johnsson, Department of Physics, Norwegian University of Science and Technology, NTNU, N-7491 Trondheim, Norway

Amitabh Joshi, Envolutionary Biology Laboratory, Evolutionary & Organismal Biology Unit, Jawaharlal Nehru Centre for Advanced Scientific Research, Jakkur P.O., Bangalore 560 064, India

Verdun M. King, Department of Anatomy, University of Cambridge, Downing Street, Cambridge CB2 3DY UK

Vinod Kumar, Department of Zoology, University of Lucknow, Lucknow 226 007, India

Peter J. Lumsden, Department of Biological Sciences, University of Central Lancashire, Preston, U.K.

Simone L. Meddle, Department of Biomedical Sciences, University of Edinburgh Medical School, George Square, Edinburgh EH8 9XD UK

Martha Merrow, Institute for Medical Psychology Chronobiology, Goethestr. 31, D-80336 München

Michael Menaker, Commonwealth Professor of Biology and NSF Center for Biological Timing University of Virginia, Charlottesville, VA 22903, USA

Hugh D. Piggins, School of Biological Sciences, 3.614 Stopford Building, Oxford Road, University of Manchester, Manchester, UK M13 9PT

Gunnar Prytz, Department of Physics, Norwegian University of Science and Technology, NTNU, N-7491 Trondheim, Norway

Shantha M.W. Rajaratnam, Centre for Chronobiology, School of Biological Sciences, University of Surrey, Guildford GU2 7XH, UK

Sangeeta Rani, Department of Zoology, University of Lucknow, Lucknow, 226 007, India

Jennifer R. Redman, Department of Psychology, Monash University, Victoria 3800, Australia

Helen E. Reed, School of Biological Sciences, 3.614 Stopford Building, Oxford Road, University of Manchester, Manchester, UK M13 9PT

Eirik Reierth, Department of Arctic Biology and Institute of Medical Biology, University of Tromso, N-9037 Tromso, Norway

Till Roenneberg, Institute for Medical Psychology Chronobiology, Goethestr. 31, D-80336 München

Vijay K. Sharma, Chronobiology Laboratory, Evolutionary & Organismal Biology Unit, Jawaharlal Nehru Centre for Advanced Scientific Research, Jakkur P.O., Bangalore 560 064, India

Sudhi Singh, Department of Zoology, Univeristy of Lucknow, Lucknow 226 007, India

Karl-Arne Stokkan, Department of Arctic Biology and Institute of Medical Biology, University of Tromso, N-9037 Tromso, Norway

Gianluca Tosini, Neuroscience Institute, Morehouse School of Medicine, 720 Westview dr., Atlanta, 30310, Georgia, U.S.A.

Mauro Zordan, Department of Biology, Univeristy of Padova, Italy

Contents

1. Biological Clocks at the End of the 20th Century

M. Menaker*[1]

Commonwealth Professor of Biology and NSF Center for Biological Timing, University of Virginia,
Charlottesville, VA 22903, USA

Biological systems regulate their internal environments as well as their relationships to the outside world. The regulatory mechanism usually involves negative feedback. This leads to the generation of oscillations because of delay in the feedback loops. Natural selection has often acted to dampen the amplitude of these oscillations but a few have been enhanced and their periods have been adjusted to match closely the periods of astronomical events that are of importance to living things (daily, tidal, lunar and annual cycles). They function as endogenous clocks built into organisms at the most fundamental level of gene transcription and translation. Best studied are the so-called circadian clocks with natural periods of about a day, normally synchronized to the day-night cycle, and used by their owners in an extraordinary variety of adaptations, many of which are discussed in this volume. In a little more than forty years, the progress has been made from superb observational natural history and ingenious behavioral experiments, to identification of clock components at organ, tissue, cellular and molecular levels, and a reasonable understanding of mechanism at each of these levels about how circadian (and other) clocks evolved and how they function in real-world ecological communities. The study of temporal programs and in particular those that generate clocks, will occupy an increasingly important place in biological research of the 21st century.

Biological systems defy, although of course they do not escape, the second law of thermodynamics largely because of their ability to regulate their internal environments and their relationships to the outside world. Regulation in biological systems usually involves negative feedback, which always generates oscillations since there is bound to be delay in the loop. As a consequence, oscillations are pervasive in living systems. Natural selection has often acted to dampen the amplitude of these oscillations but a few have been enhanced and their periods have been adjusted to match closely the periods of astronomical events that are of importance to living things (daily, tidal, lunar and annual cycles). These have been further refined and have acquired some unusual properties that enable them to function as endogenous biological clocks built into organisms at the most fundamental level of gene transcription and translation, and employed to coordinate many diverse biological processes in adaptive synchrony with the astronomically generated cycles in their environments. Such clocks have become a fundamental part of the organization of living things. Best studied are the so-called circadian clocks with natural periods of about a day, normally synchronized to the day-night cycle, and used by their owners in an extraordinary variety of adaptations, many of which are discussed in this volume.

---

*Email: mm7e@unix.mail.virginia.edu
[1]Some of this material was presented at the Royal Society Meeting, The Measurement of Time, September 20–21, 2000.

The latest of these circadian clocks to be discovered was certainly the first to evolve. Cyanobacteria provide the first fossil evidence of life on earth in three (or more) billion year-old stromatolites. Their modern descendants, modified very little as far as we can tell, have fully functional circadian clocks, formally although not mechanistically similar to the clocks of the most highly evolved vertebrates. The original selection pressure to enhance oscillations with periods close to a day in cyanobacteria may well have been the need to sequester ultraviolet-sensitive chemical reactions to the hours of darkness. The result is a temporal program that restricts the transcription of a large fraction of the entire genome of these free-living photosynthetic bacteria to the night.

The natural period of the cyanobacterial clock is under genetic control and can be artificially altered by mutation. In an elegant series of competition experiments, Carl Johnson and his colleagues (Ouyang et al. 1998) have shown that the fitness of each variant is maximal in light cycles that approximate the altered period of its clock. Although this is the only study that demonstrates a direct effect of the clock on fitness strictly defined, indirect evidence of its many important functions abounds in a wide variety of organisms.

The circadian clocks of organisms from bacteria to vertebrates share important formal properties—their natural periods are close to 24 hours and are compensated for changes in environmental temperature, and they are synchronized to the external world by similar responses to the light dark cycle–but they have probably evolved independently at least 3 or 4 times. The molecular mechanisms that generate endogenous circadian oscillations differ substantially between cyanobacteria, fungi, higher plants and metazoa although all are based on autoregulatory feedback loops involving gene transcription and translation (Dunlap 1999).

Among the metazoa there has been remarkable conservation of molecular mechanism. The molecular loop that generates circadian rhythms within cells is comprised primarily of orthologous genes in fruit flies and mice and presumably everything in between (Dunlap 1999). Among the vertebrates there has also been remarkable general conservation of circadian organization at the anatomical, physiological and behavioral levels (Menaker and Tosini 1996). On the other hand, at both the molecular and the organismic levels there is a good deal of variability in the details. The tension, so characteristic of all of biology, between commonality of basic mechanisms and the fascinating variability produced by natural selection acting to fine-tune those mechanisms, is illustrated diagrammatically by what we have already learned about the organization of the circadian systems of vertebrates.

Circadian rhythmicity in all vertebrates is driven by a central neuroendocrine loop that we have called the circadian axis. It consists of the suprachiasmatic nucleus of the hypothalamus (SCN), the pineal, the retina and the pathways, both neural and humoral, that couple these structures. Our current, though not—it should be emphasized—firmly-supported view is that their combined outputs regulate circadian rhythmicity in the rest of the body. Much of the evidence supporting this idea comes from lesion and transplantation experiments. These are compelling but only insofar as they elucidate the control of behavior. The approach can be illustrated by our early experiments with passerine birds. Removal of the pineal gland renders the perching behavior of house sparrows (and most other passerines) arrhythmic. Transplantation of the pineal of a donor bird into the anterior chamber of the eye restores rhythmicity within a day or two. The restored rhythm has the phase of the donor bird, demonstrating that this clock property was transmitted by the donor pineal (Menaker and Zimmerman 1976; Zimmerman and Menaker 1979). From these results we hypothesized that the pineal contained a clock or clocks and that its output was hormonal (probably melatonin). These speculations were confirmed by in vitro experiments which, in addition, demonstrated that the clock in the pineal could be directly entrained by light (Takahashi et al. 1980). Remarkably, a similar role is played by the pineal of lampreys, separated from sparrows by about 500 million years of evolution. Similar, though not identical, control of circadian behavior is characteristic of all non-mammalian vertebrates (Menaker and Tosini 1996). Lesion experiments implicate the SCN and, in some cases, the retina in addition to the pineal as oscillatory components of the non-mammalian circadian axis.

In mammals, the pineal has the different (but related) function of regulating reproductive responses to changes in day length. Furthermore, while non-mammalian vertebrates have multiple extra-retinal circadian photoreceptors (Menaker 1971), mammals employ specialized retinal photoreceptors and have lost all the others (Freedman et al. 1999; Lucas et al. 1999). Lesion and transplant experiments support a central role for the SCN in the control of mammalian behavior (Ralph et al. 1990). There are also circadian oscillators in peripheral tissues. Using transgenic animals in which promoter elements that normally regulate one of the clock genes are engineered to produce light when their transcription is called for by the circadian program, one can directly observe circadian oscillations in many cultured peripheral tissues. Measurement of these peripheral oscillations in transgenic rats after simulated transatlantic flights (i.e., shifting their light cycles six hours backward or forward), reveals that peripheral tissues regain normal phase only slowly. In some cases this process takes more than a week, and one can begin to appreciate the physiology that underlies the malaise of jet lag or more importantly, the serious effects of shift-work schedules on human health (Yamazaki et al. 2000). The clocks in peripheral tissues respond to different signals; experiments with isolated tissues from zebrafish indicate that the clocks they contain are light-sensitive (Whitmore et al. 2000), whereas those of mammals are not, but do respond to proximate signals such as those provided by cyclic food intake which phase-shifts the liver (Stokkan et al. 2001).

Similar variation on a common basic theme is already apparent at the molecular level. While most of what we know about the molecular mechanisms generating circadian oscillations in vertebrates comes from work with mice, we can already see differences in fish and amphibians. As we learn more about *Xenopus*, zebrafish, and eventually humans, we can expect to find the same genes used in perhaps importantly different ways.

The picture of the vertebrate circadian system that has emerged is of a central circadian axis composed of several coupled oscillators, that is synchronized to the external world by the light cycle acting on specialized circadian photoreceptors. This axis in turn synchronizes the many circadian oscillators in the periphery, employing a variety of specific signals ranging from neural impulses, to hormonal cycles and even specific behaviors. The adaptive significance of such a system is most likely to lie in the opportunities that it provides for the regulation of the phases of its components relative to the environment and to each other. Thus an organism can be thought of as an ensemble of oscillations organized by a complex temporal program. When this perspective is fully appreciated, it becomes clear that understanding the temporal organization of living systems is as important as understanding their more commonly studied spatial organization.

I have used vertebrates to illustrate briefly the progress that has been made in unraveling circadian organization, but similar and, in some cases, further advances have been made using organisms from a variety of other groups. In a little more than forty years our understanding of circadian clocks has progressed from intriguing phenomenology generated by superb observational natural history and ingenious behavioral experiments, to identification of clock components at organ, tissue, cellular and molecular levels, and a reasonable understanding of mechanism at each of these levels. We now have remarkable opportunities to build on this base. The molecular tools already at our disposal virtually guarantee that we will be able to complete our understanding of the mechanisms that generate circadian oscillations within cells. That will leave unsolved fascinating and medically important questions concerning system-level physiological and behavioral function, as well as an almost untouched panoply of questions about how circadian (and other) clocks evolved and how they function in real-world ecological communities. The study of temporal programs and in particular those that generate clocks, will occupy an increasingly important place in biological research of the 21st century.

References

Dunlap, J.C. (1999) Molecular bases for circadian clocks, Cell **96**: 271-290.

Freedman, M.S., Lucas, R.J., Soni, B., von Schantz, M., Muñoz, M., David-Gray, Z., Foster, R. (1999) Regulation of mammalian behavior by non-rod, non-cone, ocular photoreceptors. Science **284**: 502-504.

Lucas, R.J., Freedman, M.S., Muñoz, M., Garcia-Fernández, J-M., Foster, R.G. (1999) Regulation of the mammalian pineal by non-rod, no-cone, ocular photoreceptors. Science **284**: 505-507.

Menaker, M., Tosini, G. (1996) Evolution of Vertebrate Circadian Systems In: Honma, K., Honma, S. (eds..) Sixth Sapporo Symposium on Biological Rhythms: Circadian Organization and Oscillatory Coupling. Hokkaido University Press, Sapporo, pp. 37-52.

Menaker, M., Zimmerman, N. (1976) The role of the pineal in the circadian system of birds. Amer. Zool. **16**: 45-55.

Menaker, M. (1971) Rhythms, reproduction and photoreception (Invited manuscript, Third Annual Meeting of Society for Study of Reproduction) Biol. Reprod. **4**: 295-308.

Ouyang, Y., Andersson, C.R., Kondo, T., Golden, S.S., Johnson, C.H. (1998) Resonating circadian clocks enhance fitness in cyanobacteria, Proc. Natl. Acad. Sci. USA **95**: 8660-8664.

Ralph, M.R., Foster, R.G., Davis, F.C., Menaker, M. (1990) Transplanted suprachiasmatic nucleus determines circadian period. Science **247**: 975-978.

Stokkan, K-A., Yamazaki, S., Tei, H., Sakaki, Y., Menaker, M. (2001) Food directly entrains the circadian clock in the liver. Science **291**: 490-493.

Takahashi, J.S., Hamm, H., Menaker, M. (1980) Circadian rhythms of melatonin release from individual superfused chicken pineal glands *in vitro*. Proc. Natl. Acad. Sci. USA 77: 2319-2322.

Whitmore, D., Foulkes, N.S., Sassone-Corsi, P. (2000) Light acts directly on organs and cells in culture to set the vertebrate circadian clock. Nature **404**: 87-91.

Yamazaki, S., Numano, R., Abe, M., Hida, A., Takahashi, R-I., Masatsugu, U., Block, G.D., Sakaki, Y., Menaker, M., Tei, H. (2000) Resetting central and peripheral circadian oscillators in transgenic rats. Science **288**: 682-685.

Zimmerman, N.H., Menaker, M. (1979) The pineal gland: a pacemaker within the circadian system of the house sparrow. Proc. Natl. Acad. Sci. USA 76: 999-1003.

Biological Rhythms
Edited by V. Kumar
Copyright © 2002, Narosa Publishing House, New Delhi, India

2. Clocks, Genes and Evolution: The Evolution of Circadian Organization

V.K. Sharma[1] and A. Joshi[2]*

[1]Chronobiology Laboratory, [2]Evolutionary Biology Laboratory
Evolutionary & Organismal Biology Unit,
Jawaharlal Nehru Centre for Advanced Scientific Research,
Jakkur P.O., Bangalore 560 064, India

The adaptive significance of circadian rhythms has been much speculated about but has not been subjected to systematic and rigorous empirical investigation. The majority of the few empirical studies of the possible adaptive significance of circadian organisation suffer from numerous shortcomings that we can now identify based on hindsight gained through three decades of experimental work in evolutionary genetics. We present here some of the major findings from this work in evolutionary genetics that are relevant to the proper design and interpretation of experiments aiming at studying the adaptive significance of circadian organisation. We critically review past work on the adaptive significance of circadian organisation, and suggest that a long-term research programme combining selection studies on replicated sets of laboratory populations with a careful study of fitness effects of different light regimes on such populations is likely to prove far more fruitful than some of the approaches previously followed.

Introduction

> *"What might have been is an abstraction, remaining a perpetual possibility*
> *only in the realm of speculation"*   (Thomas Stearns Eliot)

Much of the information we have about the evolution of circadian organization is exemplified by the above quoted lines of Eliot. Speculation about the adaptive significance of this or that circadian rhythm in the literature far outweighs empirical data on the evolution of circadian organization. In this article, we will focus on three issues, and take a point of view that is both retrospective and prospective. We first summarize some of the recent advances in methodology and interpretation in evolutionary genetics that have yielded profound insights into the manner in which evolutionary change is constrained by the genetic architecture of various traits relevant to adaptation to a given set of environmental conditions. Next, we critically review the available empirical data on issues related to the adaptive significance of circadian rhythms and their evolution, and discuss why many of the experimental systems and approaches used in the past are unlikely to significantly enhance our understanding of the evolution of circadian organization. Finally, we outline what we feel are some of the experimental approaches that are likely to yield the best returns in terms of a thorough understanding of the genetic architecture of circadian rhythm related traits, and how they evolve.

---

*E-mail: ajoshi@incasr.ac.in

In the course of working on the evolutionary genetics of circadian organization over the past few years, we have been struck by the almost complete lack of scientific overlap between the communities of chronobiologists and evolutionary biologists, respectively (one of us is an evolutionary biologist and the other a chronobiologist). One of our hopes in writing this article is that it will be read by members of both communities, and will prove to be both informative and interesting. Consequently, before we get into the main body of the article, we will very briefly summarize what is known about behavioural expressions of circadian clocks, and about their genetic control, in order to provide the non-chronobiologist reader with some background information and references to the relevant literature. At this juncture, it is also desirable to clearly delineate the aspects of the evolution of circadian organization on which this article will be focused. Evolutionary biology today is a vast field asking many different questions, and utilizing many different methodologies. Some evolutionary biologists are primarily concerned with reconstructing past events. Into their domain would fall questions such as when in the history of life did biological rhythms evolve originally, and what might have been the selection pressures that shaped their very early evolution. We will, however, restrict our focus to questions about the evolutionary maintenance and fine-tuning of circadian rhythms. Specifically, we will concern ourselves with (i) the adaptive value of circadian rhythms to organisms possessing them, and (ii) the genetic architecture of circadian organization with regard to fitness.

Finally, before beginning our discussions of clocks and genes, we would like to clarify the sense in which some terms will be used in this article. Often different phenomena are referred to by the same name in different fields of biology, and a careful definition of such ambiguous terms right at the outset will, we hope, mitigate any misunderstanding. We will use adaptation, in the sense of a process, to refer solely to the alteration over generations, through natural selection, of some genetically controlled characteristics of a population inhabiting a particular environment, such that the mean fitness of the population in that environment is enhanced. We will also use adaptation, in the sense of a product, to refer to the characteristics that evolved during the process of adaptation. The word fitness will always be used to mean Darwinian fitness, in terms of the relative representation of an individual's offspring in the next generation. By genetic architecture, we mean the pattern of additive genetic variances and covariances between different rhythm characteristics, such as free running period and phase angle difference, and components of fitness such as pre-adult survivorship, fecundity and adult life-span. The phrase 'a tradeoff between two traits' will be used to connote a negative additive genetic correlation between them, such that it is not possible for selection to simultaneously maximize both traits.

Clocks

"O perpetual revolution of configured stars
O perpetual recurrence of determined seasons,
O world of spring and autumn, birth and dying!
The endless cycle of idea and action,
Endless invention, endless experiment . . . "

(Thomas Stearns Eliot)

Life on this planet has been subjected to various geophysical periodicities caused by the rotation of the earth around its axis, and around the sun, and the revolution of the moon around the earth, since its very origin. With the exception of organisms that live in the depths of oceans, underground caves and rivers, or any similar aperiodic environment, most organisms have evolved strategies to counteract or to exploit violent variations in their neighbouring environment. Although some of the rhythmic biological phenomena may be direct responses to environmental changes, many more are overt manifestations of endogenous periodicities. The most common of all these rhythms are the ones with periodicity close to 24 h (circadian rhythms). Organisms ranging from prokaryotic cyanobacteria to man are known to use these rhythms to "time" daily events in order to perform various important functions at the "right time of the day". Circadian rhythms in most organisms are regulated by genes, some of which are now extensively studied, for example the *period (per)* and *timeless (tim)* genes in *Drosophila*, the *frequency (frq)* gene in *Neurospora*, and the *Clock* gene in the mouse. Circadian rhythms, therefore, are neither imposed on the organisms by the environment nor are they a consequence of learning. The natural cycles of light and darkness and of temperature, however, synchronize these rhythms, in the absence of which they free-run, revealing their natural periodicity (free-running period), which is often close to but seldom equal to 24 h. It is believed that most of the biological periodicities are the outcome of millions of years of interaction between the biological and the geophysical periodicities. It may very well be that to start with organisms with all type of biological periodicities were present and only those with periodicities close to that of the geophysical environment survived.

In order for an individual organism to perform various biological functions at the right time of the day, a mere match of its periodicity with that of the geophysical counterpart is not sufficient. In addition, the organism must be able to adjust the phase of several biological functions appropriately in order to maintain a fixed phase relationship with respect to some reliable environmental cycles. The study of circadian rhythms became popular partly because of the perceived clear functional advantage for organisms that could keep in pace with the environment.

While discussing the possible role of circadian rhythms in mediating the fitness of organisms, we need to consider first a spectrum of biological periodicities and examine their possible intrinsic function, independent of the environmental cycles. Thereafter, we can proceed towards evaluating the relative advantages accruing to an individual by synchronizing its period and adjusting its phase to the environmental periodicities. It seems plausible that the adaptive advantage of circadian rhythms might stem, in principle, from any combination of two of its distinct basic properties: (i) the internal temporal order that circadian clocks can maintain in order to facilitate various physiological processes with a fixed phase relationship with the environment, and (ii) the anticipation of events occurring in the nature (Daan 1981; Pittendrigh 1993). For the first necessity the environment may play a role in ensuring the damping out of the overt rhythm due to desynchronization of the constituent oscillations originating the rhythm. The second necessity of the circadian clock is served by daily phase resetting of the clock using some reference cycle of the environment (entrainment). It remains an open question as to which of the properties of the circadian clock plays a more vital role in conferring an adaptive advantage to the organisms possessing it.

Genes

"Any variation which is not inherited is unimportant . . . "

(Charles Darwin)

Ever since the isolation of the first 'clock' mutant, *period (per)*, in *Drosophila melanogaster*, numerous mutations affecting periodicity or phasing of various rhythms have been isolated and studied in many species. Genetic studies in a number of organisms ranging from microbes to mammals have yielded much information about the molecular mechanisms regulating biological rhythms. Isolation and characterization of clock genes and the mutations that define them in *Neurospora* and *Drosophila* have, in particular, yielded remarkable insights into the functioning of circadian oscillators at the molecular level. In addition to *Drosophila* and *Neurospora*, clock mutations have been reported in *Synechococcus*, *Chlamydomonas*, *Arabidopsis*, *Mus* and *Mesocricetus*. It is not our intention here to review the molecular work on clock genes in detail; interested readers are referred to extensive reviews by Hall (1995, 1998); Young (1998) and Dunlap (1999). We will merely summarize the kind of information these studies tend to provide, and discuss their relevance to our understanding of the adaptive significance of circadian rhythms.

Any self-sustained oscillating system tends, in a regular manner, to move away from equilibrium before returning and so does the circadian clock. Several elements and loops have been identified at the molecular level which ensure such self-sustained oscillation in mRNA and protein concentration with circadian periodicity: (i) a negative element; which is involved in a process whose product feeds back to slow down the rate of the process itself, and (ii) a positive element that serves as a source of excitation or activation and thus keeps the oscillator from winding down. The genes which serve as positive and negative elements in *Synechococcus* are *KaiA* and *KaiC* respectively, whereas the role of another gene *KaiB* is still unknown (Golden et al. 1997). The positive elements for the *Neurospora* clock are *wc-1*, *wc-2* and the negative element is *frq* (Crosthwaite et al. 1997). In *D. melanogaster*, the genes *Cyc* and *Clk* serve as positive elements, *dbt* acts as a facilitating element, which is additionally required for the production of clock protein, and the genes *per* and *tim* act as negative elements (reviewed in Young 1998). In mice, *Per1*, *Per2* and *Per3* act as (putative) negative elements, *tim* as facilitating element, and *Clock* and *bmal1* as positive elements (Hogenesch et al. 1998). Very often, small changes in the clock genes produce large phenotypic effects. Among the many rhythm mutants in *Neurospora crassa*, most map to one locus, *frequency (frq)*. All *frq* mutants induced *in vivo* were found to have a single nucleotide change within the protein coding part of the gene (reviewed in Dunlap 1999). In *D. melanogaster*, there are about 10 genes known to cause alterations in rhythm when mutated. Of these, *period (per)* has been independently mutated several times by *in vivo* and *in vitro* mutagenesis, resulting in short period *(per^T, per^S, and per^{Clk})*, long period *(per^L)* or arrhythmic mutants *(per^{01}, per^{04})*. Many behavioural differences among mutant strains, geographic populations and sibling species can be traced to differences in the number of threonine-glycine repeats in the *period* gene (Costa et al. 1991; Wheeler et al. 1991).

Although the identification and characterization of clock genes raises the exhilarating possibility of a detailed understanding of some part of the 'clockworks' of certain rhythmic phenomena in living organisms, it is also worthwhile to step back and look at the impact that the molecular characterization of such genes and the signal transduction pathways to and from them is likely to have on our appreciation of how circadian organization evolves. Given the hierarchical nature of biological systems, and the often observed emergent properties at various levels of biological organization, it is quite possible that detailed molecular understanding of the functioning of the control loops of certain rhythms may not shed much light on issues of adaptive significance at the organismal level, and evolutionary change at the populational level. It should be noted that alleles at most of the clock loci identified so far cause fairly drastic changes in the period of the rhythms they affect. Moreover, it is not yet clear whether these single loci with large phenotypic effects are the only loci affecting a particular rhythm(s). It is quite possible that there may be numerous loci, other than these clock genes that have small effects on circadian parameters. For example, a recent study has reported evidence for several putative QTLs (quantitative trait loci) governing the free-running period of the wheel running activity rhythm in mice scattered over six autosomes (Hofstetter et al. 1999), even though the Clock and Wheels loci are known to have major effects on the period of this rhythm. In most organisms typically used for biological rhythms work, quantitative genetic analyses of circadian parameters have basically not been done. This, to our mind, is a serious lacuna.

If most species typically have many loci with relatively small individual effects affecting circadian rhythm parameters, then it is far more likely that such loci will be involved in the evolutionary fine-tuning of circadian organization. For example, many animal species have a locus at which mutations will give rise to dwarfs or giants, but body size typically evolves through changes at polygenic loci rather than through changes at loci of major effect. Although the observation of clines for *per* alleles and different rhythm phenotypes does suggest the involvement of this locus in the evolution of circadian organization in *Drosophila* sp., there is, as yet, no really conclusive evidence that these types of loci play a major role in the evolution of circadian organization.

Evolution

"For a biologist, the alternative to thinking in evolutionary terms is not to think at all"
(Peter B. Medawar)

Evolutionary biology as an experimental science has come a long way in the past twenty five years. Rigorous criteria have been developed for empirically establishing the adaptive significance of any trait, namely the demonstration that (i) specific environmental regimes relevant to the trait being studied have differential effects on major fitness components, and (ii) the trait in question does in fact evolve differently under those environmental regimes. It has also become increasingly apparent that careful and systematic long-term studies on systems of laboratory populations constitute what is perhaps the best approach to studying adaptive evolution in terms of the depth of understanding that they yield (reviewed by Rose et al. 1996; Mueller 1997; Joshi 1997, 2000). It has also become apparent from these studies that the evolutionary process is exceedingly subtle, and evolutionary responses to even seemingly trivial changes in the environment can be quite large and often counterintuitive. In this section, we will use examples from some of these studies to highlight some common pitfalls (methodological and inferential) confronting those interested in empirically studying the adaptive significance of any trait. We will also attempt to underscore what we feel are the advantages of a systematic laboratory based selection approach to questions in adaptive evolution, as compared to traditionally used methods, both in the field and in the laboratory.

Population replication

Individuals live, reproduce and die: as a consequence of heritable differences in reproductive output among individuals (i.e. natural selection), the genetic composition of a population changes over time in an adaptive manner. Thus, the unit of evolutionary change is the population and not the individual. The corollary to this is that the unit of replication in any study addressing evolutionary questions needs to be the population and not the individual. The problem here is that the genetic composition of populations can change in a random manner due to genetic drift. Thus, any difference observed between a control population and a population subjected to some kind of selection cannot unambiguously be assigned to selection, because drift cannot be ruled out as a cause. The observation of a consistent pattern of differentiation between a set of replicate control and selected populations, however, implicates selection as the cause, because it is unlikely that several replicate populations could undergo the same sequence of genetic changes through the operation of a random phenomenon like drift (for further discussion and empirical examples, see Joshi et al. 1997).

Population size

The size (i.e. the number of breeding adults) of a population is also an important consideration in any selection experiment. The potential number of different multi-locus genotypes in an outbreeding diploid population is extremely large, due to the reshuffling of genes through meiotic recombination and fertilization. If the population on which selection is being carried out is relatively small, the actual subset of potential genotypes that is realized each generation is very small, and the evolutionary response to selection may, therefore, be constrained by paucity of available genetic variation. For example, many attempts to select for faster egg to adult development in relatively small populations of *Drosophila* failed, leading to a view that there was no genetic variation for development time in this species, presumably due to natural selection in the wild already having maximized developmental rates. However, subsequent selection experiments on much larger populations ($N > 1000$) of *Drosophila* proved this surmise wrong (Chippindale et al. 1997).

Another serious problem in using relatively small populations for evolutionary studies is that of random genetic drift. Not only can random genetic drift retard or negate the effects of selection in a longer term study, it can also give rise to inbreeding depression, spurious positive genetic correlations between fitness components due to fixation of generally deleterious mutations, and linkage disequilibrium. Thus, even in one-generation studies aimed at examining, for example, the fitness effects of different environmental regimes, results obtained from small populations (especially if they have been maintained for many generations at low population sizes) are very likely to be misleading artefacts of the effects of random genetic drift on the genetic architecture of fitness in that population.

A very good example of a spurious picture of genetic architecture of fitness components obtained from inbred populations is provided by Rose (1984a). Inbred lines, derived from a large outbred, base population *of D. melanogaster* by three generations of full-sib mating, showed all additive genetic correlations between early and late life fecundity to be positive or zero. Yet, the base population had earlier been shown to exhibit negative additive genetic correlations (tradeoffs) between early life and late life fecundity, and subsequent selection experiments on populations derived from the same base population clearly show that these tradeoffs do exist and greatly influence the evolution of these populations under different selection regimes (Rose 1984b; Rose et al. 1996).

The effect of novel environments

A problem similar to that described above can arise when populations are moved from the field to the laboratory and studied, or more generally, if a population is studied in an environment different from that to which it is adapted. The reason for this type of artefact is the expression of loci in the new environment that had hitherto not been exposed to selection. Loci possessing alleles that pleiotropically affect more than one fitness component become fixed for alleles with overall beneficial effects on fitness over the course of selection in their familiar environment. Such loci, then, do not contribute to genetic variance or covariance for fitness traits (Falconer 1981). Other loci with alleles that have antagonistic pleiotropic effects on fitness components remain segregating, and therefore contribute to negative additive genetic covariance between fitness traits, leading to fitness tradeoffs (Falconer 1981). In a novel environment, newly expressed loci with alleles that have negative or positive effects on several fitness components are still segregating for several generations before reaching genetic equilibrium as selection in the novel environment takes effect: fitness tradeoffs, therefore, appear to vanish in a novel environment and slowly reappear as the population adapts to the new environment (Service and Rose 1985; Joshi and Thompson 1995, 1997).

Mutant and isofemale lines

Mutant lines, true breeding for certain deviant phenotypes, were traditionally quite popular for population genetics studies of adaptive evolution: evidence, perhaps, of the inertial effect of Mendel's work. The past few decades, however, have seen an increasing realization among evolutionary geneticists that the use of mutant lines (and in the case of *Drosophila*, the traditionally popular 'chromosome extraction' lines) to study the fitness consequences of particular genetic alterations is, by itself, fraught with dangers (Rose 1984a; Joshi and Mueller 2000: chapter 1). The reason is that mutant or chromosome extraction lines are typically inbred and are, therefore, likely to yield spurious positive genetic correlations between fitness components (for empirical evidence, see Mueller and Ayala 1981).

This is also a serious problem with the use of isofemale lines, or the so called 'wild type' lines (e.g. *D. melanogaster* Oregon R, or Canton S). Often, a study population is created by crossing several isofemale lines to eliminate problems relating to paucity of genetic variation: such populations, however, are subject to potentially confounding large linkage disequilibrium effects, unless they are maintained for many generations (> 25) in the laboratory after being initiated.

Ancestry, controls, and the evidence for adaptation

Contrary to much popular belief, for a conclusion of adaptedness to be drawn, a population must do better than controls (not better than its own performance in other environments) in the environment it is presumed to be adapted to. We illustate this problem with a study on larval urea tolerance of *D. melanogaster* populations selected for tolerance to toxic levels of urea (Joshi et al. 1996). If only the data from selected populations were to be examined, they would not appear to be adapted to high urea levels because they perform better in the absence of urea. It is only in comparison to the control populations that the adaptation of the selected lines becomes apparent, because the decline in fitness at toxic levels of urea is much less in the selected populations.

The above example also highlights the importance of properly matched controls in evolutionary studies. What we need, as controls, are other populations, reared in an environment differing from the selected populations in only one aspect, with all other aspects of their rearing and ancestry being identical to the selected populations. Moreover, control populations should preferably also be matched for ancestry with the selected populations, because the possibility that observed differences between control and selected populations are due to differences in initial genetic composition, rather than selection, can then be ruled out (Joshi and Mueller 1996; Rose et al. 1996; Joshi 1997). This is typically a problem when one compares wild populations from different areas that differ in some environmental variables.

Even when working in the laboratory with control populations matched for ancestry, small differences in maintenance regime can result in differential adaptive evolution in control and selected populations, quite apart from the differentiation resulting from the primary selection that one is applying. For example, Partridge and Fowler (1992) selected three populations of *Drosophila* for postponed ageing by progressively increasing the adult age at which eggs were collected to initiate the next generation, a protocol similar to that employed earlier in similar studies (e.g. Rose 1984b). Many of their results, however, were different from those seen in earlier studies; these differences were due to inadvertent larval density differences in the selected and control lines (for detailed discussion see Roper et al. 1993; Chippindale et al. 1994; Rose et al. 1996).

This is a particularly relevant example here because larval density is a variable that is typically not explicitly regulated in most *Drosophila* experiments done in chronobiology with an intention to probe issues of adaptive significance. For example, LL is known to enhance fecundity in *Drosophila*, relative to L:D 12:12 h (Sheeba et al. 2000). It is also known that larvae in populations maintained at high larval densities evolve to be less efficient in converting food to biomass (Joshi and Mueller

1996). If larval density were not to be explicitly regulated in populations kept under LL and L:D 12:12 h, respectively, then those maintained in LL would routinely experience higher levels of larval crowding potentially leading to a spurious conclusion that L:D 12:12 h led to the evolution of greater energy efficiency.

Indeed, long term laboratory studies of adaptive evolution have repeatedly shown that adaptation often occurs to seemingly trivial aspects of the maintenance regime, and the expression of these adaptations can be very environment specific due to the ubiquity of genotype × environment interactions for many fitness components (Leroi et al. 1994a; Rose et al. 1996). It is important, therefore, when comparing results from studies on fitness effects of different environmental treatments, to be mindful of possible differences in rearing conditions that could affect the observed results. In the chronobiological literature, however, details of population maintenance are rarely given even in studies explicitly aimed at addressing issues of adaptive significance, making it almost impossible to draw any meaningful conclusions from the results of different studies taken together.

We have shown that unambiguous evidence for adaptation rests upon the observation of consistent patterns of genetic differentiation between sets of control and selected populations. It is, therefore, critical to ensure that observed differences between control and selected populations are genetic, and not direct or indirect (via maternal effects) phenotypic effects of their respective maintenance regimes. Individuals from control and selected populations should, therefore, be reared for one full generation under identical environmental conditions before any phenotypic comparisons are made on their progeny. This is a point that is often ignored and can result in severely misleading conclusions.

Tradeoffs and the genetic architecture of fitness

In the past few decades, evolutionary biologists have been paying increasing attention to constraints on evolutionary change, especially those due to tradeoffs among different traits that are relevant to fitness in a particular environment (different approaches to studying such tradeoffs are reviewed in Rose et al. 1996; Sinervo and Basolo 1996; Reznick and Travis 1996). In most of the above discussion, our implicit focus has been on the use of selection experiments in the laboratory as a tool for dissecting the adaptive significance of traits. Due to constraints of space, we cannot discuss in detail the relative merits and demerits of selection approaches versus optimization arguments, phenotypic manipulations, and the comparative method (essentially correlating traits and environmental variables): the superiority of laboratory selection over these alternative approaches has been extensively argued by Leroi et al. (1994 b, c) and Rose et al. (1996).

One major aspect in which thinking in evolutionary biology today differs substantially from the 1940s, 50s and 60s is in the importance given to constraints on evolution. Natural selection is no longer seen as an all powerful force relentlessly shaping organisms to a better and better 'fit' to the environment, one trait at a time. On the contrary, empirical studies of the evolutionary process reveal a much subtler force that arises out of the complex interactions between an organism and its environment, and the resolution of these interactions into an evolutionary trajectory is often critically dependent on the genetic architecture of fitness in that population (Joshi 2000). Our ability to predict responses to selection regimes is quite poor, even in a relatively well understood ecological microcosms like a *Drosophila* culture. Thus, 'common sense' conjectures about the adaptive significance of this or that trait in real populations are likely to be wrong far more often than right.

Because components of fitness are often negatively correlated, it is also important to study multiple fitness components, and assess their relative contribution to overall fitness in a given selection regime. For example, fecundity and longevity are negatively correlated in many organisms. Therefore, if longevity is taken as the sole fitness component in a study, any environmental treatment reducing fecundity may be erroneously thought to contribute to higher fitness as a result of a correlated increase in longevity. Such an effect of LL on fecundity and longevity in *Drosophila* has recently been documented (Sheeba et al. 2000).

Understanding the pattern of tradeoffs among various traits that affect the evolutionary trajectory of a population under a given selection pressure requires some knowledge about the additive genetic correlations among traits likely to be of relevance to fitness in that population in that particular environment. Phenotypic correlations, at least among components of fitness, are not reliable indicators of either the sign or the magnitude of additive genetic correlations (Rose and Charlesworth 1981). Similarly, among-population phenotypic or genetic correlations between fitness components often do not reflect the sign or magnitude of the corresponding within-population correlations that actually affect evolutionary trajectories (Leroi et al. 1994 b, c). The latter problem is one reason why we would advocate caution in inferring adaptive significance based solely on data on clines.

Overall, what we want to emphasize is that evolutionary geneticists have gained much hindsight from empirical studies of evolution in the past three decades. If we can utilize this knowledge in studying the adaptive significance of circadian organization, we can avoid many problems that will retard the rate at which we progress toward our goal. In the following sections we will discuss some of the problems with previous studies in this area, as well as approaches that we feel will be fruitful in our quest to understand the adaptive significance of circadian organization.

Evidence for adaptive significance of circadian organization

"Evolutionary speculation . . . can be considered a relatively harmless habit, like eating peanuts, unless it assumes the form of an obsession; then it becomes a vice"

(R. Y. Stanier)

In this section we deal with the question of how studies on the adaptive significance of circadian organization stand up to critical scrutiny in light of the problematic methodological and inferential issues discussed in the previous section. These studies can be grouped into few broad categories as follows.

Persistence of circadian rhythms in aperiodic environments

Early studies on species living in caves or other such aperiodic environments suggested that these organisms either lack circadian rhythms, or exhibit rhythms that are very deviant from the circadian norm (Blume et al. 1962; Poulson and White 1969). On the other hand, several studies suggest that organisms living in aperiodic environments can exhibit circadian rhythms. For example, mice reared for six generations in continuous light showed free-running circadian rhythms throughout their life, without any symptom of impairment (Aschoff 1960). Circadian rhythms have also been reported in the cave-dwelling fishes (Trajano and Menna-Barreto 1996) and millipedes (Mead and Gilhodes 1974; Koilraj et al. 2000), and in populations of *Drosophila melanogaster* reared in LL for several hundred generations (Sheeba et al. 1999a).

The fact that at least a good proportion of individuals in populations reared in aperiodic environments for very many generations retain their capacity to be rhythmic indicates that there is some advantage for organisms to be rhythmic even in an aperiodic environment. It is possible that even in aperiodic environments circadian rhythms may contribute in various ways to the maintenance of functional integrity of the internal metabolic milieu of organisms, in a manner similar to various biological periodicities which do not have environmental correlates (Daan and Aschoff 1982).

Circadian resonance and fitness effects of rhythm disruption

Several studies have examined longevity as an indicator of fitness of organisms subjected to various periodic and aperiodic environments, and it has often been seen that longevity is greater when organisms are maintained in environments with periodicity close to their endogenous periodicity, a phenomenon termed circadian resonance by Pittendrigh and Minis (1972). *D. melanogaster* routinely maintained on L:D 12:12 h had significantly higher adult longevity in 24 h LD cycles, compared to LD cycles of other periodicity or constant light (Pittendrigh and Minis 1972). The longevity of blowflies *Phormia terraenovae* raised in L:D: 12:12 h remained unaffected when kept under L:D 14:14 h, whereas longevity decreased under L:D: 10:10 h (von Saint-Paul and Aschoff 1978). In contrast, a study of *D. melanogaster* lines fixed for different per mutations revealed lower longevity of mutant males compared to wild type homozygous males under various LD cycles, including those of periodicity closer to the free-running period of the mutants; the longevity of females, interestingly, appeared to be unaffected by the periodicity of the imposed LD cycle (Klarsfeld and Rouyer 1998).

The results of studies on the fitness consequences of LL are also conflicting (reviewed in Sheeba et al. 2000). Some reports suggest that LL is harmful for plants, while others report increased growth rates under LL in both plants and cyanobacteria. *D. melanogaster* reared for several hundred generations in LL show faster pre-adult development in LL, as compared to either L:D 12:12 h or DD, whereas dry weight at eclosion and pre-adult survivorship do not differ among the three light regimes (Sheeba et al. 1999b). The same populations *of D. melanogaster* exhibit reduced longevity in LL compared to either L:D 12:12 h or DD, but this is accompanied by increased fecundity in LL (Sheeba et al. 2000). Similarly, the longevity of *D. melanogaster*, routinely maintained in constant darkness (DD) was seen to be higher in DD, as compared to either L:D 12:12 h or LL (Allemand et al. 1973).

Conflicting results about the effects of circadian rhythm disruption on longevity are also seen in studies on mammals. Arrhythmic mutant hamsters (Menaker and Vogelbaum 1993) and mice (King et al. 1997) were not observed to undergo any reduction in longevity under laboratory conditions. Similarly, SCN lesioned golden-mantled ground squirrels, *Spermophilus lateralis* (Ruby et al. 1996), and Siberian chipmunks, *Eutamias sibiricus* (Sato and Kawamura 1984), did not suffer reduction in longevity in the laboratory. On the other hand, heterozygous tau^+/tau^s mutant golden hamsters, *Mesocricetus aureus*, showed fragmented locomotor activity and reduced longevity in L:D 14:10, as compared to either tau^s/tau^s homozygotes that free-ran in this light regime, or tau^+/tau^+ homozygotes, that entrained (Hurd and Ralph 1998). However, the longevity of the three genotypes did not differ in LL of low intensity of 20-40 lux, presumably because the heterozygotes were also able to free-run. Moreover, Hurd and Ralph (1998) also found that they could restore fragmented locomotor activity rhythms and enhance the post-transplant life expectancy of ageing individuals by means of foetal SCN grafts.

Unfortunately, interpreting the results of the effects of different light regimes on longevity is difficult unless there is also some information on effects on other fitness components, especially fecundity. Moreover, in most studies on circadian resonance, there is no population level replication, and it is hard to decide which effects are strain specific. So far, the only study to examine multiple fitness components on replicate populations are those by Sheeba et al. (1999b, 2000) and they serve to underscore the importance of looking at different components of fitness in a systematic manner.

Perhaps the cleanest evidence for circadian resonance so far obtained comes from a study of competition under different LD cycles in the cyanobacterium *Synechococcus* (Ouyang et al. 1998). Although the growth rates of various period mutant strains of *Synechococcus* in pure culture were comparable across a range of imposed LD cycles, mutant strains with periodicity closest to that of the imposed LD cycle outcompeted the others in mixed cultures.

Significance of biological clocks in the wild

Due, in part, to the ubiquity of circadian rhythms, the notion that organisms derive major benefits from the daily rhythmicity in their physiology and behaviour and by anticipating events occurring in their neighbouring environment is widely held despite the paucity of direct empirical evidence in its support. Studies on the jumping of guillemot fledglings from their nests provide an example of differential mortality based on the timing of a behaviour. Flightless young glide down the cliffs on which guillemots nest, typically during a period of a few hours in the evening, and in so doing are subject to heavy predation by gulls. Daan and Tinbergen (1980) found that predation was less severe during the daily peak of jumping activity of the fledglings, suggesting that the timing of this activity was correlated with survivorship. Eclosion in the midge *Chironomus thummi* shows a circadian rhythm with temperature dependent phasing (Kureck 1979), and provides an example of reproduction-mediated fitness importance of timing of a behaviour. At low temperatures, eclosion occurs during daytime, whereas dusk is otherwise the time of peak eclosion. It appears that at higher temperatures night swarming is more effective for reproduction, but individuals eclosing during daytime have better reproductive success at low temperatures (Kureck 1979). Studies of SCN-lesioned antelope ground squirrels, *Ammospermophilus leucurus* (DeCoursey et al. 1997) and eastern chipmunks, *Tamias striatus* (DeCoursey and Krulas 1998) kept under field conditions suggest that, relative to controls, lesioned animals suffer greater mortality, partly through increased predation.

Evidence from clines

Lankinen (1986) studied the eclosion rhythm and adult diapause in 57 European populations (mostly highly inbred in the laboratory) of *D. littoralis* collected from localities ranging from 30°N to 70°N latitude. The phase of free-running eclosion rhythms was seen to be more variable than free-running period, and the two were not correlated, although weak clines were seen for both traits, with northern populations exhibiting shorter period and earlier phase. The timing of eclosion under entrainment showed a cline, with more northern strains eclosing earlier. The amplitude of both free-running and entrained eclosion rhythms was reduced in the northern populations. The critical day length for diapause induction was most strongly correlated with latitude and was longer in more northern populations. The same relationship between latitude and phase, period and amplitude of the eclosion rhythm was seen in a study on 12 populations (again highly inbred) of *D. subobscura*, a non-photoperiodic species, from Scandinavia and the Canary Islands (Lankinen 1993). Studies on the number of threonine-glycine repeats in the per gene of 18 *D. melanogaster* populations from Europe and northern Africa revealed a strong latitudinal cline, with the frequency of the (thr-gly)$_{20}$ allele being higher in northern populations (Costa et al. 1992). Further studies on locomotor activity rhythms assayed at low and high temperatures suggested that this cline was likely related to temperature compensation. The (thr-gly)$_{20}$ allele was found to show more efficient temperature compensation than the (thr-gly)$_{17}$ allele at both 18 and 29°C, whereas the (thr-gly)$_{17}$ allele resulted in periods closer to 24 h at 29 but not 18°C (Sawyer et al. 1997). Latitudinal clines have also been observed for the oviposition rhythm in *D. melanogaster* (Allemand and David 1976).

Selection studies

A few studies have attempted to examine correlated responses of circadian parameters to selection in the laboratory. Selection for early and late eclosing strains of *D. pseudoobscura* (Pittendrigh 1967) and the of moth *Pectinophora gossypiella* (Pittendrigh and Minis 1971) entrained to 12:12 LD was successful, with mean divergence in phase angle of eclosion of 4 and 5 h after 50 and 9 generations, respectively. In both species, the early-late strain difference in phase angle of eclosion rhythm was maintained across a variety of photoperiods. Both species exhibited the same type of correlated response in free-running period of the eclosion rhythm: the early strain had longer free-running period than the parent strain, whereas the late strain had shorter free-running period than the parent strain. The phase response curves for the eclosion rhythm in the early, late and parent strains, however, did not differ, and Pittendrigh (1981) interpreted this as supporting a two-oscillator model of circadian oscillations. In *Pectinophora*, correlated responses to selection were observed for the egg hatching rhythm: the early strain eggs hatched earlier, and late strain eggs later than the parent strain. There was, however, no correlated change in the parameters of the female oviposition rhythm. In both these studies, there was no replication at the level of population within selection regime, and there is very little information provided about the handling and rearing of the experimental populations, making it difficult to assess the degree of genetic drift or inbreeding that these populations may have undergone.

In a similar study, Clayton and Paietta (1972) produced early and late eclosing strains of *D. melanogaster* starting from two base populations: an Oregon R strain, and a wild caught strain, W_2. They observed a direct response to selection over 16 generations in both sets of strains, but did not examine the selected strains for correlated changes in other circadian parameters. They observed a greater response in the strains derived from Oregon R parents, and interpret this as suggesting that laboratory strains have greater genetic variation for circadian parameters because of relaxed selection in the laboratory. Once again, no details were provided about the collection and subsequent maintenance and handling of the wild caught strain, making it impossible to unambiguously interpret the observed results.

A recent study examining correlated responses to selection on phase angle of the locomotor activity rhythm in *D. rajasekari* (Joshi 1999) exemplifies the various methodological flaws that can render even the best conceived selection studies completely uninterpretable. The base population was initiated from the progeny of a single female and reared for 'several' generations before selection was done, a recipe for ensuring inbreeding and linkage disequlibrium. One selection line each for early and late phasing of locomotor activity was maintained for 59 generations. In each line, only 40 flies were assayed for locomotor activity, and the size of the breeding populations in each selection line was 1 male and 1 female. It may be noted that breeding each generation from a pair of parents is the most severe form of inbreeding that a sexually reproducing dioecious species can be subjected to, exceeding even repeated sib-mating. A simple calculation will show that by generation 59 the inbreeding coefficient of these strains would be $F \sim 0.99999999$, even if we assume a maximal heterozygosity of $H = 1$ in the base population. The typical response half-time taken to reach a plateau in selection response is of the order of $N/2$, where N is the number of breeding adults. The predicted response half-life for these selected strains, therefore, is 1 generation and it is thus strange that significant responses to selection over 59 generations were reported. Despite the amazingly high degree of inbreeding (and therefore homozygosity) entailed by the reported selection procedure of Joshi (1999), not only did the flies survive 59 generations of inbreeding depression and continue to respond to selection, the variation in phase angle in the selected lines was actually greater than in the control (ancestral) line, even though such strong selection together with inbreeding would be expected to rapidly eliminate genetic variation. The only conclusion that can be drawn from a study suffering from so many flaws and paradoxical observations is that it is meaningless to even try and interpret any observed correlated responses, because we have no idea what we are actually looking at in genetic terms.

Studies on the melon fly *Bactrocera cucurbitae* suggest that circadian parameters can evolve as correlated responses to selection on life-history traits and that these evolved changes in the phasing of certain rhythms may lead to reproductive isolation (Miyatake and Shimizu 1999, and references therein). Two lines each of *B. cucurbitae* were selected for faster and longer pre-adult development time, respectively, starting from a mass cultured base stock. The short development lines had diverged by ~ 3 days, and the long development lines by ~ 5 days after 21 and 16 generations of selection, respectively. The base stock was quite variable in free-running period of locomotor activity, with a mean of 24.7 h (± s.d. 2.7 h). The mean free-running period was reduced in by ~ 2 h in the short development lines, and the variance was also much reduced, whereas the long development lines showed a mean increase in free-running period of ~ 3.5 h, without such a dramatic reduction in the variance. This observation parallels the finding that *D. melanogaster* individuals homozygous for the short period alleles of per have shorter development time, and vice versa (Kyriakou et al. 1990). In *D. subobscura*, too, there is a positive among-population correlation between development time and free-running period of the eclosion rhythm (Lankinen 1993). Moreover, the *Caenorhabditis elegans* protein LIN-42 that controls the temporal sequence of many developmental events has recently been shown to be similar to the Period family of *Drosophila* proteins (Jeon et al. 1999). The generality of the notion of a direct link between clock period and development time is, however, challenged by the observation that gestation length in mice entrained to LD cycles of 28 or 20 hour periodicities does not differ (Davis and Menaker 1981).

Miyatake and his colleagues also examined the inheritance of free-running period of locomotor activity in crosses between the short and long development lines and found evidence suggesting that pleiotropic control of period and development time resided in at least one major autosomal locus, showing partial dominance of short period, with other minor loci also contributing to the observed variation. Other studies on these lines revealed that short development lines tended to mate earlier in the day than long development lines. Differences in the timing of locomotion were also seen, with short development lines being diurnal, and long development lines being nocturnal. Mate choice tests on 8 lines of *B. cucurbitae*, including the short and long development lines, revealed that development time and timing of mating were correlated and that the different timing of mating in the short and long development strains resulted in significant reproductive isolation between the two types of line (Miyatake and Shimizu 1999, and references therein).

What do we know about the evolutionary genetics of circadian organization?

Surveying the various studies summarized in this section, the overwhelming impression one gets is one of conflicting and suggestive, but only occasionally conclusive, results. In light of the criteria for evidence on adaptive significance now used in experimental evolutionary genetics, most studies on the fitness effects of light regimes or other manipulations of the circadian system suffer from a lack of population replication, and inadequate control on the genetic composition of the population studied in order to avoid genetic drift and inbreeding. Similarly, multiple components of fitness are not assayed in most studies, and true controls (in the evolutionary sense) are often lacking. All these shortcomings conspire to make the drawing of unambiguous conclusions from these studies difficult. Most of the data on circadian resonance and links between circadian parameters and fitness in different light regimes are, at best, suggestive. There is, consequently, little that can be unambiguously said about some of the finer genetic issues in evolution of circadian organization. Although rigorous and well replicated studies of the kind discussed in the previous section on evolution are undoubtedly difficult to conduct on many organisms, especially mammals, the relatively slip-shod manner in which even many of the recent *Drosophila* studies have been executed is worrisome and is, in part, the motivation for this paper.

What seems to be reasonably clear is that organisms retain fairly well functioning circadian clocks even after very long spans of time in LL or DD. These observations suggest that there is likely some fitness advantage to possessing clocks of circadian periodicity that is quite independent of the ability to phase one's activities with the environment, and this notion supports the view that the maintenance of internal temporal order itself confers a fitness advantage to organisms. It is also clear that there is additive genetic variation for circadian parameters in populations of many species, and this variation may often be genetically correlated with either life history traits (such as development time or longevity) or behaviours (such as mating and courtship) that can, in turn, lead to correlations between circadian parameters and fitness. Some of the field evidence suggests that phasing of behaviours can directly be correlated with survivorship in the face of predation. In parallel with these studies, the data on clines suggest strongly that at least some of the among-population genetic variation in circadian parameters are maintained by natural selection along gradients of temperature and daylength. Strictly speaking, this is not saying much: the import of all of this is merely that circadian organization can evolve through natural selection. The evidence at hand does not really tell us anything about how circadian organization evolves, and how the evolutionary fine-tuning of this clock machinery might be achieved.

Future Paths

"Pas che bayad kard, ai aqwaam-i-sharq"
'What is to be done next, O people of the east',

(Sheikh Mohammad Iqbal)

In the beginning of this article we had stated our intention to be both retrospective and prospective in our point of view. While making the transition from retrospective to prospective mode, and having summarized what is known, it is helpful to think about the important aspects of the evolutionary genetics of circadian organization about which we have very little or no knowledge. At this time, we have hardly any information about the within-and among-population distribution of phenotypic variation for period and phase properties of most circadian rhythms. Although the phenotypic relationship between period and phase is important for precision and entrainment (Pittendrigh and Daan 1976 a, b), we know little about the genetic correlation between these two circadian parameters for most rhythms, and what is known is often contradictory. Once again, we are faced with the problem of lack of population replication. If a study shows that selection on phase in one population altered period, we cannot rule out chance factors like genetic drift or linkage disequilibrium as causes of the correlated response to selection. We have very little information about how period and phase of different rhythms are correlated, either phenotypically or genetically. The genetic correlations between circadian parameters and fitness components in different light regimes (LL, DD, LD cycles of varying period) are not known in detail for any species. Despite the isolation of numerous *Drosophila* genes affecting circadian rhythmicity, we have no idea of whether alleles at these loci interact epistatically to affect the rhythm phenotype.

In many ways, then, we are still at a very preliminary stage in our understanding of the evolutionary genetics of circadian organization. This is, no doubt, partly because chronobiology is a relatively young discipline, and even now many details of the phenomenology of circadian rhythms are still being studied. However, part of the reason for this state of affairs is also, in our opinion, due to the lack of clearly organized and focused long term studies on the issue of adaptive significance of circadian rhythms. It is here that the almost complete absence of cross-fertilization between chronobiology and evolutionary biology has had a great retarding influence on the development of the field. We are convinced that the application of methodologies refined by evolutionary geneticists over the last three decades to questions pertaining to the adaptive significance and evolution of circadian organization will yield far greater insights that the previously used methods and approaches. We outline below some areas of investigation that we feel are critical to a deeper understanding of the evolutionary genetics of circadian organization.

(i) Formal multi-locus genetic analyses of multiple clock gene mutants, to identify patterns of epistasis among clock loci.

(ii) Formal quantitative genetics and QTL mapping of loci affecting circadian parameters.

(iii) Further field studies on the among-and within-population distribution of genetic variation for circadian parameters for different circadian rhythms, with attempts to relate observed patterns to some functional benefits.

(iv) Detailed and systematic studies on fitness effects of different light regimes on populations reared for many generations in various light regimes.

(v) Artificial selection on period and phase for a variety of rhythms, to examine direct and correlated responses to selection, and fitness of selected lines under various light regimes.

(vi) Laboratory natural selection under different light regimes to examine adaptation as it occurs, and to understand the changes in circadian organization that accompany such adaptation.

We stress that in these approaches it is important to work with well characterized systems amenable to laboratory rearing, and for which we have good understanding of the laboratory ecology and the genetic architecture of fitness: *Drosophila* is a very good example of such a model system. Care must be taken to work with outbred and large populations, whose genetic architecture is much more reflective of the norm for most sexual species, and replication at the population level is an absolute must, even though it multiplies the work load many fold. Fitness component assays must, as far as possible, be done in environments similar to rearing conditions, and multiple fitness components should be assayed. The importance to evolution of tradeoffs or pleiotropic effects identified by phenotypic manipulations or formal genetic analyses must be confirmed through observing patterns of correlated responses to selection. Care must be taken in designating populations to function as evolutionary controls; matching of ancestry is particularly important here. Most importantly, a multi-pronged approach combining phenotypic manipulations, formal genetics, and selection and reverse selection approaches on a well characterized set of laboratory populations will, in the long-term, provide a far deeper understanding of the evolution of circadian organization than an equal number of scattered and isolated studies done on a varied assortment of wild populations, mutant strains and inbred lines in laboratories.

We would also like to point out at this juncture that a review of two major fields (chronobiology and evolutionary genetics) involves covering material published in hundreds of papers. In the interest of keeping the list of references to a reasonable length, we have not made citations to the literature exhaustive. We have tried, instead, to be representative, and have favoured, as far as possible, citing reviews or relatively recent articles. Omission of papers from the list of references, thus, does not imply that we consider those studies to be unimportant. Finally, we would like to say in closing that it is not our intention to fault those who have attempted to study the adaptive significance of circadian rhythms. With the benefit of hindsight which the original investigators lacked, shortcomings can be found in any empirical study. Our purpose in writing this article has been to try and provide the benefits of hindsight from evolutionary genetics to the community of chronobiologists interested in the adaptive significance of the rhythms with which they work. If this article helps even a few people to refine their experimental approaches to the issue of the adaptive significance of circadian rhythms in the future we will feel more than amply rewarded and, more importantly, the discipline of chronobiology will benefit.

Acknowledgements

Our work is supported by funds from the Indian National Science Academy (VKS) and Department of Science and Technology, Govt. of India (VKS, AJ).

References

Allemand, R., David, J.R. (1976) The circadian rhythm of oviposition in *Drosophila melanogaster*: a genetic latitudinal cline in wild populations. Experientia **32**: 1403–1405.

Allemand, R., Cohet, Y., David, J. (1973) Increase in the longevity of adult *Drosophila melanogaster* kept in permanent darkness. Exp. Gerontol. **8**: 279–283.

Aschoff, J. (1960) Exogenous and endogenous components in circadian rhythms. Cold Spr. Harb. Symp. Quant. Biol. **25**: 11–28.

Blume, J., Bünning, E., Günzler, E. (1962) Zur Aktivitätsperiodik bei Höhlentieren. Naturwissenschaften **49**: 525.

Chippindale, A.K., Hoang, D.T., Service, P.M., Rose, M.R. (1994) The evolution of development in *Drosophila melanogaster* selected for postponed senescence. Evolution **48**: 1880–1899.

Chippindale, A.K., Alipaz, J.A., Chen, H-W., Rose, M.R. (1997) Experimental evolution of accelerated development in *Drosophila*. 1. Developmental speed and larval survival. Evolution **51**: 1536–1551.

Clayton, D.L., Paietta, J.V. (1972) Selection for circadian eclosion time in *Drosophila melanogaster*. Science **178**: 994–995.

Costa, R., Peixoto, A.A., Thackeray, J.T., Dalgliesh, R., Kyriakou, C.P. (1991) Length polymorphism in a Threonine-Glycine-encoding repeat region of the per gene in *Drosophila* J. Mol. Evol. **32**: 238–246.

Costa, R., Peixoto, A.A., Barbujani, G., Kyriakou, C.P. (1992) A latitudinal cline in a *Drosophila* clock gene. Proc. R. Soc. London B **250**: 43–49.

Crosthwaite, S.C., Dunlap, J.C., Loros, J.J. (1997) *Neurospora wc*-1 and *wc*-2: transcription, photoresponses and the origin of circadian rhythmicity. Science **276**: 763–769.

Daan, S. (1981) Adaptive daily strategies in behaviour. In: Aschoff, J. (ed.) Handbook of behavioural neurobiology. Vol. 4, Biological Rhythms. Plenum, New York, pp. 275–298.

Daan, S., Tinbergen, J.M. (1980) Young guillemots (*Uria lomvia*) leaving their Arctic breeding cliffs: a daily rhythm in numbers and risk. Ardea **67**: 96–100.

Daan, S., Aschoff, J. (1982) Circadian contributions to survival. In: Aschoff, J., Daan, S., Groos, G. (eds.) Vertebrate Circadian Systems. Springer Verlag, Berlin, pp. 305–321.

Davis, F.C., Menaker, M. (1981) Development of the mouse circadian pacemaker: independence from environmental cycles. J. Comp. Physiol. A **143**: 527–539.

DeCoursey, P.J., Krulas, J.R. (1998) Behaviour of SCN lesioned chipmunks in natural habitat: a pilot study. J. Biol. Rhythms **13**: 229–244.

DeCoursey, P.J., Krulas, J.R., Mele, G., Holley, D.C. (1997) Circadian performance of suprachiasmatic nuclei (SCN)-lesioned antelope ground squirrels in a desert enclosure. Physiol. Behav. **62**: 1099–1108.

Dunlap, J.C. (1999) Molecular bases for circadian clocks. Cell **96**: 271–290.

Falconer, D.S. (1981) Introduction to quantitative genetics (2nd ed.) Longman, London.

Golden, S.S., Ishiura, M., Johnson, C.H., Kondo, T. (1997) Cyanobacterial circadian rhythms. Annu. Rev. Plant Physiol. Plant Mol. Biol. **48**: 327–354.

Hall, J.C. (1995) Tripping along the trail to the molecular mechanisms of biological clocks. Trends Neurosci. **18**: 230–240.

Hall, J.C. (1998) Genetic of biological rhythms in *Drosophila*. Adv. Genet. **33**: 135–184.

Hogenesch, J.B., Guo, Y-Z., Jain, S., Bradfield, C.A. (1998) The basic-helix-loop-helix-PAS orphan MOP3 forms transcriptionally active complexes with circadian and hypoxia factors. Proc. Natl. Acad. Sci. USA **95**: 5474–5479.

Hofstetter, J.R., Possidente, P., Mayeda, A.R. (1999) Provisional QTL for circadian period of wheel running in laboratory mice: quantitative genetics of period in R1 mice. Chronobiol. Intl. **16**: 269–279.

Hurd, M.W., Ralph, M.R. (1998) The significance of circadian organization for longevity in the golden hamster. J. Biol. Rhythms **13**: 430–436.

Jeon, M., Gardner, H.F., Miller, E.A., Deshler, J., Rougvie, A.E. (1999) Similarity of the *C. elegans* developmental timing protein LIN-42 to circadian rhythm proteins. Science **286**: 1141–1146.

Joshi, A. (1997) Laboratory studies of density-dependent selection: adaptations to crowding in *Drosophila melanogaster*. Curr. Sci. **72**: 555–562.

Joshi, A. (2000) Life-history evolution in the laboratory. J. Ind. Inst. Sci. **80**: 25–37.

Joshi, A., Thompson, J.N. (1995) Trade-offs and the evolution of host specialization. Evol. Ecol. **9**: 82–92.

Joshi, A., Mueller, L.D. (1996) Density-dependent natural selection in *Drosophila*: trade-offs between larval food acquisition and utilization. Evol. Ecol. **10**: 463–474.

Joshi, A., Thompson, J.N. (1997) Adaptation and specialization in a two-resource environment in *Drosophila* species. Evolution **51**: 846–855.

Joshi, A., Mueller, L.D. (2000) Stability in Model Populations. Princeton University Press, Princeton.

Joshi, A., Knight, C.D., Mueller, L.D. (1996) Genetics of larval urea tolerance in *Drosophila melanogaster*. Heredity **77**: 33–39.

Joshi, A., Oshiro, W.A., Shiotsugu, J., Mueller, L.D. (1997) Within—and among-population variation in oviposition preference for urea supplemented food in *Drosophila melanogaster*. J. Biosci. **22**: 325–338.

Joshi, D.S. (1999) Selection for phase angle difference of the adult locomotor activity in *Drosophila rajasekari* affects the activity pattern, free-running pattern, phase shifts and sensitivity to light. Biol. Rhythms Res. **30**: 10–28.

King, D., Zhao, Y., Sangoram, A., Wilsbacher, L., Tanaka, M., Antoch, M., Steeves, T., Vitaterna, M., Kornhauser, J., Lowrey, P., Turek, F., Takahashi, J. (1997) Positional cloning of the mouse circadian CLOCK gene. Cell **89**: 641–653.

Klarsfeld, A., Rouyer, F. (1998) Effects of circadian mutations and LD periodicity on the life span of *Drosophila melanogaster*. J. Biol. Rhythms **13**: 471–478.

Koilraj, A.J., Sharma, V.K., Marimuthu, G., Chandrashekaran, M.K. (2000) Presence of circadian rhythms in the locomotor activity of a cave-dwelling millipede *Glyphiulus cavernicolus sulu* (Cambalidae, Spirostreptida). Chronobiol. Intl. **17**: 757–765.

Kureck, A. (1979) Two circadian eclosion times in *Chironomus thummi* (Diptera) alternately selected with different temperatures. Oecologia **40**: 311–323.

Kyriacou, P., Oldroyd, M., Wood, J., Sharp, M., Hill, M. (1990) Clock mutations alter developmental timing in *Drosophila*. Heredity **64**: 395–401.

Lankinen, P. (1986) Geographical variation in circadian eclosion rhythm and photoperiodic adult diapause in *Drosophila littoralis*. J. Comp. Physiol. A **159**: 123–142.

Lankinen, P. (1993) North-south differences in circadian eclosion rhythm in European populations of *Drosophila subobscura*. Heredity **71**: 210–218.

Leroi, A.M., Chippindale, A.K., Rose, M.R. (1994a) Long-term laboratory evolution of a genetic life-history trade-off in *Drosophila melanogaster*. 1. The role of genotype by environment interaction. Evolution **48**: 1244–1257.

Leroi, A.M., Chen, W.R., Rose, M.R. (1994b) Long-term laboratory evolution of a genetic life-history trade-off in *Drosophila melanogaster*. 2. Stability of genetic correlations. Evolution **48**: 1258–1268.

Leroi, A.M., Rose, M.R., Lauder, G.V. (1994c) What does the comparative method reveal about adaptation? Amer. Natur. **143**: 381–402.

Mead, M., Gilhodes, J.C. (1974) Organization temporella de l'activité locomotrice chez un animal cavernicole *Blaniulus lichtensteini* Bröl. (Diplopoda). J. Comp. Physiol. **90**: 47–52.

Menaker, M., Vogelbaum, M.A. (1993) Mutant circadian period as a marker of suprachiasmatic nucleus function. J. Biol. Rhythms **8**: 593–598.

Miyatake, M., Shimizu, T. (1999) Genetic correlations between life-history and behavioral traits can cause reproductive isolation. Evolution **53**: 201–208.

Mueller, L.D. (1997) Theoretical and empirical examination of density dependent selection. Annu. Rev. Ecol. Syst. **28**: 269–288.

Mueller, L.D., Ayala, F.J. (1981) Trade-off between r-selection and K-selection in *Drosophila* populations. Proc. Natl. Acad. Sci. USA **78**: 1303-1305.

Ouyang, Y., Andersson, C.R., Kondo, T., Golden, S.S., Johnson, C.H. (1998) Resonating circadian clocks enhance fitness in cyanobacteria. Proc. Natl. Acad. Sci. USA **95**: 8660-8664.

Partridge, L., Fowler, K. (1992) Direct and correlated responses to selection on age at reproduction in *Drosophila melanogaster*. Evolution **46**: 76–91.

Pittendrigh, C.S. (1967) Circadian systems,. I. The driving oscillation and its assay in *Drosophila pseudoobscura*. Proc. Natl. Acad. Sci. USA **58**: 1762–1767.

Pittendrigh, C.S. (1981) Circadian organization and the photoperiodic phenomena. In: Follett, B.K., Follett, D.E. (eds.) Biological clocks in seasonal reproductive cycles. Wright, Bristol, pp. 1–35.

Pittendrigh, C.S. (1993) Temporal organization: reflections of a Darwinian clock-watcher. Annu. Rev. Physiol. **55**: 16–54.

Pittendrigh, C.S., Minis, D.H. (1971) The photoperiodic time measurement in *Pectinophora gossypiella* and its relation to the circadian system in that species. In: Menaker, M. (ed.) Biochronometry. National Academy of Sciences, Washington DC, pp. 212–250.

Pittendrigh, C.S., Minis, D.H. (1972) Circadian systems: longevity as a function of circadian resonance in *Drosophila melanogaster*. Proc. Nat. Acad. Sci. USA **69**: 1537–1539.

Pittendrigh, C.S., Daan, S. (1976a) A functional analysis of circadian pacemakers in nocturnal rodents. I. The stability and liability of spontaneous frequency. J. Comp. Physiol. **106**: 223–252.

Pittendrigh, C.S., Daan, S. (1976b) A functional analysis of circadian pacemakers in nocturnal rodents. IV. Entrainment: pacemaker as a clock. J. Comp. Physiol. **106**: 291–331.

Poulson, T.L., White, W.B. (1969) The cave environment. Science **105**: 971–981.

Reznick, D., Travis, J. (1996) The empirical study of adaptation in natural populations. In Rose, M.R., Lauder, G.V. (eds.) Adaptation. Academic Press, New York, pp. 243–289.

Roper, C., Pignatelli, P., Partridge, L. (1993) Evolutionary effects of selection on age at reproduction in larval and adult *Drosophila melanogaster*. Evolution **47**: 445–455.

Rose, M.R. (1984a) Genetic covariation in *Drosophila* life history: untangling the data. Amer. Natur. **123**: 565–569.

Rose, M.R. (1984b) Laboratory evolution of postponed senescence in *Drosophila melanogaster*. Evolution **38**: 1004–1010.

Rose, M.R, Charlesworth, B. (1981) Genetics of life history in *Drosophila melanogaster*. I. Sib analysis of adult females. Genetics **97**:173–186.

Rose, M.R., Service, P.M., Hutchinson, E.W. (1987) Three approaches to trade-offs in life-history evolution. In: Loeschke, V. (ed.) Genetic constraints on adaptive evolution. Springer Verlag, Berlin, pp. 91–105.

Rose, M.R., Nusbaum, T.J., Chippindale, A.K. (1996) Laboratory evolution: the experimental wonderland and the Cheshire Cat syndrome. In Rose, M.R., Lauder, G.V. (eds.) Adaptation. Academic Press, New York, pp. 221–241.

Ruby, N.F., Dark, J., Heller, H.C., Zucker, I. (1996) Ablation of suprachiasmatic nucleus alters timing of hibernation in ground squirrels. Proc. Natl. Acad. Sci. USA **93**: 9864–9868.

Sato, T., Kawamura, H. (1984) Effects of bilateral suprachiasmatic nucleus lesions on the circadian rhythms in a diurnal rodent, the Siberian chipmunk (*Eutamia sibiricus*). J. Comp. Physiol. A **155**: 745–752.

Sawyer, L.A., Hennessy, J.M., Peixoto, A.A., Rosato, E., Parkinson, H., Costa, R., Kyriakou, C.P. (1997) Natural variation in a *Drosophila* clock gene and temperature compensation. Science **278**: 2117–2120.

Service, P.M., Rose, M.R. (1985) Genetic covariation among life history components: the effect of novel environments. Evolution **39**: 943–945.

Sheeba, V., Sharma, V.K., Chandrashekaran, M.K., Joshi, A. (1999a) Persistence of *Drosophila* eclosion rhythm after 600 generations in an aperiodic environment. Naturwissenschaften **86**: 448–449.

Sheeba, V., Sharma, V.K., Chandrashekaran, M.K., Joshi, A. (1999b) Effect of different light regimes on pre-adult fitness in *Drosophila melanogaster* populations reared in constant light for over six hundred generations. Biol. Rhythms Res. **30**: 424–433.

Sheeba, V., Sharma, V.K., Shubha, K., Chandrashekaran, M.K., Joshi, A. (2000) The effect of different light regimes on adult lifespan *in Drosophila melanogaster* is partly mediated through reproductive output. J. Biol. Rhythms **15**: 380–392.

Sinervo, B., Basolo, A.L. (1996) Testing adaptation using phenotypic manipulations. In Rose, M.R., Lauder, G.V. (eds.) Adaptation. Academic Press, New York, pp. 221–241.

Trajano, E., Menna-Barreto, L. (1996) Freerunning locomotion activity rhythms in cave-dwelling catfishes *Trichomycterus* sp. from Brazil. Biol. Rhythms Res. **27**: 329–335.

Von Saint-Paul, U., Aschoff, J. (1978) Longevity among blowflies *Phormia terranovae* R.D. kept in non-24-hour light-dark cycles. J. Comp. Physiol. **127**: 191–195.

Wheeler, D.A., Kyriakou, C.P., Greenacre, M.L., Yu, Q., Rutila, J.E., Rosbash, M., Hall, J.C. (1991) Molecular transfer of a species-specific courtship behaviour from *Drosophila simulans* to *Drosophila melanogaster.* Science **251**: 1082–1085.

Young, M.W. (1998) The molecular control of circadian behavioural rhythms and their entrainment in *Drosophila.* Annu. Rev. Biochem. **67**: 135–152.

3. Circadian Frequency and Its Variability

S. Daan* and D.G.M Beersma

Zoological Laboratory, University of Groningen, P.O. Box 14, 9750 AA, Haren, The Netherlands

This essay is dedicated to our friend Maroli K. Chandrashekaran in recognition of his many original contributions to biological rhythm research and of his lifelong stimulation of this field, in particular in India.

The period (τ) of circadian oscillations in constant conditions typically deviates from 24 hours. This observation, first made by De Candolle (1835) on the leaf movements of the Sensitive plant, Mimosa pudica, is the key evidence of their endogenous generation, as first realized by Pfeffer (1915; see Bünning and Chandrashekaran 1975). Whether this deviation conveys any specific functional or causal meaning—beyond demonstrating that the oscillations are not directly caused by the earth's rotation—has not been settled. Functionally, the endogenous τ affects the phase angle difference (ψ) between a circadian system and its zeitgeber, and hence is expected to be subject to selection pressure in the natural periodic environment. On the causal side the fact, for instance, that genetic mutations of the circadian system are often detected by their influence on τ underscores the implications of τ for our understanding of oscillator mechanisms. In this essay, we discuss some of the issues involved in the implications of τ and its variability for both function and mechanism.

Functional aspects of τ

A question often asked by lay audiences after listening to a general lecture on human rhythms is, why should τ on average be different from 24 h, and does this deviation perhaps reflect an adaptation to the rotation of the earth in past evolutionary history ? It is well known that the earth's rotation gradually slows down. Thus any atavistic explanation of this kind would hold only for τ values shorter than 24 h. It would not explain why some species—mostly diurnal organisms but also nocturnal species such as rat and hamster—have an average τ in DD longer than 24 h. The precise rate of change of the solar day has been estimated at about 2 msec per century on the basis of the number of daily calcium deposits per annual cycle in fossil corals and bivalves (e.g., Lambeck 1978). Thus we would have to pinpoint the time of fixation for a τ of 23.5 hours in the Cretaceous around 100 million years ago, a τ of 22 h in the Devonian around 300 million years. One of the most extreme propositions in this direction was advanced by Heynick (1986), who was impressed by the fact that the endogenous human circadian period of 25 h is close to the cycle length of the lunar day, and that the lunar:solar day ratio in the earth's history has always been close to 25:24 h. Heynick suggested that this value for human τ would contribute an argument to Sir Alister Hardy's unorthodox theory (1960) of an aquatic, possibly intertidal, phase in human evolutionary history. This extravagant idea

*E-mail: s.daan@biol.rug.nl

is now buried by the more recent insight that the original τ estimate of ~ 25 h has been the result of self-selection of light-dark cycles in Wever's original experiments, and that the average endogenous human τ without self-selection is closer to ~ 24.1 h (Czeisler et al. 1995). More quintessential, however, is the general fact that the average τ is not a fixed characteristic but is dependent on the specific experimental conditions in which it is measured. The atavism approach provides no argument as to why for instance τ in DD should reflect its evolutionary history, and why τ in LL shouldn't, or *vice versa*. In speculating on the possible adaptive meaning of τ, we should focus on present day existence rather than on evolutionary history.

Is the deviation of τ from 24 hours adaptive ?

Most researchers have considered the slight deviation of τ from 24 h simply a biological imperfection. This imperfection would not lead to serious functional deficits, and thus be weeded out by natural selection, since the circadian system in nature would always be entrained to the earth's rotation. Even with a perfect τ of 24 h sensitivity to a zeitgeber would always be needed to ensure proper phase control, and such phase control simultaneously would take care of the correction of the deviation of τ.

By contrast, Colin Pittendrigh has defended the view that the deviation of τ from 24 h reflects *"a strategy, not a tolerated approximation"* (Pittendrigh 1981 p. 116). He based this argument on the stability of ψ—the phase angle difference between a circadian oscillation and its entraining zeitgeber—using the *Drosophila* light pulse PRC as a model. Under the assumption that a full photoperiod acts by eliciting a phase advance at dawn and a phase delay at dusk as dictated by the pulse PRC, Pittendrigh predicted that a small error in τ would lead to small variations in ψ when the mean τ is set away from 24 h, *e.g.*, at 23 or 25 hours, but to large ψ-instability when average τ would equal 24 h (Fig. 1). He concluded that *"... the instability of ψ .... is progressively smaller the further τ gets away from T ..."* (Pittendrigh 1981 p. 116).

This reasoning would lead to the expectation to find τ distributions for any species concentrated on either side of 24 h. If the species' average τ is close to 24 h, we would expect a bimodal distribution with peaks at larger or smaller values than 24. These predictions can be evaluated in published τ distributions for several species. Such distributions have for instance been presented by Pittendrigh and Daan (1976a fig. 2) for four species of nocturnal rodents. All four distributions straddle the 24 h mark, but none of them shows a trace of bimodality. Although these data are not supportive of the prediction, a better evaluation would make

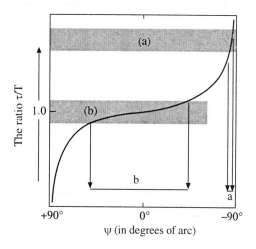

Fig. 1   **Pittendrigh's stability argument derived from the dependence of the phase-angle difference (ψ) between two oscillators on their relative periods. The period of the zeitgeber (T) is held constant and the period of the entrained oscillator (τ) is varied. When the ratio τ/T is less than 1.0 ψ is positive, when $\tau/\psi > 1.0$ ψ is negative. The figure considers the effect on ψ of a given amount of instability of τ; in case (a) the instability is around a mean τ value far from T, and the consequent ψ variation is small. The same degree of τ instability when mean $\tau = T$ is considered in case (b) where the consequent ψ instability is much greater. From Pittendrigh and Daan (1976b fig. 19)**

use of distributions where each individual contributes only a single value. In figure 2 we have restricted such an analysis to studies where τ was reported for at least 50 individuals in similar conditions. The number of such studies is astonishingly small. There must be dozens of species for which τ distributions are hidden in lab files without getting published. We encourage researchers to include such graphs, rather than mere species average τ-values in their publications. The data for house mice (Fig. 2A, Edgar et al. 1991) and humans (Fig. 2D, Wever 1975) are nearly all on one side of 24 h. By far the largest data set currently available is the study by Sharma and Chandrashekaran (1999), who measured τ in DD of 346 wild-caught rice paddy mice (*Mus booduga*). This distribution

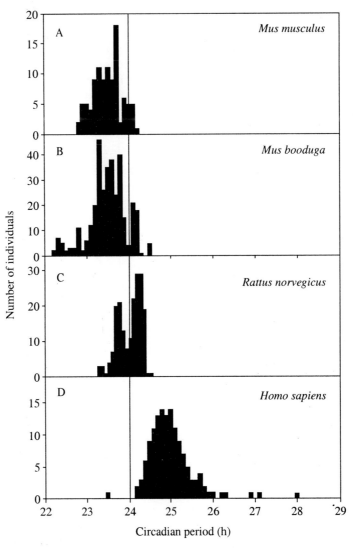

Fig. 2. Interindividual distributions of τ A. House mouse, *Mus musculus* (data from Edgar et al. 1991 fig. 4); B. Rice paddy mouse, *Mus booduga* (data from Sharma and Chandrashekaran 1999); C. Rat, *Rattus norvegicus* (data from Richter 1968); D. Human subjects, *Homo sapiens* (data from Wever 1975).

(Fig. 2B) yields no evidence for under-representation of $\tau = 24$ h. The sole case where there seems to be a bimodality in τ values with an under-representation of τ values around 24 h is Richter's data set on rats (Fig 2C). Richter himself called special attention to this distribution, but had to conclude that "*at present no explanation of this finding suggests itself*". In interpreting such a bimodal distribution as support for Pittendrigh's conjecture on the benefit of having τ unequal to T, there are at least two caveats: We must be sure that the data do not reflect two subpopulations in the data. Richter (1968) remarked that the period tends "*to be shorter than 24 hours in wild rats, longer than 24 hours in domesticated rats, particularly albinos*". If this is the case, the researcher's choice will determine whether a distribution becomes unimodal or bimodal. Also, values of (nearly) exactly 24 h may raise a researcher's suspicion of entrainment by some uncontrolled cues, and may be prone to be discarded. Given these caveats, we conclude that the published record at least does not strongly speak in favour of the idea of a selective premium on having τ unequal to T.

Pittendrigh's argument was inspired primarily by his model of non-parametric entrainment by brief light pulses, and was most obvious for zeitgebers consisting of one pulse per cycle, and oscillators with a clear "dead zone" of insensitivity to the pulse. As pointed out by Pittendrigh, in nature there are at least two zeitgeber signals available, one at dawn and one at dusk, and the reasoning that $\tau = T$ leads to phase instability is considerably weakened in a zeitgeber with two light pulses (Pittendrigh and Daan 1976b, Fig. 20).

Unlike burrowing nocturnal mammals which expose themselves naturally at most to two brief light episodes per day, most animals are exposed to complete photoperiods in nature. Under such natural circumstances the argument breaks down completely. This can be shown by computing in the non-parametric entrainment model the predicted ψ as dependent on variation in τ using PRC's for long light pulses. In Fig. 3 we have done this for the circadian system controlling pupal eclosion in the flesh fly *Sarcophaga argyrostoma*. In this species the most complete set of PRC's is available— with pulse durations varying from 1 to 20 h—from the work of Saunders (1978). In the left column of Fig. 3 three phase response curves have been plotted: (a) for a single 1-h light pulse, (b) for two 1-h pulses 12 h apart (a 'skeleton photoperiod'), as derived from (a), and (c) for a single 12-h pulse. In the middle column ψ as derived from the three PRC's is plotted as a function of τ, for τ-values of 22 to 26 h. The steepness of the slope of this dependence determines the variability of ψ as far as influenced by variation in τ around its mean. As an index of this variability, we calculated the range of ψ-values generated by assuming a range of τ-values of 1 h around the mean. The reciprocal of this range is shown in the right hand column. While Pittendrigh's argument of increased variability with mean τ near 24 h holds up for non-parametric entrainment by single pulses, and even for skeleton pulses, it evidently breaks down for long photoperiods that elicit a type-0 PRC. Here, "accuracy" of entrainment is rather maximized by having τ close to T ! (Fig. 3 lower right panel). Clearly, the complete photoperiod is doing something to the *Sarcophaga* system that is not explained by the non-parametric skeleton photoperiod approximation, and that renders much of the stability argument in favour of deviations of τ from T obsolete.

Pittendrigh was well aware of the fact that τ is not a fixed parameter of the circadian system and he was the first to describe "after-effects" on τ of entrainment (Pittendrigh 1960). These aftereffects demonstrate that exposure to a zeitgeber elicits a velocity response that brings the endogenous τ closer to T than it would have been without prior entrainment. On the basis of a simulation model we have recently reported that maximal accuracy of phase control is reached by such velocity responses which essentially maintain the pacemaker's τ at exactly 24 h in naturalistic conditions (Beersma et al. 1999a; Daan 2000). The implication is that the pacemaker in fact "learns" to run at a rate of one cycle per 24 h. Under such conditions it can ensure maximal accuracy of ψ without requiring any net

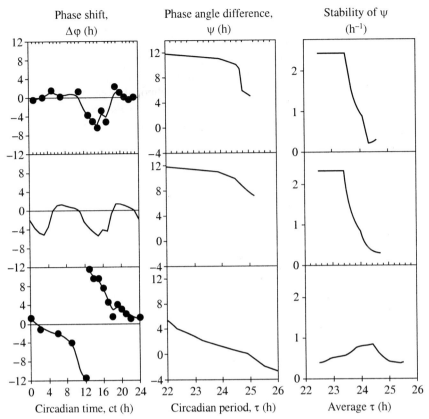

Fig. 3   Stability of entrainment predicted for the eclosion rhythm of the blowfly, *Sarcophaga argyrostoma* on the basis of non-parametric entrainment predicted from multiple phase response curves as measured by Saunders (1978). From top to bottom: A. Single light pulses (1 h); B. Skeleton photoperiod (2 1-h pulses separated by 11 h darkness), as derived from the 1-h pulse PRC; C Single light pulses (12 h). The panels in the left column show the PRC's. The middle panels present the phase angle difference ψ between the eclosion rhythm and the 24-h zeitgeber for a τ range from 22 till 26 h; the right hand panels show the stability of ψ (1/range of ψ) assuming a range of variation of τ of 1 h around the average.

phase resetting response per day. This mechanism seems to be the only one sufficient to explain entrainment in such burrowing diurnal animals that never observe the twilights (Hut et al. 1999).

In our view, then, the deviation τ-T has no specific functional meaning since it does not apply to the natural situation under which natural selection operates. It does reflect in many cases the degree to which animals are capable of maintaining the τ imposed by the zeitgeber as aftereffects in constant conditions. Variations in this capacity may of course well turn out to reflect different selection pressures.

Can variations of τ tell us something about mechanism?

As much of the further argument is based upon the variation in τ and ψ, we propose here several operational definitions, expanding on usage introduced earlier (Pittendrigh and Daan 1976a; Beersma et al. 1999a)

Precision of cycle length denotes the reciprocal of the standard deviation of single circadian cycles (t_i, $i = 1...n$). A value for precision may be obtained for instance in a 14 day segment of a freerun, which yields also a single τ estimate either as average of t_i or as the regression slope.

Stability of the endogenous period is the reciprocal of the standard deviation among a series of τ estimates, from different segments of a freerun obtained under identical conditions in the same individual.

Accuracy is the reciprocal of the day-to-day standard deviation in the phase angle difference ψ between an entrained circadian rhythm and its zeitgeber.

Clearly we expect natural selection to be concerned with accuracy, but accuracy will be affected by precision in addition to the noise generated by the entrainment process. Precision of cycle length is determined both by noise in the generating oscillation and in variations attributable to peripheral processes leading to the overt behavioural rhythm. The observed variation from cycle to cycle can be partitioned into these two sources of variance on the basis of the serial correlation between cycles (Pittendrigh and Daan 1976a). We presume that the strength of the pacemaker's oscillation will contribute both to its long-term stability and to the oscillator's component in the day-to-day precision. This presumption is supported by a comparison of individual variability among consecutive cycles (precision) and among series of τ estimates (stability), obtained from prolonged freeruns of *Mus musculus* in DD (Figure 4). The fact that we see a positive association between short-and long-term variability does not by itself provide a persuasive argument. For purely statistical reasons, one would expect that a sloppy rhythm (with large day-to-day variation or low precision) would allow a less

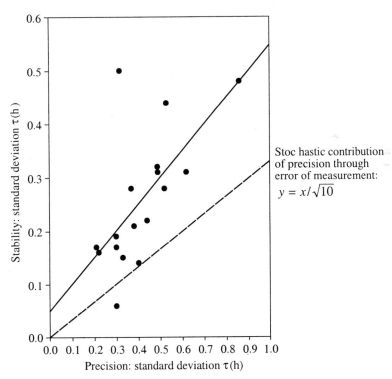

Fig. 4 The association between short-term precision and long term stability of circadian locomotor activity rhythms in *Mus musculus* in DD, based on data from Pittendrigh and Daan (1976a). Lines indicate regression through individual data points (solid) and the estimated stochastic contribution of precision to stability through error of measurement: $y = x/\sqrt{10}$.

precise determination of τ in each segment, and thereby a larger variation between τ estimates. This error of measurement would lead to the expected relationship s.d.τ = s.d.t/$\sqrt{n}$, where n = the number of cycles per segment. In the data evaluated in figure 4, segments were about 10 cycles long, and the line s.d.τ = s.d. t/$\sqrt{10}$ has been indicated. Clearly, the observed relationship is much steeper than solely attributable to error of measurement. This qualitatively supports our contention that both day-to-day precision and long-term stability of circadian cycles reflect a property of the underlying pacemaker.

In all three measures, accuracy, precision and stability, a curious relationship has been noted by several researchers: variability increases when the average deviates from the norm. This holds in the first place for the accuracy of ψ measured under natural daylight. Such data have been extensively reported for diurnal as well as nocturnal homeotherm species exposed year-round to natural light conditions at different latitudes (Daan and Aschoff 1975). One result from these studies was that ψ was most accurately controlled when the mean activity onset coincided with civil twilight, i.e., with the maximal logarithmic change in light intensity (Fig. 5A). As the authors noted, this might be

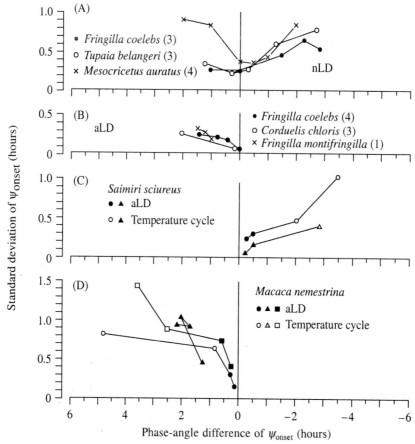

Fig. 5   Accuracy (expressed as s.d. ψ) of entrained circadian locomotor activity rhythms as a function of mean ψ. (A) Finch (*Fringilla coelebs*), Tree shrew (*Tupaja belangeri*) and hamster (*Mesocricetus auratus*) in natural LD cycles (from Daan and Aschoff 1975). (B) Finch, Greenfinch (*Carduelis chloris*) and Brambling (*Fringilla montifringilla*) in artificial LD cycle with interposed twilights (from Wever 1967). (C) Squirrel monkey (*Saimiri sciureus*) and (D) Rhesus monkey (*Macaca nemestrina*) in artificial LD and temperature cycles (data from Aschoff and Tokura, unpublished).

attributed either to more accurate entrainment, or to stronger masking effects of light when $\psi = 0$. In other experiments, by Wever (1967), Aschoff and Tokura (unpublished) the masking explanation is excluded since mean activity onset occurred well before any change in light intensity (Fig. 5B-D). It is thus likely that indeed the underlying oscillator is most precise when $\psi = 0$, *i.e.*, under conditions of maximal resonance between pacemaker and zeitgeber.

A related and more illuminating fact is that long-term *stability* of circadian oscillators is reduced when their average cycle length τ is further away from 24 h. This tendency for reduced stability has been documented for τ-values shorter than 24 h in four nocturnal rodent species by Pittendrigh and Daan (1976b fig. 12). Cycle-to-cycle *precision* appears to be subject to the same pattern, as was first documented by Aschoff (1960) for chaffinches. Results from his study have been summarized in figure 6, along with those of Eskin (1969) obtained in sparrows: Both are from diurnal birds, where in LL, with $\tau > 24$ h, the variation is negatively associated with τ, and in DD, with $\tau > 24$ h, there is a positive association.

In both cases (Fig. 6), the distributions are restricted to one side of the "norm" value of 24 h. More powerful results are those where the spontaneous variation of τ between individuals straddles the 24 h mark. Four such data sets, all concerning nocturnal rodents, are summarized in figure 7. One of these refers to the stability of circadian period in the Kangaroo rat, studied by Natalini (1972) and plotted as the standard deviation of τ (Fig. 7A). In the other three panels the inverse of precision of circadian cycles (s.d.t.) as based on the onset of activity has been plotted. In all distributions a quadratic regression (cubic regression in Fig. 7D) has been fitted. The most extensive data set is from Chandrashekaran's group on the rice paddy mouse, *Mus booduga*, in which 346 individual mice were evaluated (Fig. 7D; Sharma and Chandrashekaran 1999). Two smaller data sets refer to mice (Fig. 7C; Pittendrigh and Daan 1976a) and Syrian hamsters (Fig. 7B; Aschoff unpublished). All cases

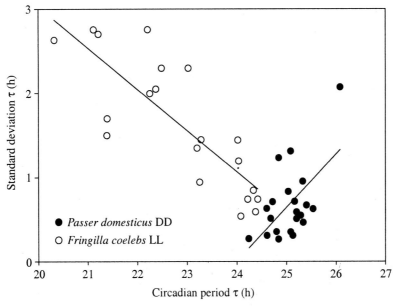

Fig. 6   **Precision (expressed as day-to-day standard deviation in cycle length) of freerunning circadian rhythms in birds as a function of their mean τ in DD (*Passer domesticus*; from Eskin 1969) and LL (*Fringilla coelebs*; from Aschoff 1960)**

provide clear evidence for maximal stability or precision of the rhythm in midrange, i.e., when τ approaches 24 h.

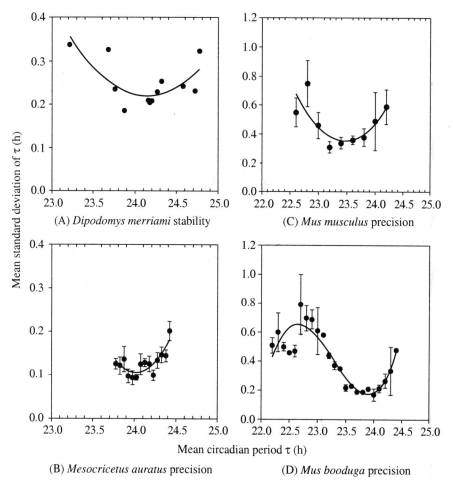

Fig. 7   **Stability and precision as a function of mean τ: the standard deviation increases towards the extreme τ values in DD. Stability in the *(A)* Kangaroo rat (Natalini 1972). Precision in *(B)* Syrian hamster (data J. Aschoff, unpublished), *(C)* house mice (data from Pittendrigh and Daan 1976a) and *(D)* rice paddy mice (*Mus booduga*, data from Sharma and Chandrashekaran 1999).**

This is a remarkable phenomenon. Pittendrigh interpreted it as indicating that circadian pacemakers running at a frequency close to 24 h require tighter homeostatic control of frequency because of the inherent instability of phase when τ equals T (Pittendrigh and Daan 1976b). As explained above, we now have reservations about this functional argument. But what remains is that somehow the circadian pacemaker must be able to 'recognize' its own frequency and becomes more precise as it's τ approaches the earth's rotation. In the absence of any external time cue, there must be some internal time cue 'telling the pacemaker how fast it is running'. So far there has been no theoretical approach attempting to explain this phenomenon. Aschoff et al. (1971) have provided a general explanation for changes in the precision with τ-change. Their approach describes the change in waveform of a van der Pol oscillator that accompanies τ changes with variations in one parameter. Although helpful to understand why with lengthening τ the onset of activity becomes less precise, while the offset becomes more precise, this approach does not explain the nonmonotonic changes in precision around 24 h shown in Figure 7. Intuitively, it seems to us that a second timing device would be needed to explain such changes, and we have been curious to find out whether a system of two circadian oscillators, as proposed long ago (Pittendrigh and Daan 1976c) would be sufficient to generate the pattern.

Model simulations

This E/M model postulated a "Morning oscillator" locking on to dawn, and internally coupled to an "Evening oscillator" locking on to dusk. Although until recently there has not been much concrete evidence for a system with two functionally distinct oscillators, it has remained popular among some researchers to explain different behavioural and physiological responses at the beginning and end of the subjective night (*e.g.,* Illnerova 1986; Wehr 1996). The current renewed interest in the model is primarily due to findings at the neurophysiological and molecular levels. Jagota et al. (2000) observed two clear peaks in multi-unit electrical activity of horizontal SCN slice preparations, that behave in response to light in the manner postulated for the Evening and Morning oscillators. At the molecular level Van der Horst et al. (1999) identified circadian phenotypes for mice where either the cry1 or the cry2 genes have undergone targeted manipulations. These developments have led to the postulation of a molecular substrate for the *E/M* model (Daan et al. 2000). Independent of molecular or physiological detail, we address here the question whether such a two-oscillator model may generate the dependence of precision on τ

In exploring this proposition by simulation, we have used a simplistic 'phase-only' model for each of the two oscillators like that employed by Daan and Berde (1978). Only the interaction between the oscillators was not restricted to a single event in each oscillator's cycle, but was introduced as a continuous change in velocity depending both on its own phase and on the phase of the other oscillator. This allowed us to study the accumulation of noise randomly introduced in each time step eventually to the precision of each oscillator's period. In practice we programmed a computer such that there were two oscillators, E and M, with $\tau_E = 23$ h and $\tau_M = 25$ h. Per time unit i of 15 minutes, the phases of E and M were calculated from two recursive equations:

$$\varphi_{Ei} = \varphi_{Ei-1} + 1/(4 \cdot \tau_E) + c_1 \cdot \sin(2 \cdot \pi \cdot \varphi_{Mi-1}) \cdot \sin(2 \cdot \pi \cdot \varphi_{Ei-1}) + \varepsilon_1$$

$$\varphi_{Mi} = \varphi_{Mi-1} + 1/(4 \cdot \tau_M) + c_2 \cdot \sin(2 \cdot \pi \cdot \varphi_{Ei}) \cdot \sin(2 \cdot \pi \cdot \varphi_{Mi-1}) + \varepsilon_2,$$

where φ_{xi} is the phase of oscillator X at time i, and ε_1 and ε_2 are each time randomly drawn from a uniform distribution between -0.005 and $+0.005$. For each combination of a series of 50 values of c_1 and c_2 between -0.05 and $+0.05$ (representing different strengths of coupling in both directions, and restricted to the condition that $c_1 \cdot c_2 < 0$) the program computed the standard deviation of the last 50 cycle durations in a run of 100. Clearly, when $c_1 = 0$, φ_E is unaffected by φ_M, and hence there is unilateral coupling with the coupled system expressing the period (23 h) of the E oscillator. Similarly, when $c_2 = 0$ there is unilateral coupling with the M oscillator dominantly controlling the system ($\tau \sim 25$ h). Thus over the range of varying coupling strength the coupled system displays periods varying between ~ 23 and ~ 25 hours.

Using this simple model we have evaluated how precision of cycle length (1/s.d.t.) varies with τ as long as the two oscillators remain coupled, when τ varies due to varying strength of coupling. The result is shown in figure 8. The simulations clearly demonstrate that under these conditions two coupled oscillators attain maximal precision when they run at an intermediate frequency, *i.e.,* when the coupling forces are about equal in both directions.

In spite of the interesting result in Figure 8, we should be cautious in interpreting the intermediate maximum in precision around $\tau = 24$ h as evidence for the *E/M* model. Enright (1980b) has shown in a complex neuronal model that the noise in the coupled system decreases with increasing numbers of component oscillators. Intuitively, this is likely to be a general result. In our simulations, dominance of one of the two oscillators at the extreme τ-values in figure 8 simply reduces the effective number of oscillators from two to one, and hence is expected to reduce precision. Presumably, a similar result would be obtained with any number of coupled oscillators, if their intrinsic frequencies vary around $1/24$ h^{-1}, and if their mutual coupling varies in strength.

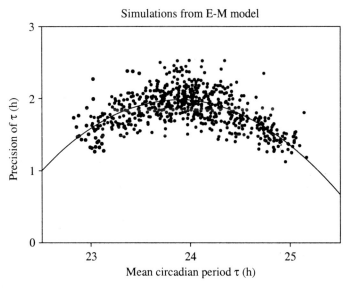

Fig. 8   **Precision of circadian rhythms in a simple two-oscillator model, generating variance in mean τ by varying mutual coupling strengths between the two component oscillators, each with equal and constant noise (see text).**

It is now well established that there is indeed a multitude of circadian oscillators active in most organisms. There are multiple oscillators within unicellulars (Roenneberg and Morse 1993), there are multiple oscillating neurons (mutually synchronized) within the SCN of mammals (Enright 1980a; Welsh et al. 1995; Honma et al. 1998), there are two SCN's with possibly redundant function left and right (De la Iglesia et al. 2000), and there are multiple circadian rhythm generators in different organs (e.g., Menaker and Tosini 1996; Yamazaki et al. 2000). At all these levels there is probably mutual interaction with varying strength of coupling and mutual feedback. At all these levels we would expect that selection has aimed for circadian periods not too distant from 24 hours. When the system displays τ further away from 24 h, this might be due to assymetry in the coupling forces and hence to a smaller number of effectively controlling oscillators, leading to less resonance, and larger noise. The inverse U shape of precision versus τ may thus simply reflect variability across the τ range in the internal resonance among oscillators. In these highly complex systems, the two-oscillator (E/M) model may be regarded as the simplest form of a multioscillatory system. It is of course of theoretical interest that already this minimal model displays the remarkable characteristic of maximal precision close to the intermediate frequency.

There is one specific aspect of the *E/M* model that makes it of particular interest in this respect beyond simply being the minimal multioscillator model. On functional grounds the endogenous frequencies of E and M are each expected to differ. In order to lock on to dawn, M is expected to be accelerated by light, and have a long τ_M in constant darkness (DD). In order to lock on to dusk, E is expected to be decelerated by light, and thus to have a short period (τ_E) in DD. Thus, in contrast to the multiple circadian oscillators elsewhere in the body, we theoretically expect that selection has set these two putative component oscillators in the SCN at frequencies away from 1/24 h^{-1}. Therefore, we expect in particular in the coupled E/M system that the inverted U shape of precision against τ is expressed. Whether it is the primary cause of the phenomenon can be investigated if the cry1 and cry2 knockout mice, characterized by short and long τ, respectively (Van der Horst et al. 1999), indeed represent SCN's where either the M or the E oscillator has been genetically suppressed as was recently postulated (Daan et al. 2000). The prediction is that in both the inverted U-shape should be gone.

In several recent publications we have argued that under natural conditions circadian systems reach highest accuracy of entrainment by adjusting τ itself in response to prior light conditions to 24 h exactly (Beersma et al. 1999a, b; Daan 2000). The data summarized in Figures 6 and 7 indeed provide evidence that circadian systems are built to function with maximal precision at a τ of 24 h. Deviations from this frequency are probably the consequence of disbalance between the constitutive oscillators—regardless of whether or not these are distributed in two or more functional subunits. This fact can be considered independent evidence that circadian systems are evolved to run at a frequency of 1/24 h, in contrast with Pittendrigh's view of strategic deviations of τ from 24 h.

Conclusion

In summary, we feel that there is neither a strong theoretical argument for natural selection to have set circadian periods for the coupled system away from 24 h in order to maintain maximal accuracy of phase, nor that maximal precision at intermediate τ (at 24 h) reflects the need for tighter homeostatic control when τ is close to 24 h. This maximal precision is a remarkable circadian phenomenon that may reflect maximal resonance between constituent oscillators in general. It may also reflect the symmetric influence of two major component oscillators tuned by natural selection to frequencies at both sides of 24 h in particular.

Acknowledgements

Part of the work reported here was originally conceived as a joint contribution with Professor Jürgen Aschoff on precision and accuracy. With gratitude we have made use of some of the analyses done by him before his death (Figures 5 and 6). We are grateful to Dr. V K Sharma for providing us with the original data on *Mus booduga* used in Figures 2 and 7.

References

Aschoff, J. (1960) Exogenous and endogenous components in circadian rhythms. Cold Spr. Harb. Symp. Quant. Biol. **25**: 11–28.

Aschoff, J., Gerecke, U., Kureck, A., Pohl, H., Rieger, P., Von Saint Paul, U., Wever, R. (1971) Interdependent parameters of circadian activity rhythm in birds and man. In: Menaker, M. (ed). Biochronometry. Nat. Acad. Sciences, Washington D.C., pp. 3–29.

Beersma, D.G.M., Daan, S., Hut, R.A. (1999a) Accuracy of circadian entrainment under fluctuating light conditions: contributions of phase and period responses. J. Biol. Rhythms **14**: 320–329.

Beersma, D.G.M., Spoelstra, K., Daan, S. (1999b) Accuracy of human circadian entrainment under natural light conditions: model simulations. J. Biol. Rhythms **14**: 524–531.

Bünning, E., Chandrashekaran, M.K. (1975) Pfeffer's views on rhythms. Chronobiologia **2**: 160–167.

Czeisler, C.A., Duffy, J.F., Shanahan, T.L., Brown, E.N., Mitchell, J.F., Dijk, D.J., Rimmer D.W., Ronda, J.M., Allan, J.S., Emens, J.S., Kronauer, R.E. (1995) Reassessment of the intrinsic period (τ) of the human circadian pacemaker in young and older subjects. Sleep Res. **24A**: 505.

Daan, S. (2000) Colin Pittendrigh, Jürgen Aschoff, and the natural entrainment of circadian systems. J. Biol. Rhythms **15**: 195–207.

Daan, S., Albrecht, U., Van der Horst, G.T.J., Illenrova, H., Roenneberg, T., Wehr, T.A., Schwartz, W.J., (2001) Assembling a clock for all seasons: Are there M and E Oscillators in the genes? *J. Biol. Rhythms* **16**: 105–116.

Daan, S., Aschoff, J. (1975) Circadian rhythms of locomotor activity in captive birds and mammals: their variations with season and latitude. Oecologia **18**: 269–316.

Daan, S., Berde, C. (1978) Two coupled oscillators: simulations of the circadian pacemaker in mammalian activity rhythms. J. Theor.Biol. **70**: 297–313.

De Candolle AP (1835) Physiologie végétale. Paris.

De la Iglesia, H.O., Meyer, J., Carpino, A., Schwartz, W.J. (2000) Antiphase oscillation of the left and right suprachiasmatic nuclei. Science **290**: 799–801.

Edgar, D.M., Martin, C.E., Dement, W.C. (1991) Activity feedback to the mammalian circadian pacemaker: Influence on observed measures of rhythm period length. J. Biol. Rhythms **6**:185–199.

Enright, J.T. (1980a) The timing of sleep and wakefulness. Springer, Berlin.

Enright, J.T. (1980b) Temporal precision in circadian systems: a reliable neuronal clock from unreliable components? Science **209**: 1542–1544.

Eskin, A. (1969) The sparrow clock: behavior of the free running rhythm and entrainment analysis. Ph.D. thesis, University of Texas at Austin.

Hardy, A.C. (1960) Was man more aquatic in the past? New Scientist **7**: 642–645.

Heynick, F. (1986) A geophysical note on man's free-running circadian rhythm. J. Interdiscipl. Cycle Res. **17**: 113–119 .

Honma, S., Shirakawa, T., Katsuno, Y., Namihira, M., Honma, K. (1998) Circadian periods of single suprachiasmatic neurons in rats. Neurosci. Lett. **250**: 157–160.

Hut, R.A., van Oort, B.E.H., Daan, S. (1999) Natural entrainment without dawn and dusk: the case of the European Ground squirrel *(Spermophilus citellus)* J. Biol. Rhythms **14**: 290–299.

Illnerova, H.. (1986) Circadian rhythms in the mammalian pineal gland. Academia, Praha.

Jagota, A., de la Iglesia, H.O., Schwartz, W.J. (2000) Morning and evening circadian oscillations in the suprachiasmatic nucleus *in vitro*. Nature Neuroscience **3**: 372–376.

Lambeck, K. (1978) The earth's palaeorotation. In: Brosche P, Suendermann, J. (eds.) Tidal friction and the earth's rotation. Heidelberg, Springer Verlag, pp. 145–153.

Menaker, M., Tosini, G. (1996) The evolution of vertebrate circadian systems. In: Honma, K., Honma, S. (eds.) Circadian organization and oscillatory coupling. Hokkaido Univ. Press, Sapporo, pp. 39–52.

Natalini, J.J. (1972) Relationship of the phase-response curve for light to the free-running period of the Kangaroo rat *Dipodomys merriami*. Physiol. Zool. **45**: 153–166.

Pfeffer, W. (1915) Beiträge zur Kenntnis der Entstehung der Schlafbewegungen. Ber. d. math.-phys. Kl. d. Koenigl. Sächs. Gesellsch. d. Wissensch. 34(I-VI): 1-154.

Pittendrigh CS (1960) Circadian rhythms and the circadian organization of living systems. Cold Spr. Harb. Symp. Quant. Biol. **25**: 159-184.

Pittendrigh, C.S. (1981) Circadian Systems: Entrainment. In: Aschoff, J. (ed.) Handbook of Behavioral Neurobiology. New York, Plenum Press. **4**: 95–124.

Pittendrigh, C.S., Daan, S. (1976a) A functional analysis of circadian pacemakers in nocturnal rodents, I. The stability and lability of spontaneous frequency. J. Comp. Physiol. **106**: 223–252.

Pittendrigh, C.S., Daan, S. (1976b) A functional analysis of circadian pacemakers in nocturnal rodents IV, Entrainment: pacemaker as clock. J. Comp. Physiol. **106**: 291–331.

Pittendrigh, C.S., Daan, S. (1976c) A functional analysis of circadian pacemakers in nocturnal rodents, V. Pacemaker structure: a clock for all seasons. J. Comp. Physiol. **106**: 333–355.

Richter, C.P. (1968) Inherent twenty-four hour and lunar clocks of a primate—the squirrel monkey. Commun. Behav. Biol. A 1: 305–332.

Roenneberg, T., Morse, D. (1993) Two circadian oscillators in one cell. Nature **362**: 362–364.

Saunders, D. (1978) An experimental and theoretical analysis of photoperiodic induction in the flesh-fly, *Sarcophaga argyrostoma*. J. Comp. Physiol. **124**: 75–95.

Sharma, V.K., Chandrashekaran M.K. (1999) Precision of a mammalian circadian clock. Naturwissenschaften **86**: 333–335.

Van der Horst, G.T., Muijtjens, M., Kobayashi, K., Takano, R., Kanno, S., Takao, M., de Wit, J., Verkerk, A., Eker, A.P.M., van Leenen, D., Buijs, R., Bootsma, D., Hoeijmakers, J., Yasui A (1999) Mammalian Cry1 and Cry2 are essential for maintenance of circadian rhythms. Nature **398**: 627–630.

Wehr, T.A. (1996) A 'clock for all seasons' in the human brain. In: Buijs, R.M., Kalsbeek, A., Romijn, J.H., Pennartz, C.M.A., Mirmiran, M. (eds.) Progress In Brain Research 111, Elsevier, Amsterdam, pp. 319–340.

Welsh, D.K., Logothetis, D.E., Meister, M., Reppert, S.M. (1995) Individual neurons dissociated from rat suprachiasmatic nucleus express independently phased circadian firing rhythms. Neuron **14**: 697–706.

Wever, R. (1967) Zum Einflus der Daemmerung auf die circadiane Periodik. Z. vergl. Physiol. **55**: 255–277.

Wever, R. (1975) The circadian multi-oscillator system of man. Internat. J. Chronobiol. **3**: 19–55.

Yamazaki, S., Numano, R., Abe, M., Hida, A., Takahashi, R., Ueda, M., Block, G.D., Sakaki, Y., Menaker, M., Tei, H. (2000) Resetting central and peripheral circadian oscillators in transgenic rats. Science **288**: 682–685.

4. Period Doubling of Rhythmic Water Regulation in Plants

A. Johnsson* and G. Prytz

Department of Physics, Norwegian University of Science and Technology,
Trondheim, Norway NTNU, N-7491

Under certain conditions rhythmic water regulation in plants can be found, and both circadian and ultradian components can be recorded. Short term oscillations in the primary leaf of the oat shoot are discussed in this contribution. The period time of the oscillations is about 40 min. The mechanisms underlying the rhythms are discussed with an electric model of the water flow in the system as the starting point. In experiments, period doubling and other phenomena have been found to exist in the rhythmic transpiration and are reported here. Model simulations can also show period doubling in the water regulation. The system is non-linear and the period doubling can indicate that it is chaotic in character, at least under some conditions. It is emphasized that in order to treat properly several features of the oscillatory water regulation of plants, refinements of existing models are needed.

Introduction

In the present contribution we will study rhythmic water transpiration from primary plant leaves. The rhythms have a period of about 40 minutes and are thus conventionally termed as *ultradian*. The transpiration rhythm is the measurable indication of synchronized opening and closing of leaf *stomata*, i.e. the openings on the leaf surface that allow water vapor transport to the atmosphere and uptake of carbon dioxide.

An example of such a rhythm is seen in Fig. 1a, where the transpiration from an oat leaf (*Avena sativa* L.) is recorded as a function of time. The rhythmic water loss from the leaf surface is accompanied by simultaneous oscillations in water uptake by the plant and under some conditions oscillations in the uptake of carbon dioxide can likewise be recorded. Experimental studies of this oscillating system have mainly been published by Johnsson and collaborators (see references). Theoretical treatments of oscillations in this and other species have been given by Cowan (1972); Johnsson (1976) and Gumowski (1981, 1983) and others.

Under some circumstances the sinusoidal transpiration curve can change into a curve form in which every second peak is higher with respect to its neighbors (every second period will also be different from the neighboring periods). An example of such a recording is seen in Fig. 1(b). It is seen that the transpiration curve now shows a period component that is doubled — the rhythm shows a *period doubling*. Other patterns can also occur (Fig. 1c, 1d) and will be discussed later.

We would like to discuss the phenomenon of period doubling in this rhythmic system, describe some new and unpublished results and relate them to theoretical models and predictions.

---

*E-mail: Anders.Johnsson@phys.ntnu.no

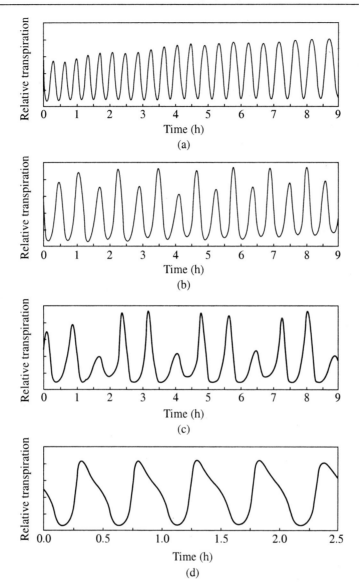

Fig. 1 (a) Transpiration oscillations in an intact primary leaf of *Avena*. The curve form is slightly asymmetric but sinusoidal, and the period of the rhythm is typically 40 min. (b) Recording of period doubling in the transpiration oscillations of a primary leaf of an intact *Avena* shoot. (c) Recording in which every third peak has a lower amplitude. (d) Example of recording in which the curve form has a more pronounced asymmetry.

Models of the water regulatory system

Figure 2 can conveniently describe the structure of the water regulatory system in plants. The model description originates from Cowan (1972). The water flux in the plant system is represented by electric currents, i, and water flow resistances by electric resistances, R. Cells capable of taking up water are depicted as capacitors, denoted by C in the figure.

The model has three explicit types of cells, *mesophyll* cells in the leaves, index m, and *subsidiary* and *guard* cells of the stomata, indices s and g, respectively. The water fluxes into these cells are then denoted i_m, i_g and i_s respectively, and they pass through the resistances R_m, R_s and R_g. Water moves according to differences in the water potential ψ, i.e. differences in the electrochemical potential of water (see e.g. Nobel 1974). The volume of the cells will correspond to the electric charges on the capacitors.

In the model the water flux resistances R_1 and R_2 denote the root and xylem resistances, while ψ_0 denotes the root medium water potential.

The stomata control the water transpiration i_a indicated by the current generator in Fig. 2. An increase of the volume of the subsidiary cells (W_s) in *Avena decreases* the stomatal opening while an increase of the guard cell volume (W_g) *increases* it.

The mesophyll cells (with their volume W_m corresponding to the charge on the capacitor C_m) of the leaves participate in the water regulation by forming a tissue in which water can be stored. However, they do not directly regulate the stomatal opening as the other two cell types. The water transpiration stream through the stomatal opening, i_a, is controlled by the subsidiary and guard cells. The functional dependence of i_a on the W_s and W_g can be written as $i_a = f(W_s(t), W_g(t))$, where the function f is an unknown function of time, t.

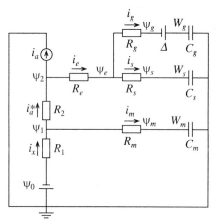

Fig. 2   Model to describe and simulate oscillatory water regulation. The rhythmic water fluxes can be modeled in electric terms as depicted (from Cowan 1972). In the electric analogue, ψ represents water potentials, R the water flux resistances, i water fluxes, C the ability of a cell to "store" water and Δ the light induced osmotic changes in the guard cells. Further explanations are given in the text.

Cowan (1972) proposed the function f to have a specific non-linear form. The resulting transpiration control system will, under certain conditions, allow oscillations to be generated. This would mean that the transpiration stream, i_a, from the plant, as well as the water fluxes and the water charges, can display self-sustained oscillations. Johnsson (1976) suggested a functional relationship that took into account certain results from stability studies of the transpiration dynamics (Johnsson et al. 1979). Another version of this model, however in hydraulic terms, has been published by Delwiche and Cooke (1977). The similarities with the Cowan model are pronounced (cf. Prytz et al. to be published).

The figure (and the model discussed above) contains, although simplified, the main elements of the versions that have been presented and used for numerical simulations. In many of the simulations, the parameter data, the curve forms etc. have been tested against experimental results for oscillatory water transpiration in the primary leaf of the *Avena* shoot (see Johnsson and coworkers 1973–1979; Brogårdh and Johnsson 1974, 1975; Klockare et al. 1978).

In the next section we focus on the mathematical aspects of the model. It contains *three* dynamic elements, viz. the water transport of the subsidiary cells, the guard cells and the mesophyll cells, respectively.

A mathematical description

The physical picture in Fig. 2 corresponds to a three state variable model (as discussed e.g. by Gumowski (1983)). With slightly modified notation the model can thus be described as follows:

$$\frac{dW_g}{dt} = \alpha_1 f(W_g, W_s) + \alpha_{11}W_g + \alpha_{12}W_s + \alpha_{13}W_m$$

$$\frac{dW_s}{dt} = \alpha_2 f(W_g, W_s) + \alpha_{21}W_g + \alpha_{22}W_s + \alpha_{23}W_m \tag{1}$$

$$\frac{dW_m}{dt} = \alpha_3 f(W_g, W_s) + \alpha_{31}W_g + \alpha_{32}W_s + \alpha_{33}W_m$$

Here the variables and abbreviations are as follows: W's represent the cell volumes of the guard, subsidiary and mesophyll cells. The α_i's are constants (analogue to flux resistances) and the α_{ij}'s are also constants (with dimension time^{-1}).

The non-linear function f describing the transpiration stream is of great importance as it determines the behavior, e.g. stability, of the overall model. In general, it can be described as

$$f = b_g \int h_g(u)W_g(u)du + b_s \int h_s(u)W_s(u)du \tag{2}$$

The b's denote constants and the *weight functions* h_g and h_s denote the accumulated and time weighted influences of the cell volumes W_g and W_s. Integration is performed over time.

Assuming simple weight functions with no time delays in the regulation simulations have been performed with the following equation (see Gumowski 1983):

$$\frac{d^3x}{dt^3} + \alpha\frac{d^2x}{dt^2} + \frac{dx}{dt} + \alpha x + \mu(\mu\alpha x - \beta x^2) = 0 \tag{3}$$

In this third order differential equation α, β and μ are constants.

With certain parameter values, simulations gave oscillations as shown in Fig. 3. Besides the basic sinusoidal rhythms, *subharmonic frequencies*, or "period doubling" were also generated, (Fig. 3a). Different curve forms could be seen in the time series (one example is given in Fig. 3b). Such features are typical for non-linear systems. With three system variables as in equation (1), one can anticipate that the present model of the water regulatory system will be able to describe chaotic behavior. The period doubling could be an indicator of a route to chaos (not simulated).

Atypical waveforms

The transpiration of primary *Avena* leaves was recorded following methods described in the literature (e.g. Johnsson 1973). Dry air was used as a carrier gas and sent into a cuvette where the leaf transpiration increased the water content of the air in a rhythmic way. The relative humidity of the outgoing air was monitored. Light levels (approx. 100 μmol/m^{2}s) and temperature (approx. 24°C) were kept constant. In some experiments a physical pressure was applied across the leaf base, thereby increasing the resistance to the xylem water flux. By doing so we succeeded in achieving period doubling in the water regulatory system of *Avena* plants. A few relevant examples are shown in Figs. 1 and 4.

Period doubling. Under certain conditions every second peak in the rhythmic transpiration has a

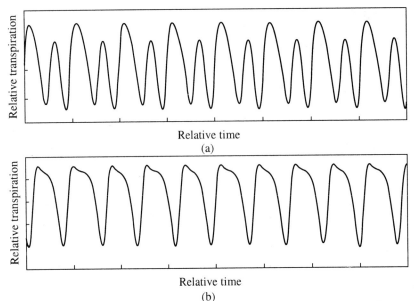

Fig. 3   **Model simulations. Simulations performed on the model discussed, equation (3), can show period doubling (a) and curve forms that deviate from sinusoidal ones (b). Parameters in equation (3): (a) $\alpha = 1.2$, $\beta = 0.6$, $\mu = 0.8$, (b) $\alpha = 1.2$, $\beta = 1.0$, $\mu = 1.03$.**

different amplitude than those in between, as exemplified in Fig. 1b. This pattern can arise spontaneously in some plants or can be induced. The pattern may continue for a number of periods.

More complicated waveforms. In a few rare cases even more complicated waveforms were observed. An example is shown in Fig. 1(c), where every third peak has higher amplitude than the others.

Inducing period doubling. Period doubling could be initiated in experiments on the transient response of the water control system to temperature changes and to a physical increase of the xylem resistance (R_1 in Figure 2). Figure 4a demonstrates period doubling after step changes of the xylem resistance (physical compression of the xylem).

A recording of period doubling after a temporary temperature change in the root medium (from 25 to 9°C). is shown in Fig. 4(b). As can be seen, a peak which should normally be lower than the neighboring peaks has a higher amplitude after the perturbation.

Figure 4(c) shows that a light pulse can reverse the phase of the asymmetric rhythm by 180 degrees.

The concentration of *calcium* ions in the growth medium seems to be important for the induction of period doubling (Prytz et al, to be published). Plants grown in approx. 0.3 mM of Ca^{2+} solution showed period doubling to a higher extent than did plants grown at lower or higher calcium concentration.

Atypical waveforms after exposure to different ions etc. In some cases, very asymmetric waveforms were achieved. Interestingly, Gumowski (1983) stresses the importance of the waveform in the analysis of his model, and also presents some oscillatory curves that have great similarities to those found experimentally.

Some experimental results on changes in the period length or the curve form can be mentioned: the prolonging effects of lithium ions (Brogårdh and Johnsson 1974), the influence of ATP, kinetin and theophylline on the curve form (see examples in Johnsson 1976).

Johnsson and Skaar (1979) reported an *alternating light pulse* response in the primary *Avena* leaves. They studied leaves without a root system but dipped into the medium with the cut end and thus with free water access to the xylem system. The leaves were exposed to repetitive light pulses. Successive responses to the pulses were different while every second response had the same shape. Several similarities to the period doubling can be mentioned, e.g. the period length, phase shifting by light etc. The alternating pulse response was affected by increased concentration of ions, particularly sodium ions (see Johnsson and Skaar 1979). This alternating light pulse response seems to reveal features of the mechanism behind the period doubling of the control system.

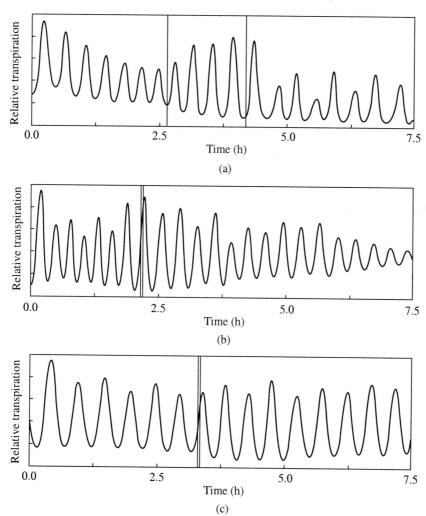

Fig. 4   Recordings of transpiration from primary leaves of *Avena* shoots. (a) demonstrates that compression of the xylem affects the transpiration pattern. The first compression (vertical line) affects the amplitude of the transpiration while the second one induces period doubling in the oscillations. (b) a temporary temperature lowering of the root medium (vertical lines) phase shifts the rhythm with 180 degrees. (c) A dark pulse can likewise phase shift the rhythm.

Discussion of the model and period doubling

The present example of an ultradian plant system adds to the number of oscillatory systems that can show period doubling under certain experimental conditions. Another example is the Belousov-Zhabotinsky reaction (see e.g. Strogatz 1994). The period doubling of the rhythmic transpiration of the *Avena* shoot has previously been described (Johnsson and Skaar 1979) but the present report is the first to discuss period doubling as found in the *selfsustained* oscillations.

The transition to period doubling may be a step on the way to chaotic behavior, but further experiments must be performed to answer this question. Changing the light intensity changes the period of oscillations (Klockare et al. 1978) and the period abruptly goes through a transition to new values. However, the light intensity cannot be varied over a very large range without the oscillations being damped out. It is thus an unsuitable parameter for revealing a possible period doubling route to chaos.

The period doubling can to a certain extent be "explained" by the physical model used in the present contribution. However, we would like to stress that several "cavities" are present in the model and that several experimental results are not in accordance with it. Some points to this discussion can be (cf. Johnsson 1976, who lists "Pro et Contra" with respect to the Cowan model):

—very long periods can be experimentally achieved (e.g. in kinetin treatment of plants in darkness). This seems incompatible with the changes of the physical parameters of the model at its present stage.
—the model is basically only a hydraulic one, i.e. it does not take explicit cellular processes into account.
—the model does not consider the CO_2 regulation of the plant although this influences the stomatal regulation.

The existence of period doubling in ultradian rhythms seems to be a fairly common feature. As is well known it also exists in many circadian rhythms. It might, therefore, be tempting to ascribe the phenomenon to the general behavior of non-linear, oscillatory biological systems. However, studies of the detailed physiological mechanisms during period doubling and analysis of the system still remain to be carried out.

Acknowledgements

The authors would like to thank Tor Egil Rødahl, M. Sc., at the present Department for his participation in the experimental work. Associate Prof. C. Futsæther, Norwegian Agricultural University, Ås, is thanked for project collaboration and reading of the manuscript. Financial support from the Norwegian Research Council is gratefully acknowledged.

References

Brogårdh, T., Johnsson, A. (1974) Effect of lithium on stomatal regulation. Zeitschr. *f.* Naturforsch. **29c**: 298–300.
Brogårdh, T., Johnsson, A. (1975) Effects of magnesium, calcium and lanthanum ions on stomatal oscillations in *Avena* sativa, L. Planta **124**: 99–103.
Cowan, I.R. (1972) Oscillations in stomatal conductance and plant functioning associated with stomatal conductance: Observations and a model. Planta (Berl.) **106**: 185–219.
Delwiche, M.J., Cooke, J.R. (1977) An analytical model of the hydraulic aspects of stomatal dynamics. J. Theor. Biol. **69**: 113–141.
Gumowski, L. (1981) Analysis of oscillatory plant transpiration. J. Interdiscipl. Cycle Res. **12**: 273–291.
Gumowski, I. (1983) Analysis of oscillatory plant transpiration. II. J. Interdiscipl. Cycle Res. **14**: 33–41.

Johnsson, A. (1973) Oscillatory transpiration and water uptake of *Avena* plants. I. Preliminary observations. Physiol. Plantarum **28**: 40–50.

Johnsson, A. (1976) Oscillatory water regulation in plants.—Bull. Inst. Mathematics and its Applications. **12**: 22–26.

Johnsson, A., Brogårdh, T., Holje Ø. (1979) Oscillatory transpiration of *Avena* plants: perturbation experiments provide evidence for a stable point of singularity. Physiol. Plantarum **45**: 393–398.

Johnsson, M., Brogårdh, T., Johnsson, A., Engelmann, W. (1975) Period lengthening effect of theophylline on oscillatory water regulation in *Avena*. Zeitschr. *f.* Pflanzenphysiol. **76**: 238–242.

Johnsson, A., Skaar, H. (1979) Alternating perturbation response of the water regulatory system in *Avena* leaves. Physiol. Plantarum **46**: 218–220.

Klockare, R., Johnsson, A., Brogårdh, T., Hellgren, M. (1978) Oscillatory transpiration and water uptake of *Avena* plants. VI. Influence of the irradiation level. Physiol. Plantarum **42**: 379–386.

Nobel, P.S. (1974) Introduction to biophysical plant physiology. Freeman and Company. San Francisco. ISBN 0–7167–0593–3.

Strogatz, S.H. (1994) Nonlinear dynamics and chaos. Addison-Wesley Publishing Company. ISBN 0–201–54344–3.

Biological Rhythms
Edited by V. Kumar

5. The Genetics and Molecular Biology of Circadian Clocks

M. Zordan*, R. Costa*, C. Fukuhara[#] and G. Tosini[#]*

*Department of Biology, University of Padova, Italy

[#]Neuroscience Institute, Morehouse School of Medicine, 720 Westview dr., Atlanta, 30310, Georgia, USA

For many years it was believed that daily rhythms in behaviour and physiology were imposed by rhythms in the physical environment to which organisms responded passively. The demonstration that plants and animals do not respond passively to environmental stimuli, but they possess accurate endogenous time-measuring systems (i.e., circadian clocks) engendered renewed interest in the study of biological rhythms. During the last 30 years, thousands of studies have demonstrated that most organisms have an (or more than one) internal circadian clock(s) which oscillates with a period close to the astronomical day (i.e., 24 hours). Currently we are aware of many of the physiological mechanisms underlying the control of circadian rhythmicity, while the improvement of molecular biology techniques has fostered dramatic advancements in our knowledge of the molecular workings of circadian clocks in many different organisms, man included. The present review is an attempt to summarize our current understanding of the genetic and molecular biology of circadian clocks.

Introduction

In nature, with few exceptions, organisms dwell in environments, which change cyclically over the course of the 24 hours. Under the circumstances, it is reasonable to assume that the ability to anticipate such changes, through the coordinated regulation of physiology and behaviour, should afford a selective advantage to the organism. The capability to anticipate daily cycles is provided by endogenous biological clocks with an intrinsic circadian (circa = approximately; dian = day) period of approximately 24 hours. Currently much work is being devoted to understanding the mechanisms underlying the biological oscillators in organisms ranging from cyanobacteria to humans. To date, much experimental evidence suggests multiple independent origins for the intracellular clocks in living organisms (Dunlap 1999). Clock research in the last two decades has witnessed a steady increase in the publication of reports, and has seen a remarkable advancement in the last few years. The following, by no means exhaustive account, will attempt to summarize the concepts and information, available at the moment, in this rapidly evolving field, for key model organisms including cyanobacteria, fungi, insects and mammals.

The circadian clock in bacteria

Circadian clocks have so far been extensively described only in Eukaryotes. While they seem to be

---

*E-mail: tosini@msm.edu

ubiquitous in Plantae, Animalia and Fungi, none has been detected in Archaebacteria, and among Eubacteria, only seem to exist in some cyanobacteria. In cyanobacteria, clock mutants were produced by EMS (Ethyl Methane Sulphonate) mutagenesis and subsequently identified in *Synechococcus sp.* (Kondo et al. 1994). More than 100 mutants were obtained, from which a cluster of three clock genes, namely *kaiA*, *kaiB* and *kaiC* (kai is a Japanese word meaning "cycle") were identified, cloned and characterized (Ishiura et al. 1998). Most of the mutants analyzed so far (more than 50) showed complementation with the *kai* gene cluster and only one additional gene, *pex* (which does not seem to have an important function *per se* in the clock but acts as a modifier of the circadian clock) has been discovered in this way (Kutsuna et al. 1998). The *kai* genes appeared as completely "new" clock genes, related to those of eukaryotes only by virtue of having similar regulatory mechanisms. This suggests an independent origin of this clock system from those of eukaryotes, among which a further polyphyletic origin can be hypothesized for the clock of fungi (i.e. *Neurospora*) in relation to those of insects and mammals (Dunlap 1999). The circadian clock of *Synechococcus* controls rhythms which have a periodicity of about 24h, can be reset by light and temperature pulses and show temperature compensation (Golden et al. 1997). The *kaiA* gene is under the control of its own promoter and encodes for an activator, which positively regulates the transcription, through interaction with a different single promoter, of both the *kaiB* and *kaiC* genes. Transcription of the three genes occurs in near synchrony, with a more pronounced peak in the case of *kaiB* and *kaiC* at about ZT12. KaiC protein seems to play the role of a negative element in the cyanobacteria pacemaker, via inhibition of its own transcription, thus generating a negative feedback loop, which is an essential component of the circadian oscillator. Furthermore, in heterologous experimental systems, KaiC appears to enhance KaiA-KaiB interactions, suggesting that the three Kai proteins may interact to form a heteromultimeric complex (Iwasaki et al. 1999), although this complex has yet to be assigned a function. No known structural motifs have yet been identified in the KaiA and KaiB putative clock proteins, while only inferred ATP/GTP binding domains have been detected in KaiC. Figure 1 shows a model illustrating the transcription/translation feedback loop involved in the *Synechococcus* circadian clock.

The expression of almost all the genome of *Synechococcus* is under control of the circadian clock (Liu et al. 1995). Consequently a significant amount of genes are expressed in phase or antiphase with respect to the *kai* genes, suggesting the existence of a "general mechanism" for the cis-regulation of specific genes (Golden et al. 1997). While one might gain the impression that most bacteria lack a circadian clock, the information available so far on *Synechococcus sp.* proves that a circadian pacemaker can operate also in a prokaryotic cell. The presence of a circadian clock in cyanobacteria suggests that it serves an important biological function, with possible adaptative significance. Results of competition experiments with *Synechococcus* mutant strains, support this hypothesis, indicating that an improved Darwinian fitness can be associated to those cells with a free running period which "resonates" with the environmental cycles of light and temperature (Ouyang et al. 1998).

From an evolutionary perspective, it may be of interest to point out that while *kai* homologues have not been found in Eubacteria genome databases yet, putative homologues have been identified in Archaebacteria databases (Kondo and Ishiura 1999). Moreover, the hypothesized evolutionary relationships between cyanobacteria and chloroplasts suggest the possibility that the circadian clock, which was first "invented" within these prokaryotes, was subsequently transferred to plants via endosymbiosis. If this hypothesis were to prove correct, then at least some molecular components of the core of the circadian pacemaker of plants (which is still mostly undescribed) would be expected to have maintained some similarity to that of cyanobacteria (Kondo and Ishiura 1999).

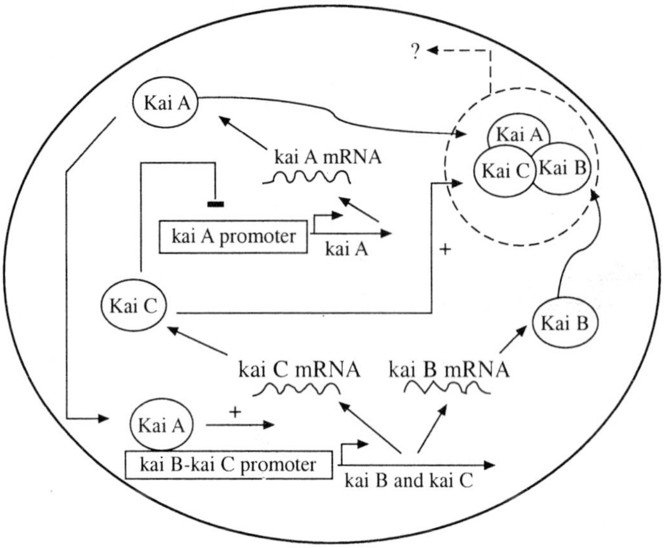

Fig. 1   The circadian clock in bacteria. The *kai A*, *B*, *C* cluster constitutes the core of the circadian oscillator in *Synechococcus*. For ease of representation, the *kai A* gene has been drawn separately from the *kai B* and *C* genes. Kai A protein functions as a positive effector of the transcription of *kai B* and *C*, while Kai C is known to negatively regulate transcription of *kai A*. Kai C protein promotes the formation of a ternary complex involving the three *proteins in vitro*. However, whether this occurs *in vivo* or what function the complex might have is unknown. In the figure "plus" signs indicate positive effectors and lines terminating in perpendicular dashes indicate negative effectors.

The circadian clock in fungi

The circadian clock of fungi, together with that of the insects, was one of the first for which the analysis at the molecular level identified *bona fide* components of the circadian machinery. Almost all of the knowledge available regarding this issue in fungi concerns the species *Neurospora crassa,* which has become one of the most important model organisms for this kind of studies. The isolation of the first clock mutants, in which the circadian rhythm of conidiation was altered (Feldman and Hoyle 1973), led to the cloning of the clock gene *frequency* (*frq*) (McClung et al. 1989). *frq* is characterized by the presence of two (in frame) start codons which allow translation of two variants of the FREQUENCY (FRQ) protein, which differ in length (FRQ$_s$). The two forms are translated in a temperature-dependent quantitative ratio (Garceau et al. 1997; Liu et al. 1997) and are both required for robust circadian conidiation rhythms. *frq* mRNA and FRQ$_s$ oscillate in abundance with a 4h lag between the two peaks (Garceau et al. 1997). FRQ$_s$ act as negative elements depressing *frq* transcription via interaction with a heterodimer, the White Collar Complex (WCC). WCC consists of WHITE COLLAR-1 (WC 1) and WHITE COLLAR-2 (WC 2), which are PAS-containing positive transcription factors (Aronson et al. 1994). PAS multifunctional domains are signaling modules (see Taylor and Zhulin 1999, for a comprehensive review on this issue) which also mediate homo- and heterodimer formation and have been found in other clock proteins (CLOCK and BMAL1) acting as positive elements in the molecular pacemaker of insects and mammals (Dunlap 1999). In *Neurospora* low levels of *frq* mRNA and FRQ$_s$ are present during the night; *frq* mRNA peaks at about mid morning, while total FRQ$_s$ (nuclear plus cytoplasmatic), with about a 4 h lag, peaks in the early afternoon. FRQ$_s$ enter the nucleus where they interact with the WCC (Luo et al. 1998). While inhibition of *frq* mRNA transcription and the successive *frq* mRNA disappearance are probably relatively fast processes (requiring about 6 h), FRQ$_s$ turnover seems a time demanding step (about 14h), which requires progressive phosphorylation leading to the degradation and consequent decline of FRQ$_s$ during the early night (Merrow et al. 1997). The FRQ$_s$-WCC interactions generate a negative autoregulatory feedback loop with an approximately 24h period.

WC1 and WC2 are also involved in blue light photoresponses in *N. crassa* (Ballario and Macino 1997; Linden and Macino 1997; Talora et al. 1999). WCC regulates the expression of light induced genes and is necessary for light resetting of the circadian clock. In particular, blue-light exposure produces an increase in phosphorylation of WC1, which in turn causes transcription of *frq* mRNA and other clock controlled genes (ccg$_s$). Hyperphosphorylated WC1 is then degraded and "substituted" in the WCC by newly translated WC1, which is however, no longer able to sustain light-induced transcription of *frq* (Talora et al. 1999). The *frq*-FRQ$_s$-WCC system constitutes a light-entrainable feedback loop, in which FRQ$_s$ are essential for response to light. This feedback loop has long been considered the core of the circadian clock in *Neurospora crassa*, but recent results obtained by Merrow et al. (1999), suggest the existence of another independent oscillator which can interact with the one characterized by the "*frq*-FRQ$_s$-WCC" feedback loop. Figure 2 illustrates a model for the

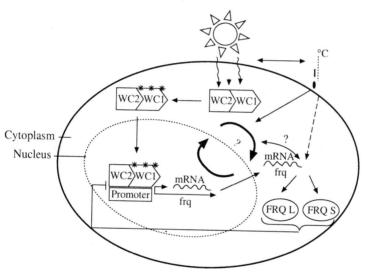

Fig. 2   **The circadian clock in *Neurospora crassa*. In *Neurospora*, the core of the light-entrainable clock elements consists in the products of the *wc1*, *wc2* and *frq* genes. The WC1 and WC2 proteins dimerize and, following exposure to blue light, WC1 is phosphorylated (symbolized by the small black spheres), thus the dimer enters the nucleus and acts as a positive effector of the transcription of *frq*. In the cytoplasm *frq* mRNA can be translated from alternative start codons giving rise to 2 length variants (FRQ S and FRQ L). FRQ protein acts as an inhibitor of its own transcription. Recently, it has been shown that temperature cycles can entrain the *Neurospora* conidiation rhythms in the absence of functional FRQ protein, suggesting the existence of an as yet unidentified "*frq*-less" oscillator (symbolized by the thick-lined cycling arrows enclosing a question mark). The possibility of feedback between the classical oscillator and the "*frq*-less" oscillator has been depicted by a double arrow connecting the two with a question mark. In the figure "plus" signs indicate positive effectors and lines terminating in perpendicular dashes indicate negative effectors.**

molecular clock operating in this organism. The new finding comes from experiments in which entrainment was imposed by using temperature as the zeitgeber instead of light. Intriguingly, *frq* mutants and in particular a null mutant (*frq*[9]), turned out to be entrainable by temperature cycles, as is expected to occur in *frq*[+]. On the basis of these observations, it has been suggested that *N. crassa* possesses an independent additional oscillator which can work even in the absence of FRQ$_s$ (FRQ$_s$-less oscillator) and mediate entrainment (Merrow et al. 1999; Mc Watters et al. 1999). Moreover, Merrow et al. (1999), demonstrated the existence of feedback between the two oscillators. It seems worthwhile to recall that many years ago, Pittendrigh (1960) also hypothesized the existence of two independent but coupled pacemakers, to explain the effects of light and temperature on the pupal eclosion rhythm of *Drosophila*. The recent sophisticated analysis of *N. crassa* circadian rhythmicity, which allowed the discovery of an additional *frq*-less oscillator in this organism, underlines the importance of studies in which the integrated entraining action of light and temperature are considered.

The circadian clock in insects

Among insects, the fruit fly *Drosophila melanogaster* has so far provided the most detailed description of the workings of the central oscillator. In fact, mutations in a clock gene (the *period* gene) were first identified in this organism (Konopka and Benzer 1971). In particular, the null mutant *per*[01] abolished free-running circadian periodicity in (i) the eclosion of imagoes from the pupal case and (ii) the rhythmicity of daily locomotor activity patterns. Furthermore, with reference to the wild type 24h period, missense mutations shortened (*per*[s]) to 19h or lengthened (*per*[L]) to 29h the free-running periods of both rhythms (Konopka and Benzer 1971). A second clock gene, *timeless* (*tim*), is also required for circadian rhythmicity in *Drosophila* (Sehgal et al. 1994). In the *tim*[01] mutant, loss of behavioural circadian rhythms is accompanied by a loss of daily fluctuations of *per* mRNA, as well as of abundance and phosphorylation of PER protein. These effects are accompanied by a failure of PER protein to accumulate in the nucleus. In particular in *Drosophila*, rhythmic expression of the clock genes *per* and *tim*, accompanied by the temporal nuclear localization of the respective protein products, is detectable in cells throughout the body. Whereas these cells could be pacemakers for circadian rhythms of as yet unspecified physiological processes, groups of cells (the lateral neurons) located in the central/optic lobe area of the brain, probably contain the relevant pacemakers, thought to be responsible for circadian behaviour (Kaneko 1998). Furthermore, detailed experimental work showed that a PER-reporter fusion protein cycles in the Malpighian tubules of the fruit fly, showing accumulation in the cytoplasm, followed by translocation into the nucleus. In the same study it was also shown, in decapitated flies (assayed for 3 days after decapitation), that this rhythm can occur independently of the brain (Hege et al. 1997).

At the cellular level, *per* and *tim* mRNA oscillations are interdependent, and are related to the formation of a PER-TIM complex in the cytoplasm, which stabilizes the monomers and allows nuclear entry of both proteins. In this way PER and TIM act as negative regulators of their own transcription, by participating in a negative feedback autoregulatory loop. The lag of 4–6h between mRNA and protein peaks, together with posttranscriptional regulatory mechanisms, such as protein degradation and phosphorylation, introduce temporal delays in the molecular oscillations necessary to generate a 24 h clock (Price et al. 1998). Figure 3 represents a simplified model of the circadian clock in *Drosophila*. In particular, the role of phosphorylation in the destabilization of PER has been clearly established by the recent discovery and characterization of a further clock gene called *doubletime* (*dbt*). *dbt* encodes for a specific kinase, closely related to human casein kinase type 1 (Kloss et al. 1998; Price et al. 1998). Various alleles of *dbt* produce phenotypes similar to *per* and *tim* mutants.

Unlike the former two genes, however, the *dbt* mRNA is not rhythmic (though DBT is essential for rhythmicity) and null mutations are lethal. In addition, gene dosage studies indicate that *dbt* function negatively correlates with periodicity (Young 1998).

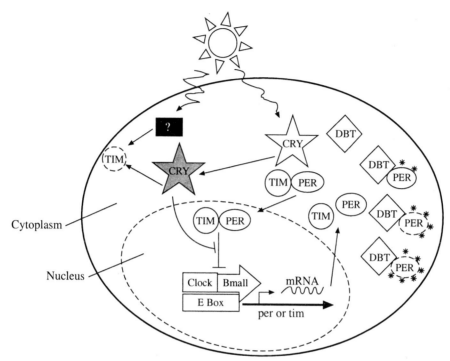

Fig. 3 **The circadian clock in *Drosophila melanogaster*. In this organism transcription of the clock genes *per* and *tim*, is positively regulated by the heterodimer formed by the protein products of the *clock* and *bmal1* (*dcycle*) genes. PER and TIM form heterodimers which can then translocate into the nucleus, thereby negatively regulating their own transcription by interfering with the positively acting Clock-Bmal1 dimers. Moreover, the inhibiting action of the PER-TIM dimers is contrasted by the light-activated product (gray star) of the *cry* gene. The action of a constitutive homologue of human casein kinase type 1, encoded by the *dbt* gene, delays the formation of TIM-PER dimers by subtracting PER monomers, which are degraded via the ubiquitin pathway, following targeting via DBT-dependent phosphorylation (depicted from top to bottom along the right hand side of the cell). Clock resetting by light is effected by light-dependent degradation of TIM protein, although the precise mechanism by which light leads to this degradation has still to be described. In the figure "plus" signs indicate positive effectors and lines terminating in perpendicular dashes indicate negative effectors.**

The identification of a short enhancing sequence (E-box, CACGTG) that was shown to be responsible for the robust rhythmic transcription of *per* (Hao et al. 1997) led to the search for the corresponding basic-helix-loop-helix (bHLH) transcription factor(s), known to bind to E-boxes, usually in the form of heterodimers with a partner protein. Two recently identified clock mutants in *Drosophila*, *cycle* (Rutila et al. 1998) and *dclock* (also known as *jerk*) (Allada et al. 1998), identified the sought after genes. The CYCLE and dCLOCK proteins bind as heterodimers to the E-box of the *Drosophila per* and *tim* promoters (Darlington et al. 1998), consequently activating transcription. *In vitro* experiments using promoters of *Drosophila per* and *tim*, showed that binding of the CYCLE/dCLOCK heterodimer is necessary as the activating element for *per* and *tim* mRNA rhythmicity (Darlington et al. 1998; Gekakis et al. 1998), while interactions of PER/TIM heterodimers with CYCLE/dCLOCK are inhibitory.

It is worth noting that both the inhibitory and the activating heterodimers are expressed rhythmically; CYCLE/dCLOCK is rhythmic owing to the cycling of dCLOCK, while cycling of PER/TIM is determined by rhythmicity of both elements. Interestingly, dCLOCK is absent in tim^0 and per^0 flies (Lee et al. 1998) thus providing evidence suggesting the existence of another feedback loop.

A question of central importance is whether *Drosophila* circadian clock genes and their molecular mechanisms are evolutionarily conserved. Work addressing this issue has led to the cloning of *per* from various other Dipteran and non-Dipteran insects (Costa and Kyriacou 1998). In this respect, the cloning of *per* from the giant silkmoth *Antherea pernyi*, an insect that diverged from Dipterans about 240 million years ago, is of particular interest (Reppert et al. 1994). The silkmoth *per* cDNA encodes for a protein that shares extensive sequence identity with *Drosophila* PER. *Per* mRNA levels exhibit a prominent circadian oscillation in silkmoth heads, and PER levels show robust daily variations in *Antherea* photoreceptor nuclei. Expression of the silkmoth *per* cDNA in per^0 transgenic *Drosophila*, showed that the silkmoth homologue can rescue circadian clock function in *Drosophila* (Levine et al. 1995). As occurs in *Drosophila* brain and eye cells, in the eye cells of *A. pernyi*, *per* mRNA and protein levels are expressed rhythmically with a 4–6h delay between the two rhythms. This is accompanied by temporally regulated appearance of PER in the nuclei of the photoreceptor cells (Sauman and Reppert 1996). Remarkably, however, the dynamics of PER regulation in the silkmoth brain are notably different from what occurs in the eye. PER is expressed rhythmically in the cytoplasm of eight brain cells (the presumptive pacemaker cells), with no evidence of temporally regulated nuclear translocation. In addition, there is evidence for the existence of an antisense *per* transcript in these cells, raising the intriguing possibility that this might be involved in the regulation of *per* sense mRNA and protein oscillations through the temporal sequestration of sense *per* transcripts (Sauman and Reppert 1996).

The endogenous generation of circadian rhythmicity is a determining characteristic of circadian systems. Synchronization to 24h environmental cycles is also of utmost importance. In *Drosophila*, photic entrainment is mediated via rapid light-dependent degradation of TIM, thereby also destabilizing PER (Hunter Ensor et al. 1996; Myers et al. 1996). Recent experimental work in *Drosophila* has shown that photic stimuli can be transduced to the clock mechanism through the visual and/or extraocular transduction pathways (Stanewsky et al. 1998). In particular, photic entrainment in plants has been shown to depend upon blue-light-specific circadian photoreceptors of the cryptochrome family. The identification of such receptors in plants was paralleled by the isolation of homologous genes in animals including, recently, *Drosophila*. Cryptochromes are flavin binding, redox-sensitive, soluble proteins (Cashmore et al. 1999) and were first identified as structural homologues of light-activated DNA repair enzymes called photolyases. Like photolyases the cryptochromes contain FAD and a pterin as chromophores, but they do not possess DNA repair activity.

Recent work in *Drosophila* points to the existence of a single cryptochrome gene, *cry* (Stanewsky et al. 1998) or DCry (Egan et al. 1999; Ishikawa et al. 1999; Selby and Sancar 1999) in this organism. Detailed experimental work on cry^b, an apparent null mutation in the *Drosophila* cryptochrome gene, shows that in this organism both the classical opsin/retinal-based visual pathway, as well as that based on pterin/flavin (cryptochrome), provide a light input pathway to the central clock mechanism (Stanewsky et al. 1998). Evidence at the molecular level further suggests that, in *Drosophila*, cryptochrome may also play a role in the central clock mechanism. This is supported by the finding that in cry^b mutants the circadian oscillations of TIM are blocked in most tissues of the organism, even in the absence of light-dark entrainment, whereas in wild-type flies TIM oscillations occur even in the absence of photic entrainment. TIM oscillations do, however, continue to occur in *cry*[b] mutants in a very small group of neurons (the lateral neurons) which are thought to comprise the CNS pacemaker in *Drosophila* (Hall 1995). Besides the lateral neurons, circadian oscillators are present throughout the entire fly (Plautz et al. 1997) and these appear to become arrhythmic in the *cry* mutant. Although biochemical photoreception remains to be demonstrated for the animal crytpochromes, *Drosophila cry* plays some important role in circadian light reception. Its overexpression results in stronger responses to brief light pulses compared to wild type, while these responses are absent in *cry* mutants (Emery et al. 1998). The activity rhythm of the mutant, however, remains entrainable to light/dark cycles, possibly via feedback from zeitgeber-driven activity. These results also lend support to the idea that *Drosophila* activity rhythms are controlled essentially by the lateral neurons and not by any of the other oscillators.

Drosophila, has so far been invested with a central role on the scenes of the "circadian clocks in animals" play. The genetic tractability of this organism, the abundance of unique molecular tools available and the accessibility to a wealth of genetic and molecular information have characterized *Drosophila melanogaster* as a choice model in many fields of research. In this particular case, it may be said that what is seen to occur in the *Drosophila* circadian clock, may be broadly generalized to the animal world, although the fine details of the processes involved may not turn out be strictly shared across the evolutionary tree. Nonetheless, important contributions regarding the elucidation of the central clock mechanism and the way this is interfaced to the input (i.e. reception of zeitgeber stimuli) and the output elements (i.e. the clock controlled genes) are probably still to come from studies conducted in this model organism.

The circadian clock in vertebrates

Circadian clocks in vertebrates have been localized to few neural structures such as: the suprachiasmatic nucleus of the hypothalamus (SCN), the retina and the pineal (Moore and Eichler 1972; Takahashi and Menaker 1984; Tosini and Menaker 1996). However, the role that these structures play in the circadian organization may vary among classes and even among similar species (Menaker and Tosini 1996). In general, in non-mammalian vertebrates, the pineal and the retina have a predominant role, while in mammals the SCN is considered to be the master circadian pacemaker, which controls most of the body circadian rhythms. In mammals the role of the pineal seems marginal, since it does not contain a circadian oscillator and its removal does not influence any circadian rhythms, while the retinal clock is believed to control some of the circadian rhythms present in the eye.

In the late eighties, Ralph and Menaker, while working with the golden hamster, discovered the first genetic mutation affecting the biological clock in mammals (Ralph and Menaker 1988). The identification of this spontaneously occurring mutation, which was called *tau*, was a fortunate event, since the mutant turned up in a shipment of animals sent to Menaker's laboratory from Charles River's Laboratories. *Tau* has since been shown to be a semidominant autosomal mutation, characterized by a short period in the rhythm of circadian running-wheel activity (22 hours in the heterozygous mutant and 20 hours in the homozygote). The second mutation affecting the circadian system in mammals was discovered in 1994, and the corresponding gene was named *clock* (Circadian Locomotor Output Cycles Kaput). *Clock* is also an autosomal, semidominant mutation. Differently, from *tau,* the discovery of the *clock* mutant was not a serendipitous event, but it was isolated in Takahashi's laboratory in a deliberate ENU (N-ethyl-N-nitrosourea) mutagenesis screen for mutations affecting the circadian rhythmicity phenotype (Vitaterna et al. 1994). The *Clock* mutation abolishes the locomotor activity rhythms in animals kept in constant darkness (Vitaterna et al. 1994). The mouse *clock* gene was later successfully cloned (King et al. 1997), and shown to be a rather large gene (100 Kb) located on chromosome 5. *Clock* is a member of the bHLH-PAS family of transcription factors (i.e., it contains a PAS protein-protein interaction domain, along with a bHLH DNA binding domain). Interestingly, *clock* transcription does not appear to be regulated in a circadian manner (i.e. it is not expressed in an oscillatory fashion).

A few months after *clock* was cloned, two different research groups finally identified a mammalian homologue of the *Drosophila period* gene (Sun et al. 1997; Tei et al. 1997). Like the *Drosophila period* gene, this mammalian *period* gene possesses a PAS domain but not a functional bHLH domain or any other known DNA binding sites. In the mouse SCN, the transcript levels of this gene begin to rise in the late night before dawn, and peak around mid-day, while in the retina and in other tissues its expression appears to lag four hours behind that observed in the SCN. Additional studies (Zylka et al. 1998) went on to show that, in mammals, in addition to the first *period* gene just described (*period* 1), there are two other genes encoding *period* products (*period* 2 and *period* 3), these also contain a PAS domain, but no functional DNA-binding domain. Further studies, using yeast two-hybrid assays to identify protein partners of mammalian CLOCK, led to the identification of BMAL1 (Gekakis et al. 1998) (the mammalian homologue of the *Drosophila cycle* gene). Finally, the mammalian homologue of the *Drosophila timeless* gene was also recently cloned (Sangoram et al. 1998).

Very recently two new genes, involved in the generation of rhythmicity in circadian locomotor activity, have been identified. In particular, a Dutch-Japanese research group discovered, using a reverse genetic approach, that two other genes (*cryptochrome 1* and *2*) are necessary for the generation of the circadian rhythm. In fact, mice in which both *cryptochrome* (*cry*1 and *cry*2) genes have been removed (i.e., a double knock out), are completely arrhythmic when placed in constant darkness (van der Horst et al. 1999). Although most of the research on the genetics of circadian rhythms focuses on the mouse and rat as model organisms, many of the clock genes mentioned so far have been cloned for other vertebrate species as well (Whitmore et al. 1998).

The most widely accepted model describing the molecular mechanisms leading to the generation of the circadian rhythm-associated phenotypes in mammals is similar to that already described for *Drosophila* (Figure 3), albeit with some important differences. Firstly, in mammals the system may be redundant, as is suggested by the presence of three *period* genes, and two *cryptochrome* genes. A further important difference in the mammalian mechanism lies in the role played by *timeless*. Specifically, mammalian *timeless* shows no evidence of rhythmic circadian regulation, TIM protein is not degraded on exposure to light (as in *Drosophila*) and, furthermore, TIM may not dimerize with PER as is known to occur in *Drosophila*.

Our current knowledge regarding the workings of the mammalian system is summarized in Fig. 4. The autoregulatory feedback loop begins with the heterodimer CLOCK/BMAL1 binding to the E-box of the *period* 1 gene promoter, initiating the transcription of the gene. This results in the rise in the levels of the PERIOD 1 protein in the cytoplasm. Soon after that, PERIOD 3 and later PERIOD 2 also start to accumulate, with each of these proteins reaching peak levels at different circadian times. PERIOD 1 peaks at between circadian time (CT) 4–6, PERIOD 3 between CT 4–8 and lastly, PERIOD 2 at around CT 8. The difference in the phases of the peak levels of proteins expressed by these genes suggests that this might depend upon the existence of other (yet to be discovered) transcriptional activators besides the CLOCK/BMAL1 heterodimer. CLOCK/BMAL1 also acts as a transcriptional activator of the two mammalian *cryptochrome* genes. Once the *per* proteins (PER1, PER2 and PER3) and the *cry* proteins (CRY1 and CRY2) have reached determined levels, they form heterodimers with each of the PER proteins, and the heterodimers then translocate to the nucleus. Having reached the nucleus, the PER/CRY heterodimers inhibit the transcription of their own genes through inactivation of the activating CLOCK/BMAL1 complex. At this point, the level of the PER1, PER2, PER3, CRY1 and CRY2 proteins in the cytoplasm begins to decrease, leading to a parallel decrease in the formation of the inhibiting heterodimers. Ultimately, the level of heterodimers will be insufficient to inhibit the transcription of the *per* and *cry* genes, which will once again fall under the positive control of the CLOCK/BMAL1 complex.

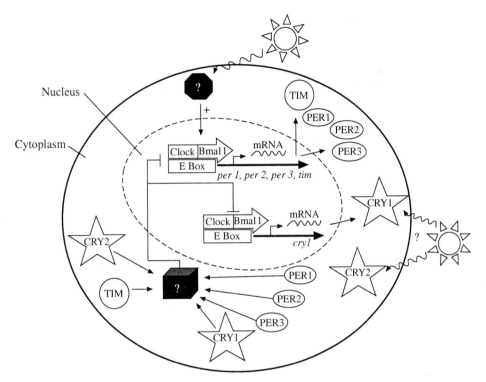

Fig. 4   The circadian clock in mammals. In this model, transcription of the 3 *per* genes, *tim* and *cry*1 is positively regulated by the heterodimer formed by the protein products of the *clock* and *bmal1* genes. In the case of the *per* genes, light plays the role of a positive effector through an as yet uncharacterized mechanism. From *in vitro* experiments it has been shown that the products of the 3 *per* genes, TIM, CRY1 and CRY2 are all able to participate in the formation of complexes in various ways. However, it is not known which complexes are actually formed *in vivo*. In addition, some or all of these proteins may participate as negative regulators of the transcription of the 3 *per* genes, *tim* and *cry1*, although recent evidence implies that PER-CRY dimers may be the most important in this respect. The actual role of CRY1 and CRY2 in photoreception has also been diagrammed as uncertain. In the figure "plus" signs indicate positive effectors and lines terminating in perpendicular dashes indicate negative effectors.

When *cry1* and *cry2* genes were discovered in mice, it was immediately thought that the corresponding proteins could be involved in blue light reception (Cashmore et al. 1999). Subsequent experiments clearly suggested a crucial role for CRYs as negative elements in the circadian feedback loop. But a very recent report by Okamura et al. (1999), strongly implies that CRY1 and CRY2 are not required for light-induced phase shifting, raising the possibility that other photoreceptor molecules could be involved in photic entrainment. Nevertheless, at the moment, it remains impossible to exclude that CRYs might still play some role in the transduction of the photic signals.

With the cloning of the genes believed to be responsible for the functioning of the circadian pacemaker, new tools have become available for the investigation of the expression of these genes in tissues or organs, other than the anatomical site known to be the seat of the central circadian pacemaker. Currently, the picture that is emerging is that "clock" genes are in fact expressed in many tissues. The three *period* genes are expressed in several peripheral non-neural tissues, and the same is true for *clock*, *bmal1*, and *timeless*. In addition, a self-sustained circadian oscillator in cultured mammalian cells (fibroblast) has recently been described (Balsalobre et al. 1998). These new data suggest that in vertebrates, circadian oscillators may be present in many tissues and cells and may not be restricted exclusively to the retina, SCN and pineal (Whitmore et al. 1998). The role that these circadian oscillators play in these cells and/or tissues is not clear to date, but undoubtedly constitutes a new and intriguing aspect of vertebrate circadian organization that is worthwhile investigating.

Conclusion

Much of the work described in this overview shows that, although the molecules involved in the clock machinery show heterogeneity across taxa, the mechanisms underlying the basic functioning of biological clocks show substantial conservation at various levels in the phylogeny of living organisms. This has proven to be basically true, insofar as all known circadian oscillators use intracellular loops that are based on positive and negative elements. Such elements act in oscillators, in which transcription of clock genes yields proteins (negative elements), which act to block the action of positive element(s), whose role is to activate the clock genes encoding the negative element(s). However, the details of such processes have been shown to vary even among organisms of the same Class (i.e. Insects). Accordingly, this underlines the importance of comparative molecular and functional studies, in unveiling the evolutionary mechanisms and steps involved in the origin(s) and elaboration(s) of the molecular elements implicated in the construction of biological circadian time-keeping machinery.

References

Allada, R., White, N.E., So, W.V., Hall, J.C., Rosbash, M. (1998) A mutant *Drosophila* homolog of mammalian Clock disrupts circadian rhythms and transcription of period and timeless. Cell **93**: 791–804.

Aronson, B., Johnson, K., Loros, J.J., Dunlap, J.C. (1994) Negative feedback defining a circadian clock: autoregulation in the clock gene frequency. Science **263**: 1578–1584.

Ballario, P., Macino, G. (1997) White collar proteins: Passing the light signal in *Neurospora crassa*. Trends Microbiol. **5**: 458–462.

Balsalobre, A., Damiola, F., Schibler, U. (1998) A serum shock induces gene expression in mammalian tissue culture cells. Cell **93**: 929–937.

Cashmore, A.R., Jarillo, J.A., Wu, Y.J., Liu, D. (1999) Cryptochromes: blue light receptors for plants and animals. Science **284**: 760–765.

Costa, R., Kyriacou, C.P. (1998) Functional and evolutionary implications of natural variation in clock genes. Curr.Opin. in Neurobiol. **8**: 659–664.

Darlington, T.K., Wager-Smith, K., Ceriani, M.F., Staknis, D., Gekakis, N., Steeves, T.D.L., Weitz, C.J., Takahashi, J.S., Kay, S.A. (1998) Closing the circadian loop: CLOCK-induced transcription of its own inhibitors *per* and *tim*. Science **280**: 1599–1603.

Dunlap, J.C. (1999) Molecular bases for circadian clocks, 1999. Cell **96**: 271–290.

Egan, E.S., Franklin, T.M., Hilderbrand-Chae, M.J., McNeil, G.P., Roberts, M.A., Schroeder A.J., Zhang, X., Jackson, F.R. (1999) An extraretinally expressed insect cryptochrome with similarity to the blue light photoreceptors of mammals and plants. J. Neurosci. **19**: 3665–3673.

Emery, P., So, W.V., Kaneko, M., Hall, J.C., Rosbash, M. (1998) CRY, a Drosophila clock and light-regulated cryptochrome, is a major contributor to circadian rhythm resetting and photosensitivity. Cell **95**: 669–679.

Feldman, J.F., Hoyle, M., (1973) Isolation of circadian clock mutants of *Neurospora crassa*. Genetics **75**: 605–613.

Garceau, N., Liu, Y., Loros, J.J., Dunlap, J.C., (1997) Alternative initiation of translationand time-specific phosphorylation yield multiple forms of the essential clock protein FREQUENCY. Cell **89**: 469–476.

Gekakis, N., Staknis, D., Nguyen, H.B., Davis, F.C., Wilsbacher, I.D., King, D.P., Takahashi J.S., Weitz, C.J. (1998) Role of the CLOCK protein in the mammalian circadian mechanism. Science **280**: 1564–1569.

Golden, S., Ishiura, M., Johnson, C.H., Kondo, T. (1997) Cyanobacterial circadian rhythms. Annu. Rev. Plant Physiol. Plant. Mol. Biol. **48**: 327–354.

Hall, J.C. (1995) Tripping along the trail to the molecular mechanisms of biological clocks. Trends Neurosci. **18**: 230–240.

Hao, H., Allen, D.L., Hardin, P.E. (1997) A circadian enhancer mediates PER-dependent mRNA cycling in *Drosophila melanogaster*. Mol. Cell Biol. **17**: 3687–3693.

Hege, D.M., Stanewsky, R., Hall, J.C., Giebultowicz, J.M. (1997) Rhythmic Expression of a PER-reporter in the Malpighian tubules of decapitated *Drosophila*: evidence for a brain-independent circadian clock. J. Biol. Rhythms **12**: 300–308.

Hunter Ensor, M., Ousley, A., Sehgal, A. (1996) Regulation of the *Drosophila* protein Timeless suggests a mechanism for resetting the circadian clock by light. Cell **84**: 677–685.

Ishikawa, T., Matsumoto, A., Kato, T. Jr., Togashi, S., Ryo, H., Ikenaga, M., Todo, T., Ueda R., Tanimura, T. (1999) DCRY is a *Drosophila* photoreceptor protein implicated in light entrainment of circadian rhythm. Genes Cells **4**: 57–65.

Ishiura, M., Kutsuna, S., Aoki, S., Iwasaki, H., Andersson, C.R., Tanabe, A., Golden, S.S., Johnson, C.H., Kondo, T. (1998) Expression of a gene cluster kaiABC as a circadian feedback process in Cyanobacteria. Science **281**: 1519–1523.

Iwasaki, H., Taniguchi, Y., Ishiura, M., Kondo, T. (1999) Physical interactions among circadian clock proteins kaiA, kaiB and kaiC in cyanobacteria. The EMBO J. **18**: 1137–1145.

Kaneko, M. (1998) Neural substrates of *Drosophila* rhythms revealed by mutants and molecular manipulations. Curr. Opin. Neurobiol. **8**: 652–658.

King, D.P., Zhao, Y., Sangoram, A.M., Wilsbacher, L.D., Tanaka, M., Antoch, M.P., Steeves, T.D., Vitaterna, M.H., Kornhauser, J.M., Lowrey, P.L., Turek, F.W., Takahashi, J.S. (1997) Positional cloning of the mouse circadian *Clock* gene. Cell **89**, 641–653.

Kloss, B., Price, J.L., Saez, L., Blau, J., Rothenfluh, A., Wesley, C.S., Young, M.W., (1998) The *Drosophila* clock gene *double-time* encodes a protein closely related to human casein kinase I epsilon. Cell **94**: 97–107.

Kondo, T., Ishiura, M. (1999) The circadian clocks of plants and cyanobacteria. Trends Plant Sci. **4**: 171–176.

Kondo, T., Tsinoremas, N.F., Golden, S.S., Johnson, C.H., Kutsuna, S., Ishiura, M. (1994) Circadian clock mutants of cyanobacteria. Science **266**: 1233–1236.

Konopka, R.J., Benzer, S. (1971) Clock mutants of *Drosophila melanogaster*. Proc. Natl. Acad. Sci. USA **68**: 2112–2116.

Kutsuna, S., Kondo, T., Aoki, S., Ishiura, M. (1998) A period extender gene pex that extends the period of the circadian clock in the cyanobacterium *Synechococcus*. J. Bacteriol. **180**: 2167–2174.

Lee, C., Bae, K., Edery, I. (1998) The Drosophila CLOCK protein undergoes daily rhythms in abundance, phosphorylation, and interactions with the PER-TIM complex. Neuron **21**: 857–867.

Levine, J.D., Sauman, I., Imbalzano, M., Reppert, S.M., Jackson, F.R. (1995) Period protein from the giant silkmoth *Antheraea pernyi* functions as a circadian clock element in *Drosophila melanogaster*. Neuron **15**: 147–157.

Linden, H., Macino, G. (1997) White collar 2, a partner in blue light signal transduction, controlling expression of light regulated genes in *Neurospora crassa*. EMBO J. **16**: 98–109.

Liu, Y., Garceau, N., Loros, J.J., Dunlap, J.C. (1997) Thermally regulated translational control mediates an aspect of temperature compensation in the *Neurospora* circadian clock. Cell **89**: 477–486.

Liu, Y., Tsinoremas, N., Johnson, C., Lebdeva, N., Golden, S., Ishiura, M., Kondo, T. (1995) Circadian orchestration of gene expression in cyanobacteria. Genes Dev. **9**: 1469–1478.

Luo, C., Loros, J.J., Dunlap, J.C. (1998) Nuclear localization is required for function of the essential clock protein FREQUENCY. EMBO J. **17**: 1228–1235.

McClung, C.R., Fox, B.A., Dunlap, J.C. (1989) The *Neurospora* clock gene frequency shares a sequence element with the *Drosophila* clock gene *period*. Nature **339**: 558–562.

McWatters, H., Dunlap, J.C., Millar, A. (1999) Circadian biology: Clocks for the real world. Curr. Biol. **9**: 633–635.

Menaker, M., Tosini, G. (1996) The Evolution of Vertebrate Circadian Systems. In: Honma K., Honma, S. (eds) Circadian Organization and Oscillatory Coupling, Hokkaido University Press, Sapporo, pp. 39–52.

Merrow, M.W., Garceau, N.Y., Dunlap, J.C. (1997) Dissection of a circadian oscillator into the discrete domains. Proc. Nat. Acad. Sci. USA **94**: 3877–3882.

Merrow, M., Brunner, M., Roenneberg, M. (1999) Assignment of circadian function for the *Neurospora* clock gene frequency. Nature 1999: **399**: 584–586.

Moore, R.Y., Eichler, V.B. (1972) Loss of a circadian adrenal corticosterone rhythm following suprachiasmatic lesion in the rat. Brain Res. **42**: 201–206.

Myers, M.P., Wager, Smith, K., Rothenfluh., H.A., Young, M.W. (1996) Light-induced degradation of TIMELESS and entrainment of the *Drosophila* circadian clock. Science **271**: 1736–1740.

Okamura, H., Miyake, S., Sumi, Y., Yamaguchi, S., Yasui, A., Muijtjens, M., Hoeijmakers, J.H.J., van der Horst, G.T.J. (1999) Photic induction of mPer1 and mPer2 in Cry-deficient mice lacking a biological clock. Science **286**: 2531–2534.

Ouyang, Y., Andersson, C.R., Kondo, T., Golden, S.S., Johnson, C.H. (1998) Resonating circadian clocks enhance fitness in cyanobacteria. Proc. Natl. Acad. Sci. USA **95**: 8660–8664.

Pittendrigh, C.S. (1960) Circadian rhythms and the circadian organisation of living systems. Cold Spr. Harb. Symp. Quant. Biol. **25**: 159–184.

Plautz, J.D., Kaneko, M., Hall, J.C., Kay, S.A. (1997) Independent photoreceptive circadian clocks throughout *Drosophila*. Science **278**: 1632–1635.

Price, J.L., Blau, J., Rothenfluh, A., Abodeely, M., Kloss, B., Young, M.W. (1998) double-time is a novel Drosophila clock gene that regulates PERIOD protein accumulation. Cell **94**: 83–95.

Ralph, M.R., Menaker, M. (1988) A mutation of the circadian system in golden hamsters. Science **241**: 1225–1227.

Reppert, S.M., Tsai, T., Roca, A.L., Sauman, I. (1994) Cloning of a structural and functional homolog of the circadian clock gene period from the giant silkmoth *Antheraea pernyi*. Neuron **13**: 1167–1176.

Rutila, J.E., Suri, V., Le M, So, W.V., Rosbash, M., Hall, J.C. (1998) CYCLE is a second bHLH-PAS clock protein essential for circadian rhythmicity and transcription of *Drosophila period* and *timeless*. Cell **93**: 805–814.

Sangoram, A.M., Saez, L., Antoch, M.P., Gekakis, N., Staknis, D., Whiteley, A., Fruechte E.M., Vitaterna, M.H., Shinomura, K., King, D.P., Young, M.W., Weitz, C.J., Takahashi, J.S. (1998) Mammalian circadian autoregulatory loop: a timeless ortholog and mPer1 interact and negatively regulate CLOCK-BMAL1-induced transcription. Neuron **5**: 1101–1113.

Sauman, I., Reppert, S.M. (1996) Circadian clock neurons in the silkmoth *Antheraea pernyi*: novel mechanisms of Period protein regulation. Neuron **17**: 889–900.

Sehgal, A., Price, J.L., Man, B., Young, M.W. (1994) Loss of circadian behavioral rhythms and per RNA oscillations in the *Drosophila* mutant *timeless*. Science **263**: 1603–1606.

Selby, C.P., Sancar, A. (1999) A third member of the photolyase/blue-light photoreceptor family in *Drosophila*: a putative circadian photoreceptor. Photochem. Photobiol. **69**: 105–107.

Stanewsky, R., Kaneko, M., Emery, P., Beretta, B., Wager-Smith, K., Kay, S.A., Rosbash, M., Hall, J.C. (1998) The *cryb* mutation identifies cryptochrome as a circadian photoreceptor in *Drosophila*. Cell **95**: 681–692.

Sun, Z.S., Albrecht, U., Zhuchenko, O., Bailey, J., Eichele, G., Lee, C.C. (1997) Rigui, a putative mammalian ortholog of the *Drosophila period* gene. Cell **90**: 1003–1011.

Talora, C., Franchi, L., Linden, H., Ballario, P., Macino, G. (1999) Role of a white collar-1-white collar-2 complex in blue-light signal transduction. EMBO J. **18**: 4961–4968.

Takahashi, J.S., Menaker, M. (1984) Multiple redundant circadian oscillators within the isolated avian pineal gland. J. Comp. Physiol. **170**: 479–489.

Tei, H., Okamura, H., Shigeyoshi, Y., Fukuhara, C., Ozawa, R., Hirose, M., Sakaki, Y. (1997) Circadian oscillation of a mammalian homologue of the *Drosophila period* gene. Nature **389**: 512–516.

Tosini, G., Menaker, M. (1996) Circadian rhythms in cultured mammalian retina. Science **272**: 419–421.

van der Horst, G., Muijtjens, M., Kobayashi, T., Takano, R., Kanno, A., Takao, M., de Wit J, Verkerk, A., Eker, A., van Leenen, D., Buijs, R., Bootssma, D., Hoeijmakers. J.H.J., Yasui, A. (1999) Mammalian Cry1 and Cry2 are essential for maintenance of circadian rhythms. Nature **398**: 627–630.

Vitaterna, M.H., King, D.P., Chang, A.M., Kornhauser, J.M., Lowrey, P.L., McDonald, J.D., Dove, W.F., Pinto, L.H., Turek, F.W., Takahashi, J.S. (1994) Mutagenesis and mapping of a mouse gene, *Clock*, essential for circadian behaviour. Science **264**: 719–725.

Whitmore, D., Foulkes, N., Strahle, U., Sassone-Corsi, P. (1998) Zebrafish Clock rhythmic expression reveals independent peripheral circadian oscillators. Nature Neurosci. **1**: 701–707.

Young, M.W. (1998) The molecular control of circadian behavioural rhythms and their entrainment in *Drosophila*. Ann. Rev. Biochem. **67**: 135–152.

Zylka, M.J., Shearman, L.P., Weaver, D.R., Reppert, S.M. (1998) Three *period* homologs in mammals: differential light responses in the suprachiasmatic circadian clock and oscillating transcripts outside of brain. Neuron **20**: 1–20.

Biological Rhythms
Edited by V. Kumar

6. The Circadian Systems of Cells

T. Roenneberg* and M. Merrow

Institute for Medical Psychology Chronobiology, Goethestr. 31, D-80336 München, Germany

It has become clear over the past years that the circadian system in all organisms is based on molecular processes within single cells, and an unprecedented surge of research has revealed the molecular details of a transcriptional/translational feedback loop that is considered as the molecular clock works. In spite of this excellent progress, central questions are still unanswered: what makes the loop turn over with the long circadian period? How is the loop compensated for extra-and intracellular noise and different levels of metabolism (e.g., temperature compensation)? How do different clocks communicate and couple and how does the system adapt to changing conditions in development, in the environment, in life history or in seasons. Although the complexity of the feedback loop is increasing with every new insight, it still represents a fairly simple molecular network. Many decades of physiological and biochemical research in unicellular organisms shows, however, that the circadian system is highly complex on one hand and exquisitely compensated on the other—qualities that are as yet not entirely explained by the molecular loop. The results gained in unicellular and simple organisms can be used to understand the functioning and the constraints of single cell clocks within complex organisms. A comparison between single cell clocks, be it unicells or in cells of complex organisms can be used to outline general circadian principles that apply to both.

Introduction

For more than half a century, circadian systems have been investigated at the level of the organism. These efforts included the search for circadian centers and the biological mechanisms which generate circadian rhythmicity at the organismal level. Over the course of this approach, it has become clear that single cells of multicellular organisms are capable of generating circadian rhythmicity on their own (Michel et al. 1993; Takahashi 1987; Welsh et al. 1995), a fact that has been known since 1948 for unicellular organisms (Edmunds 1988; Hastings and Sweeney 1958; Pohl 1948; Sweeney 1987) and has more recently been extended beyond the eukaryotic world (Johnson et al. 1996).

Of course, there is a categorical difference between unicells and single cells; the former are entire organisms, the latter are building blocks of individual fungi, plants, or animals. So how are their circadian systems comparable? An organism is the ultimate expression of its genome reflecting the current state of its evolution. It is the platform at which the combined 'action' of the genes is put to a test. Yet, the only way the results of this test can feed back on the gene pool is through the comparative success by which genes are passed on and multiply. Thus, evolution only acts through genes (Dawkins 1976), and not on organisms or ecological systems (although the latter contribute to the pressure that drives selection). In this respect, single intra-organismal cells and organs within organisms resemble organisms and populations within ecosystems; they form a network of units

*E-mail: till.roenneberg@imp.med.uni-muenchen.de

which communicate and "shape" the system without being themselves directly subject to evolution and without "knowing" that they form a necessary platform for the expression of genes.

Single cells are the most rudimentary elements of biology, they have metabolism, they grow and reproduce. Like organisms, intra-organismal cells are surrounded by a structured environment, have receptors and signal transduction pathways for sensing changes in their exterior, and like in ecosystems, many of the intra-organismal changes occur in a diurnal fashion. It has always been an intriguing question why organisms have developed an autonomous circadian program in spite of the fact that most of them live in a strict daily regime. One of the arguments for this evolution has been the necessity to anticipate and pre-program the daily changes with the help of an endogenous clock. If this argument is correct for clocks of organisms, it should also be correct for clocks of individual cells within an organism. In spite of the similarities between organisms and cells, there also must be distinct differences between the "tasks" of their circadian systems. One approach to find the similarities and differences is to compare the circadian system of single cells within organisms with that of a unicellular organism.

Gonyaulax, an example for a single cell clock

The complexity of circadian systems of single cells has been well documented in the marine dinoflagellate *Gonyaulax polyedra* (Fig. 1). Endogenous "ecology" (metabolism) and exogenous ecology are closely linked. Energy (from light) and carbon sources are spatially separated in the water column from nitrogen and other inorganic sources. During the day, the cells aggregate in the top layers of the ocean and photosynthesize; during the night, they sink more than 30 m (and near the shore to the bottom) and exploit the higher concentrations of fixed nitrogen. Our experiments showed that environmental factors can be both resources for the organisms' metabolism as well as *zeitgeber* for its circadian system.

Light reaches the *Gonyaulax* clock via two separate transduction pathways, and both are apparently under circadian control (Roenneberg and Deng 1997; Roenneberg and Hastings 1988; Roenneberg and Taylor 1994). One of them is predominantly blue light sensitive and is activated only during the subjective night. The other responds both to red and blue light and is possibly associated with photosynthesis which is also clock-controlled. Yet, photosynthesis is not only a clock output but feeds back into the circadian system, demonstrated by physiological and pharmacological experiments (Johnson and Hastings 1989; Roenneberg 1994). Thus, light, photosynthesis, and the clock form a metabolic feedback. This is also true for nitrogen metabolism in *Gonyaulax*. Nitrate reductase activity and abundance is circadianly regulated (Ramalho et al. 1995) as is the rate of NO_3^- uptake (Roenneberg and Merrow, unpublished results). Along the pathway of nitrogen assimilation, NO_3^- acts also as an input to the *Gonyaulax* clock (Roenneberg and Rehman 1996). Finally, a metabolic feedback loop is also apparent in the cells' regulation of pH, which undergoes circadian changes both in light:dark cycles (Hastings 1960) and in constant light (Eisensamer and Roenneberg, unpublished). The changes in H^+, which are measured in the culture medium, correspond to proton translocations in the millimolar range when corrected to the total volume of the cells. When the pH of the culture medium is changed experimentally it can shift the circadian phase.

In addition to the complexity of the *Gonyaulax* daily temporal program with multiple feedbacks closing between the clock, metabolism and its environment, its circadian system involves two circadian oscillators which control different output rhythms and show different responsiveness to environmental stimuli (nitrate, different qualities of light, etc.) (Roenneberg and Morse 1993).

The *Gonyaulax* clock is one of the traditional model systems for circadian research. Compared to other classical model systems, e.g., *Drosophila* and *Neurospora*, it is not amenable for molecular

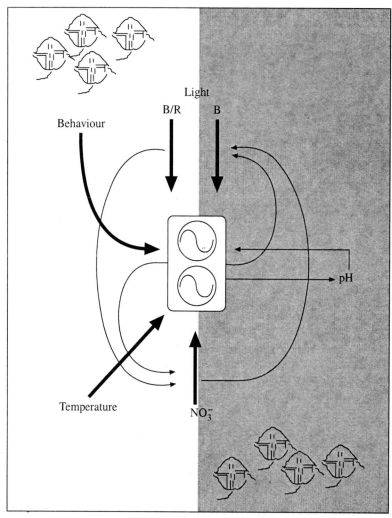

Fig. 1   The circadian system of *Gonyaulax* is highly complex. It controls numerous biochemical functions as well as behavior (vertical migration, formation of swarms, and orientation in light gradients) and consists of at least two separable oscillators. Each of these is influenced specifically by light via two distinct input pathways (one is both blue and red sensitive, BR, the other senses only blue light, B), by temperature, as well as by nitrogen sources. Inputs and outputs form feedbacks with the oscillator(s). Redrawn from (Roenneberg and Merrow 2001).

genetics. Yet even without the possibilities of crossing or transforming *Gonyaulax*, great progress has been made in understanding the molecular mechanisms of circadian regulation (Mittag et al. 1995; Morse et al. 1989). While circadian control in the model systems allowing the traditional approaches of molecular genetics was found mostly at the level of transcription, circadian control in *Gonyaulax* appears to be predominantly at the level of translation. It may well be that these differences reflect phylogenetic divergence of the different systems, but there is also the possibility that the methods and possibilities of experimental approaches partly determine the results. The strength of the *Gonyaulax* system is the possibility to investigate physiological mechanisms in an ecological context. Many of the results concern the interactions of a single cell's circadian system with its environment. On a conceptional level, *Gonyaulax* is therefore an ideal model to generate ideas how the circadian system of intra-organismal cells interacts with its immediate environment.

Entrainment of cellular clocks

Entrainability is essential for the synchronization of circadian systems with their environment, be it the ecosystem or the rest of the organism. Basically, it is not important which *zeitgeber* signal is used to ensure this synchronization as long it offers a reliable measure for the outside rhythmicity. The type of *zeitgeber* used for entrainment may be one of the main differences between organisms and intra-organismal cells. While light is the most prominent *zeitgeber* for practically all organisms, its role for the circadian system of intra-organismal cells is questionable. This is apparently not due to the fact that cells are commonly embedded in tissue that cannot easily be penetrated by light. In many animals, specialized photoreceptor cells are buried deep in the brain, mostly using opsin based receptors (Foster et al. 1994). It has been claimed that the human circadian system is light entrainable through the skin via the blood stream (Campbell and Murphy 1998). There is, however, overwhelming evidence for the fact that the mammalian circadian system is exclusively light entrainable via the retina—once the eyes are removed, light entrainability is lost (Lucas and Foster 1999a; Lucas and Foster 1999b; Lucas and Foster 1999c; Roenneberg and Foster 1997). In contrast, there is no doubt that most arthropods and lower vertebrates use multiple possibilities for circadian light detection outside of their eyes. Recent work in zebra fish has shown that circadian rhythmicity is light entrainable even in individual organs, e.g., heart, kidney, and cell lines in culture, (Whitmore et al. 2000). The same is also true for many body parts of *Drosophila* when isolated from the rest of the animal (Hege et al. 1997; Plautz et al. 1997). The phylogenetic fact that different species chose very different channels for their circadian light input carries a message for understanding the strategies of circadian light entrainment.

That light is the predominant *zeitgeber* for organisms reflects its reliability but also the fact that all other potential *zeitgeber* signals (temperature, food, or social cues) are derived from the light rhythm ruling our daily structure. It is, therefore, not surprising that the circadian pacemaker centers in all phyla are morphologically closely linked to light input pathways (Roenneberg and Foster 1997). The same is true for the molecular components of the circadian system (Crosthwaite et al. 1997; Crosthwaite et al. 1995; Emery et al. 1998; Hunter-Ensor et al. 1996; Stanewsky et al. 1998). A good example is the *Neurospora* circadian system, where the clock genes *white collar*-1, *white collar*-2, and *frequency*, are intimately involved in light perception (Ballario et al. 1996; Crosthwaite et al. 1997; Linden et al. 1999; Linden and Macino 1997; Merrow et al 2001). Without the *white collar* genes, all light responses appear to be absent.

It could be argued that light always is the main *zeitgeber*, even if its action is only indirect. SCN cells do not respond to light per se but to glutamate (Ebling 1996; Honma et al. 1996), yet its excretion from collaterals of the optic tract is a direct translation of retinal light responses. The light-correlated activation of *mper*[1] and *mper*[2] in the SCN is, therefore, often coined as light responsiveness (Shearman et al. 1997; Shigeyoshi et al. 1997; Zylka et al. 1998). The intracellular transduction pathways that transmit light information to the circadian oscillator in *Drosophila* involve *cryptochrome* (d*cry*) (Emery et al. 1998; Stanewsky et al. 1998), and the dCRY protein has been postulated as a circadian light receptor in flies (Ceriani et al. 1999). Mouse *cryptochrome* (m*cry*) is also intimately involved in the molecular mechanisms of the mammalian circadian loop (Kume et al. 1999), but in this case, it is neither directly nor indirectly light-inducible (Field et al. 2000; Okamura et al. 1999).

The differences in the specific function of orthologue clock genes across species indicate a common origin of the molecular components used to construct a cellular circadian pathway initially involved in intracellular light transduction.

The existence of a circadian clock in peripheral cells, e.g. in fibroblasts, has been discovered because it was possible to synchronize the individual cellular clocks within an immortalized culture. The synchronizing agents in this case are components of the serum. Schibler and coworkers exposed fibroblast cultures to a serum shock and found that the RNA of *per* and of other genes known to be rhythmic in the intact animal displayed circadian rhythmicity (Balsalobre et al. 1998). The specific *zeitgeber* substances present in the serum are now being identified. Again, one could argue that these internal non-photic *zeitgebers* represent light or darkness, similarly to melatonin (Vondrasova-Jelinkova et al. 1999).

As described above, the *Gonyaulax* circadian system can also be entrained by non-photic, chemical *zeitgebers*. In this case, nitrate is largely independent of the solar light changes and it is also not strictly a *zeitgeber* because its concentration in the deeper layers of the ocean is very unlikely to vary in a daily rhythm. However, *Gonyaulax*' exposure to nitrogen is diurnal due to the cells' vertical migration, which again is under control of the light entrainable circadian clock. So, eventually all *zeitgebers* are derivatives of the solar light signals.

Coupling between oscillators

Circadian clocks of individual animals can synchronize each other (Chandrashekaran and Marimuthu 1985) but in some cases, e.g., a male and a female hamster, they do not lead to mutual entrainment but influence each other by masking (Aschoff and Goetz 1988). Whether clocks of individual organisms, be it within a population or as part of the food chain, entrain each other depends on the specific case. While it may not be important for a male to alter its circadian program in response to a non-estrous female, it apparently is advantageous for social animals to synchronize their activities both in the circadian (Bovet and Oertli 1974; Crowley and Bovet 1980; Gwinner 1966; Marimuthu et al. 1978; Marimuthu et al. 1981; Menaker and Eskin 1966) and the ultradian range (Aschoff and Gerkema 1985).

In constant conditions (e.g., in constant light), circadian rhythms can split into different components, which continue to be rhythmic and, therefore, are thought to be controlled by separate circadian oscillators. This splitting or internal desynchronization has initially been associated with different neuronal pacemakers in complex organisms (Aschoff 1967; Aschoff et al. 1967; Ebihara and Gwinner 1992; Gwinner 1974; Hoffman 1971; Lovegrove and Papenfus 1995; Meijer et al. 1990; Schardt et al. 1989) but has subsequently also been demonstrated in a unicellular organism (Roenneberg and Morse 1993). The bioluminescence rhythm (together with the vertical migration) and the behavioral swarming rhythm in *Gonyaulax* can run with different periods, leading to internal desynchronization and complex crossing patterns (Morse et al. 1994; Roenneberg et al. 1998). Thus, the circadian system within single cells can also be constructed of multiple circadian oscillators. Under normal conditions, these oscillators communicate with each other and couple themselves to produce a uniform output. Because the two oscillators respond differently to light and metabolites and control different outputs, they provide the organism with a flexibility concerning specific phase angles for different biological functions under entrained conditions.

There are examples of independence between different oscillators within an organism, e.g., the circadian and the estrus cycle in hamsters (Refinetti and Menaker 1992; Yamazaki et al. 1998) and in humans (Chandrashekaran et al. 1991). However, there are no examples (at least in wild type animals without experimental manipulations) of independence when the different oscillators within an organism are circadian—even in constant conditions, their rhythmicities show at least relative coordination (Holst 1939) proving residual communication or coupling among them. Although communication among circadian oscillators can only be demonstrated in the absence of external *zeitgeber*, it is very likely that communication also exists in their presence.

But even in constant conditions, circadian clocks create, via their outputs, a cyclic environment both within the organism and outside of it. When *Gonyaulax* cells are kept, for example, in small vesicles, the cells impose systematic changes on the medium due to their circadian physiology. Throughout the cycle, they take up nitrate at different rates (Roenneberg and Merrow unpublished), and circadianly gated cell death makes more nitrate available at the end of each subjective night. Thus, the endogenous system creates a nitrate oscillation in the exogenous medium. As described above, this is also true for the medium's pH. Since these factors are at the same time signals for the algae's circadian clock, they feed back and, thereby, synchronize the cells amongst each other. These feedbacks contribute to the remarkable precision of the cell population rhythm over many weeks in constant conditions (Njus et al. 1981). They also offer an explanation of how different *Gonyaulax* cultures can entrain each other. When two cultures, raised in light-dark cycles out of phase, are mixed, their rhythms slowly synchronize (Broda et al. 1985; Hastings et al. 1985), but only do so if the medium is not exchanged. Do similar feedbacks exist between cells and their immediate environment within organisms?

Communication or coupling between circadian oscillators within an organism can suberve many different functions. Isolated SCN cells continue to generate circadian rhythmicity *in vitro*, but the cells which were previously part of the same neuronal structure oscillate on the multi-electrode plate with very different periods (Welsh et al. 1995). Within the intact SCN, the different neuronal pacemaker cells are strongly coupled to produce a robust central signal for the rest of the body. So, one reason for coupling multiple oscillators lies in making a circadian signal precise and robust (Enright 1980).

Unlike in birds, reptiles, and several fish species, the mammalian pineal is not light responsive, and the circadian melatonin production appears to be purely driven by a multi-synaptic, noradrenergic projection from the SCN. Yet, SCN cells contain melatonin receptors and can, thus, receive a feedback from the pineal (Gwinner et al. 1997). The eyes, which are essential for light entrainment of the mammalian SCN, are able to produce an independent circadian rhythm (Tosini and Menaker 1996), and, again they apparently receive feedback from the pineal (Steinlechner 1989). In some non-mammalian vertebrates, the pineal is both light responsive and self-sustained, and in addition SCN-like thalamic structures are capable of self-sustained circadian rhythmicity. The degree to which the pineal and the avian SCN contribute to the circadian program appears to correlate with their circannual behavior, whether they are sessile or only local migrators or whether they seasonally migrate over long distances (Gwinner et al. 1997). We still know very little about the interdependencies between different circadian oscillators in the brain, and some of the circadian centers outside of the SCN have only been shown to exist by physiological experiments but have not yet been identified (e.g., the methamphetamine dependent oscillator; Honma et al. 1987), but the communication of these oscillators with each other appears to be an important prerequisite for fine-tuning daily and annual behavior.

Besides communication between circadian oscillators within the brain, we now know that the periphery also contains millions of cellular circadian clocks which have to be specifically embedded into the temporal program of the whole organism. Communication between central pacemakers and peripheral oscillators, as judged by the mRNA oscillations of rhythmic genes, results in a phase angle difference by several hours (Balsalobre et al. 1998; Yamazaki et al. 2000; Zylka et al. 1998). We do not yet understand the reasons for the substantial phase delay between the brain and the periphery, but oscillator theory tells us that this phase relationship can only be established accurately if it involves coupling of oscillators. So one of the reasons for the existence of these peripheral cellular clocks, besides the advantage of a systematic temporal organization of cellular metabolism, may lie in the fact that they need to oscillate with a different phase than the central pacemaker (Yamazaki et al. 2000).

Open questions

The question whether all cells within an organism contain their own circadian clock is still open but there are many indications that this is not the case (e.g., in situ hybridization using probes against rhythmic clock genes, shows them to be rhythmic only in specific parts of the brain, predominantly in the SCN). So, why have some cells and tissues retained an endogenous circadian program and others not. Or why is there plasticity with the option of turning rhythmicity on and off in oscillators, depending on the situation? These questions can be studied in the pineal gland which appears to play very specific roles in different vertebrate species. Many experiments investigating possible self-sustained rhythmicity in the mammalian pineal have been unsuccessful (Paul Pevet, personal communication). Yet, the circadian output used in these experiments was melatonin production, which clearly is driven by SCN outputs. Because other outputs, such as a panel of rhythmic genes, were not investigated, the question whether the mammalian pineal is completely arhythmic in the circadian range is, therefore, still not conclusively settled, and recent experiments indicate that the pineal like many other tissues does show circadian rhythmicity of clock gene RNA in *vitro* (Yamazaki, Abe, Herzog, and Menaker, as well as Tosini, personal communication). Similar to birds (see above), communication between the pineal and the thalamus plays an important role in controlling circadian behavior in reptiles (Foà et al. 1994; Innocenti et al. 1996). Again, their interplay varies with the season, supporting the hypothesis that this concerns fine tuning of seasonal physiology. In the lamprey, physiology, morphology, and behavior drastically change across different life stages, and the self-sustained, rhythmicity of the pineal is either robust or shut off (Menaker and Tosini 1996). Finally, circadian rhythmicity is greatly dampened when hibernators arise from hibernation underground so that the circadian system can be easily reset. Thus, there are obviously advantages for, as well as against rhythmicity, depending on season and/or developmental stage.

We still have to go a long way to fully understand the circadian system. We have found concrete building blocks that are essential for circadian behavior. We have defined morphological clock centers within the organism and clock genes within the cell. An understanding of how the different oscillators which exist at the ecological, the organismal, and potentially also at the cellular level interact is a prerequisite for understanding the ubiquitous phenomenon of circadian temporal programs.

Acknowledgements

This paper is dedicated to M. K. Chandrashekaran, the pioneer and champion of Indian chronobiology and the first holder of "Aschoff's Rule" (1991). We are grateful to Drs. Serge Daan and Mike Menaker for helpful comments on the manuscript. Our work is supported by the DFG and the Meyer-Struckmann-Stiftung.

References

Aschoff, J. (1967) Human circadian rhythms in activity, body temperature and other functions. Life Science and Space Research. 159–173.

Aschoff, J., Gerecke, U., Wever, R. (1967) Desynchronization of human circadian rhythms. Jap. J. Physiol. **17**: 450–457.

Aschoff, J., Gerkema, M. (1985) On diversity and uniformity of ultradian rhythms. In: Schulz, H., Lavie, P. (eds.) Ultradian rhythms in physiology and behavior, pp. 321–334.

Aschoff, J., Goetz Cv (1988) Masking of circadian activity rhythms in male golden hamsters by the presence of females. Behav. Ecol. Sociobiol. **22**: 409–412.

Ballario, P., Vittorioso, P., Magrelli, A., Talore, C., Cabibbo, A., Macino, G. (1996) *White collar-1*, a central regulator of blue light responses in *Neurospora*, is a zinc finger protein. EMBO J. **15**: 1650–1657.

Balsalobre, A., Damiola, F., Schibler, U. (1998) A serum shock induces gene expression in mammalian tissue culture cells. Cell. **93**: 929–937.

Bovet, J., Oertli, E.F. (1974) Free-running circadian activity rhythms in free-living beaver (*Castor canadensis*). J. Comp. Physiol. **92**: 1–10.

Broda, H., Brugge, D., Homma, K., Hastings, J.W. (1985) Circadian communication between unicells? Effects on period by cell-conditioning of medium. Cell Biophys. **8**: 47–67.

Campbell, S.S., Murphy, P.J. (1998) Extraocular circadian phototransduction in humans. Science. **279**: 396–399.

Ceriani, M.F., Darlington, T.K., Staknis, D., Mas, P., Petti, A.A., Weitz, C.J., Kay, S.A. (1999) Light-dependent sequestration of TIMELESS by CRYPTOCHROME. Science. **285**: 553–556.

Chandrashekaran, M.K., Geetha, L., Marimuthu, G., Subbaraj, R., Kumarasamy, P., Ramkumar, M.S. (1991) The menstrual cycle in a human female under social and temporal isolation is not coupled to the circadian rhythm in sleep-wakefulness. Curr. Sci. **60**: 703–705.

Chandrashekaran, M.K., Marimuthu, G. (1985) Communication and synchronization of circadian rhythms in insectivorous bats. Proc. Indian Acad. Sci. **94**: 655–665.

Crosthwaite, S.K., Dunlap, J.C., Loros, J.J. (1997) *Neurospora wc-1* and *wc-2*: Transcription, photoresponses, and the origin of circadian rhythmicity. Science. **276**: 763–769.

Crosthwaite, S.K., Loros, J.J., Dunlap, J.C. (1995) Light-induced resetting of a circadian clock is mediated by a rapid increase in *frequency* transcript. Cell. **81**: 1003–1012.

Crowley, M., Bovet, J. (1980) Social synchronization of circadian rhythms in deer mice (*Peromycus maniculatus*). Behav. Ecol. Sociobiol. **7**: 99–105.

Dawkins, R. (1976) The selfish gene. Oxford University Press, Oxford.

Ebihara, S., Gwinner, E. (1992) Different circadian pacemakers control feeding and locomotor activity in European starlings. J. Comp. Physiol. A. **171**: 63–67.

Ebling, F.J. (1996) The role of glutamate in the photic regulation of the suprachiasmatic nucleus. Prog. Neurobiol. **50**: 109–132.

Edmunds, L.N. Jr (1988) Cellular and molecular bases of biological clocks: Models and Mechanisms of circadian timekeeping. Springer, New York, Heidelberg.

Emery, P., So, W.V., Kaneko, M., Hall, J.C., Rosbash, M. (1998) CRY, a *Drosophila* clock and light-regulated cryptochrome, is a major contributor to circadian rhythm resetting and photosensitivity. Cell. **95**: 669–679.

Enright, J.T. (1980) Temporal precision in circadian systems: a reliable neuronal clock from unreliable components? Science. **209**: 1542–1545.

Field, M.D., Maywood, E.S., O'Brien, J.A., Weaver, D.R., Reppert, S.M., Hastings, M.H. (2000) Analysis of clock proteins in mouse SCN demonstrates phylogenetic divergence of the circadian clockwork and resetting mechanisms. Neuron. **25**: 437–447.

Foà, A., Tosini, G., Minutini, L., Innocenti, A., Quaglieri, C., Flamini, M. (1994) Seasonal changes of locomotor activity patterns in ruin lizards *Podarcis sicula*. I, Endogenous control by the circadian system. Behav. Ecol. Sociobiol. **34**: 267–274.

Foster, R.G., Grace, M.S., Provencio, I., Degrip, W.J., Garcia-Fernandez, J.M. (1994) Identification of vertebrate deep brain photoreceptors. Neurosci. Biobehav. Rev. **18**: 541–546.

Gwinner, E. (1966) Entrainment of a circadian rhythm in birds by species-specific song cycles (Aves, Fringillidae; *Carduelis spinus, Serinus serinus*). Experientia. **22**: 765.

Gwinner, E. (1974) Testosterone induces "splitting" of circadian locomotor activity rhythms in birds. Science. **185**: 72–74.

Gwinner, E., Hau, M., Heigl, S. (1997) Melatonin: generation and modulation of avian circadian rhythms. Brain Res. Bull. **44**: 439–444.

Hastings, J.W. (1960) Biochemical aspects of rhythms: phase shifting by chemicals. Cold Spr. Harb. Symp. Quant. Biol. **25**: 131–148.

Hastings, J.W., Broda, H., Johnson, C.H. (1985) Phase and period effects of physical and chemical factors. Do cells communicate? In: Rensing, L., Jaeger, N.I., (eds.) Temporal Order, Springer Verlag, Berlin, Heidelberg, pp. 213–221.

Hastings, J.W., Sweeney, B.M. (1958) A persistent diurnal rhythm of luminescence in *Gonyaulax polyedra*. Biol. Bull. **115**: 440–458.

Hege, D.M., Stanewsky, R., Hall, J.C., Giebultowicz, J.M. (1997) Rhythmic expression of a PER-reporter in the Malpighian tubules of decapitated Drosophila: evidence for a brain-independent circadian clock. J. Biol. Rhythms. **12**: 300–308.

Hoffman, K. (1971) Splitting of the circadian rhythm as a function of light intensity. In: Menaker, M. (ed) Biochronometry, National Academy of Sciences, Washington, D.C., pp. 134–151.

Holst, E.V. (1939) Die relative Koordination als Phänomen und als Methode zentralnervöser Funktionsanalyse. Ergebn. Physiol. **42**: 228–306.

Honma, K-I, Honma, S., Hiroshige, T. (1987) Activity rhythms in the circadian domain appear in suprachiasmatic nuclei lesioned rats given methamphetamine. Physiol. Behav. **40**: 767–774.

Honma, S., Katsuno, Y., Shinohara, K., Abe, H., Honma, K. (1996) Circadian rhythm and response to light of extracellular glutamate and aspartate in rat suprachiasmatic nucleus. Am. J. Physiol. **271**: R579–R585.

Hunter-Ensor, M., Ousley, A., Sehgal A (1996) Regulation of the *Drosophila* protein Timeless suggests a mechanism for resetting the circadian clock by light. Cell. **84**: 677–685.

Innocenti, A., Bertolucci, C., Minutini, L., Foà A (1996) Seasonal variations of pineal involvement in the circadian organization of the ruin lizard *Podarcis sicula*. J. Exp. Biol. **199**: 1189–1194.

Johnson, C.H., Golden, S.S., Ishiura, M., Kondo, T. (1996) Circadian clocks in prokaryotes. Mol. Microbiol. **21**: 5–11.

Johnson, C.H., Hastings, J.W. (1989) Circadian phototransduction: phase resetting and frequency of the circadian clock of *Gonyaulax* cells in red light. J. Biol. Rhythms. **4**: 417–437.

Kume, K., Zylka, M.J., Sriram, S., Shearman, L.P., Weaver, D.R., Jin, X., Maywood, E.S., Hastings, M.H., Reppert, S.M. (1999) mCRY1 and mCRY2 are essential components of the negative limb of the circadian clock feedback loop. Cell. **98**: 193–205.

Linden, H., Ballario, P., Arpaia, G., Macino, G. (1999) Seeing the light: news in *Neurospora* blue light signal transduction. Adv. Genet. **41**: 35–54.

Linden, H.G. Macino (1997) *White collar* 2, a partner in blue-light signal transduction controlling expression of light-regulated genes in *Neurospora crassa*. EMBO J. **16**: 98–109.

Lovegrove, B.G., Papenfus, M.E. (1995) Circadian activity rhythms in the solitary cape molerat (*Georychus capensis*: Bathvergidae) with some evidence of splitting. Physiol. Behav. **58**: 679–685.

Lucas, R.J., Foster, R.G. (1999a) Circadian clocks: A *cry* in the dark? Curr. Biol. **9**: R825–828.

Lucas, R.J., Foster, R.G. (1999b) Circadian rhythms: Something to cry about? Curr. Biol. **9**: R214–217.

Lucas, R.J., Foster, R.G. (1999c) Photoentrainment in mammals: a role for *cryptochrome*? J. Biol. Rhythms. **14**: 4–10.

Marimuthu, G., Subbaraj, R., Chandrashekaran, M.K. (1978) Social synchronization of the activity rhythm in cave-dwelling bat. Naturwissenschaften **65**: 6000.

Marimuthu, G., Subbaraj, R., Chandrashekaran, M.K. (1981) Social entrainment of the circadian rhythm in the flight activity of the microchiropteran bat *Hipposideros speoris*. Behav. Ecol. Sociobiol. **8**: 147–150.

Meijer, J.H., Daan, S., Overkamp, G.J.F., Hermann, P.M. (1990) The two-oscillator circadian system of tree shrews (*Tupaia belangeri*) and its response to light and dark pulses. J. Biol. Rhythms. **5**: 1–16.

Menaker, M., Eskin, A. (1966) Entrainment of circadian rhythms by sound in *Passer domesticus*. Science. **154**: 1579–1581.

Menaker, M.G. (1996) The evolution of vertebrate circadian systems. In: Honma K-I, Honma, S. (eds.) Circadian Organization and Oscillatory Coupling. Hokkaido University Press, Sapporo, pp. 39–52.

Merrow, M., Franchi, L., Dragovic, Z., Görl, M., Johnson, J., Brunner, M., Macino, G., Roenneberg, T. (2001) Circadian regulation of the light input pathway in *Neurospora Crasa*. EMBO. J. **20**; 307–315.

Michel, S., Geusz, M.E., Zaritsky, J.J., Block, G.D. (1993) Circadian rhythm in membrane conductance expressed in isolated neurons. Science. **259**: 239–241.

Mittag, M., Lee, D-H, Hastings, J.W. (1995) Are mRNA binding proteins involved in the translational control of the circadian regulated synthesis of luminescence proteins in *Gonyaulax polyedra*. In: Horoshige, T., Honma, K. (eds.) Evolution of Circadian Clocks from Cell to Human. Hokkaido University Press, Sapporo, pp.

Morse, D., Hastings, J.W., Roenneberg, T. (1994) Different phase responses of two circadian oscillators in *Gonyaulax*. J. Biol. Rhythms. **9**: 263–274.

Morse, D., Milos, P.M., Roux, E., Hastings, J.W. (1989) Circadian regulation of bioluminsecence in *Gonyaulax* involves translational control. Proc. Natl. Acad. Sci. USA **86**: 172–176.

Njus, D., Gooch, V.D., Hastings, J.W. (1981) Precision of the *Gonyaulax* circadian clock. Cell Biophys. **3**: 223–231.

Okamura, H., Miyake, S., Sumi, Y., Yamaguchi, S., Yasui, A., Muijtjens, M., Hoeijmakers, J.H.J., Horst, G.T.J.v.d. (1999) Photic induction of *mPer1* and *mPer2* in *Cry*-defcient mice lacking a biological clock. Science. **286**: 2531–2534.

Plautz, J.D., Kaneko, M., Hall, J.C., Kay, S.A. (1997) Independent photoreceptive circadian clocks throughout *Drosophila*. Science **278**: 1632–1635.

Pohl, R. (1948) Tagesrhythmus in phototaktischem Verhalten der *Euglena gracilis*. Z. Naturforsch. **3**: 367–374.

Ramalho, C.B., Hastings, J.W., Colepicolo, P. (1995) Circadian oscillation of nitrate reductase activity in *Gonyaulax polyedra* is due to changes in cellular protein levels. Plant Physiol. **107**: 225–231.

Refinetti, R., Menaker, M. (1992) Evidence for separate control of estrous and circadian periodicity in the golden hamster. Behavioral and Neural Biology **58**: 27–36.

Roenneberg, T. (1994) The *Gonyaulax* circadian system: Evidence for two input pathways and two oscillators. In: Hiroshige, T., Honma K-I (eds.) Evolution of circadian clock. Hokkaido University Press, Sapporo, pp. 3–20.

Roenneberg, T., Deng, T-S. (1997) Photobiology of the *Gonyaulax* circadian system: I Different phase response curves for red and blue light. Planta. **202**: 494–501.

Roenneberg, T., Foster, R.G. (1997) Twilight Times-Light and the circadian system. Photochem. Photobiol. **66**: 549–561.

Roenneberg, T., Hastings, J.W. (1988) Two photoreceptors influence the circadian clock of a unicellular alga. Naturwissenschaften **75**: 206–207.

Roenneberg, T., Merrow, M. (2001) The role of transcriptional feedbacks in circadian systems. In: Hiroshige, T., Honma, K. (eds.) The brain that keeps time. Hokkaido Univ. Press, Sapporo. (in press)

Roenneberg, T., Merrow, M., Eisensamer, B. (1998) Cellular mechanisms of circadian systems. Zoology **100**: 273–286.

Roenneberg, T., Morse, D. (1993) Two circadian oscillators in one cell. Nature **362**: 362–364.

Roenneberg, T., Rehman, J. (1996) Nitrate, a nonphotic signal for the circadian system. FASEB J. **10**: 1443-1447.

Roenneberg, T., Taylor, W. (1994) Light induced phase responses in *Gonyaulax* are drastically altered by creatine. J. Biol. Rhythms **9**: 1–12.

Schardt, U., Wilhelm, I., Erkert, H.G. (1989) Splitting of the circadian activity rhythm in common marmosets (*Callithrix j.jacchus*; Primates). Experientia **45**: 1112–1115.

Shearman, L.P., Zylka, M.J., Weaver, D.R., Kolakowski, L.F., Reppert, S.M. (1997) Two *period* homologs: circadian expression and photic regulation in the suprachiasmatic nuclei. Neuron **19**: 1261–1269.

Shigeyoshi, Y., Taguchi, K., Yamamoto, S., Takekida, S., Yan, L., Tei, H., Moriya, T., Shibata, S., Loros, J., Dunlap, J.C., Okamura, H. (1997) Light-induced resetting of a mammalian clock is associated with rapid induction of the *mPer1* transcript. Cell **91**: 1043–1053.

Stanewsky, R., Kaneko, M., Emery, P., Beretta, B., Wagner-Smith, K., Kay, S.A., Rosbash, M., Hall, J.C. (1998) The *cry^b* mutation identifies cryptochrome as a circadian photoreceptor in *Drosophila*. Cell **95**: 681–692.

Steinlechner, S. (1989) Is there is a feedback loop between the pineal gland and the retina? Adv. Pineal Res. **3**: 175–180.

Sweeney, B.M. (1987) Rhythmic phenomena in plants. Academic Press, San Diego.

Takahashi, J.S. (1987) Cellular basis of circadian rhythms in the avian pineal. In: Hiroshige T, Honma, K. (eds.) Comparative aspects of circadian clocks. Hokkaido University Press, Sapporo, pp. 3–15.

Tosini, G., Menaker, M. (1996) Circadian rhythms in cultured mammalian retina. Science **272**: 419–421.

Vondrasova-Jelinkova, D., Hajek, I., Illnerova, H. (1999) Adjustment of the human melatonin and cortisol rhythms to shortening of the natural summer photoperiod. Brain Res. **816**: 249–253.

Welsh, D.K., Logothetis, D.E., Meister, M., Reppert, S.M. (1995) Individual neurons dissociated from rat suprachiasmatic nucleus express independently phased circadian firing rhythms. Neuron **14**: 697–706.

Whitmore, D., Foulkes, N.S., Sassone-Corsi, P. (2000) Light acts directly on organs and cells in culture to set the vertebrate circadian clock. Nature **404**: 87–91.

Yamazaki, S., Kerbeshian, M.C., Hocker, C.G., Block, G.D., Menaker, M. (1998) Rhythmic properties of the hamster suprachiasmatic nucleus *in vivo*. J. Neurosci. **18**: 10709–10723.

Yamazaki, S., Numano, R., Abe, M., Hida, A., Takahashi R-i, Ueda, M., Block, G.D., Sakaki, Y., Menaker, M., Tei, H. (2000) Resetting central and peripheral circadian oscillators in transgenic rats. Science **288**: 682–685.

Zylka, M.J., Shearman, L.P., Weaver, D.R., Reppert, S.M. (1998) Three *period* homologs in mammals: differential light responses in the suprachiasmatic circadian clock and oscillating transcripts outside of brain. Neuron **20**: 1103–1110.

Biological Rhythms
Edited by V. Kumar
Copyright © 2002, Narosa Publishing House, New Delhi, India

7. Retinal Circadian Rhythms

G. Fleissner* and G. Fleissner

Zoologisches Institut, J. W. Goethe-Universität, Siesmayerstr. 70, D-60054 Frankfurt a.M., Germany

In addition to light and darkness, endogenous signals control the sensitivity of the eyes, most pronounced in night active animals. Endogenous signals from the circadian clock anticipate the daily changes of environmental light and protect the sensitive eyes against possible damage by daylight – even in constant darkness. Oscillators driving these eye rhythms could be shown to be located in the retina itself in several vertebrates (e.g., *Xenopus*, chicken, quail, rat and several rodents) and in the snails *Bulla* and *Aplysia*. These retinal oscillators are in use for studying the cellular mechanisms of pacemaker neurons. In arthropods, however, the pacemaker of retinal rhythmicity is centrally located and bilaterally symmetrical – in insects most likely in neurons of the medulla accessoria. As mutually coupled oscillators they function as the master clock for circadian behavioural and developmental rhythms. The retinal rhythms of arthropods can be used as direct monitors of pacemaker time. In this paper we mainly review the basic mechanisms of adaptation, especially the signals and signalling pathways of the endogenous control of retinal rhythmicity in arthropod models. We use these results to demonstrate their impact on the analysis of neuronal mechanisms of circadian clock systems and on current paradigms to interpret results on the cellular and subcellular level.

Introduction

In nearly all textbooks, the visual sensitivity of photoreceptors is described as controlled directly by light and darkness, eyes are termed "light adapted" or "dark adapted", according to the change of sensitivity (= adaptation) induced by the surrounding light conditions. But additionally, endogenous processes influence the sensitivity of the eyes, mainly controlled by circadian clocks. This circadian retinal rhythm is known since long, and it may even be named one of the oldest reports on endogenously controlled circadian rhythms in animals, while prior in the history of chronobiology observations of leaf and petal movements of plants and periodicities in human body functions attracted the interest of scientists. While already Leydig (1864) and Exner (1881) described day/night changes of the screening pigment in insect eyes, it was Kiesel (1894) to show these changes to occur rhythmically, by means of describing the nocturnal eye shine of the moth *Plusia gamma* under constant darkness. The persistence of this rhythm despite constant experimental conditions can only be explained by an endogenous drive (Fig. 1). Similar results could be obtained by Demoll (1917); Parker (1932) and Welsh (1938). Jahn and Crescitelli (1940) were the first to confirm diurnal rhythms in beetle eyes by electrophysiological methods. Descriptions of circadian retinal rhythms in other arthropod groups followed: e.g., crustaceae (histologically: Welsh 1930; electrophysiologically: Arechiga and Wiersma 1969); chelicerates (histologically and electrophysiologically: Fleissner 1971, 1972). Retinal circadian

*E-mail fleissner@zoology.uni-frankfurt.de

rhythms could also be demonstrated in the eyes of a mollusc by the rhythmic bursts of compound action potentials, first in *Aplysia californica* (Jacklet 1969) and later in several other mollusc species: *Bulla* (Block and Wallace 1982), *Bursatella* (Block and Roberts 1981), *Haemonea* (McMahon and Block 1982), *Navanax* (Eskin and Harcombe 1977). Only little later the first retinal rhythms could be shown in the eyes of vertebrates mainly by rhythmic melatonin production: in chicken (Hamm and Menaker 1980), quail (Underwood and Siopes 1985) and the frog *Xenopus* (Besharse and Iuvone 1983). In rats, additionally a discsshedding rhythm could be demonstrated (Reme et al. 1986).

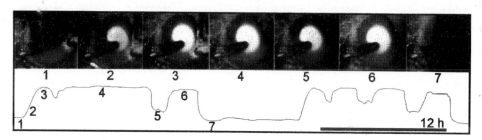

Fig. 1 **Circadian rhythm of eye shine in the compound eye of the beetle *Dynastes hercules*. At night the tapetum at the base of the retina is exposed and reflects incident light. (upper row: serial infrared video-recordings, lower row: rel. Intensity of the reflected light. DD conditions; the numbers indicate corresponding frames in the recordings). (changed from Fleissner and Fleissner 1988).**

Already Exner (1881) posed the questions, how can endogenous factors regulate adaptational processes in the eyes, and he suggested the three alternatives of neural, hormonal and metabolic drives. But it took several decades to experimentally show first examples of mechanisms controlling the state of retinal sensitivity in arthropods: e.g., Day 1941; hormonal control in crustacean eye stalk: Fingerman et al. 1959; efferent terminals in the scorpion median and lateral eyes: Fleissner and Schliwa, 1977b; and their role as circadian signalling pathways: Fleissner and Fleissner 1978; efferent terminals in the lateral eye of Limulus: Fahrenbach 1981; and their control function: Barlow et al. 1977; Chamberlain and Barlow 1979.

Other questions were addressed little later: Where and what is the origin of these endogenous periodic signals? In vertebrates and molluscs the retina houses its own oscillator. Therefore, these model systems were used for the cellular analysis of pacemaker cells, which lead to a detailed understanding of pacemaker synchronisation and the mechanisms involved in the expression of cellular rhythms (for review: Block and Michel 1997).

It turned out that in all known examples of arthropod retinal rhythm, the circadian oscillator was not situated inside the retina, but between the second and third optic neuropil (e.g., beetles: Fleissner 1982; cockroach: Wills et al. 1985). In these animal models, the neurobiological analysis of the circadian pathways gained relevance for identifying circadian pacemaker neurons, as well as the input and output mechanisms. In this paper we will mainly review the results with our arthropod model systems, scorpion and beetle. They are ideal chronobiological research subjects, as they live in arid environments and even in nature, they experience constant conditions, often for many weeks, when hidden in their burrow during the desert summer. In this paper we will show the basic adaptational mechanisms involved in retinal rhythmicity and the known neuronal components of the underlying circadian clock system.

Mechanisms controlling the sensitivity of compound eyes are governed by the circadian clock

The sensitivity of a photoreceptor can be controlled by changes of the photon flux through in the dioptric apparatus, the shielding pigment and the photoreceptive membranes, the rhabdomes themselves, and additionally, by changes along the molecular receptor cascade. The relative contribution of the different mechanisms mirrors both, the phylogenetic background and the species-specific ecological demands.

In principle, light-adaptation reduces the light input to photoreceptors, dark-adaptation increases this light flux. Changes of the acceptance angle and spectral sensitivity may arise as "side-effects". The *dioptric apparatus* contributes to light/dark adaptation by changing the length of the crystalline cone and the crystalline tract for regulating the distance between corneal surface and rhabdom. The *shielding pigment* can be used like a sunglass and placed in front of the rhabdomeres, it can generate a wide or narrow pupil and it can optically isolate the different light channels, the ommatidia, from each other, or allow common exploitation of light. The *rhabdoms* may massively increase or decrease their volume, thus changing the size of the photoreceptive membrane. Additionally, various principles of *mirror optics* can be found to achieve the highest possible photon-capture, e.g., in the eyes of night-active animals. Reflecting layers (tapetum) behind the retina may be exposed or concealed by the shielding pigment. Peri-rhabdomeric ER-cisterns can be found so close to the microvilli that they make the rhabdom a fibre-light guide.

In the beetle, as can be seen here, the basic framework for photoreceptor sensitivity is set by the clock-controlled day/night adaptation, it is only superimposed by light/dark adaptation, yielding 4 adaptational states clearly different in all involved structural components (Fig. 2) (Dube and Fleissner 1986). This figure demonstrates that even in constant darkness, there is a prominent difference between the day eye and the night eye. This is reflected in a rhythmic change of visual sensitivity as can be monitored in long-term recordings of the electroretinogram (ERG) (Fig. 3) (Fleissner 1986).

Intensity response functions of day and night eyes show that the night eye is up to 10,000 fold (beetles: Fleissner and Fleissner 1986; scorpions: Fleissner 1971; 1972) more sensitive than the day eye, thus rhythmically pre-adapting the eyes to the lower environmental light intensities during night, and, in the end of the night, functioning as a precaution against the high daylight levels sheltering these sensitive eyes by a clock controlled sunglass rhythm. The putatively damaging effect of high light intensities applied to sensitive eyes has been shown several times (e.g., Meyer-Rochow 1994; for review see Eguchi 1999) and has even been used to analyse the spectral sensitivity in certain photoreceptor cells (e.g., Langer et al. 1986). Restoration processes of damaged photoreceptive membranes can last several days, leaving the animals functionally blind. Therefore, in all nocturnal animals, studied so far, such a protective circadian rhythm of sensory adaptation could be demonstrated by electrophysiological and histological methods (e.g., cockroach: Wills et al. 1985; for review see Fleissner and Fleissner 1988; or Meyer-Rochow 1999). In diurnal animals, which do not need precise and very sensitive night-time vision, these adaptational rhythms are less pronounced, but still discernible (fly: Chen et al. 1992) .

The circadian pacemaker controlling these retinal rhythms lies in the photic input pathways next to the higher optic neuropils

By means of series of lesion experiments and backfillings it could be shown that the circadian clock behind these ERG-rhythms is located neither in the retina nor in the brain, but in the optic lobe: In

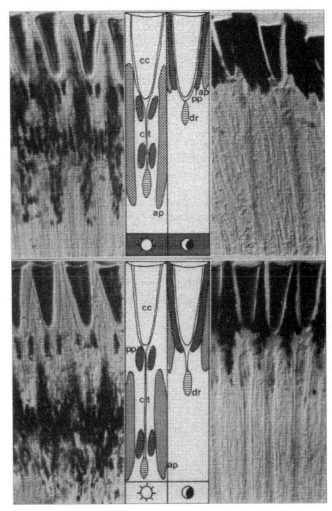

Fig. 2   Histology of retino-motor mechanisms in the retina of the compound eye of the beetle *Pachymorpha sexguttata*. Efferent signals from the circadian clock during day (left side) and during night (right side) - set the limits for the migration of shielding pigment and the length of the crystalline tract as caused by light (lower row) and dark adaptation (upper row). (longitudinal sections of the distal retina; unstained resin-embedded; cornea and proximal ends of receptor cells not shown; ap accessory pigment cells; cc crystalline cone; ct crystalline tract; dr distal rhabdom; pp primary pigment cells) (after Dube and Fleissner 1986).

beetles, this pacemaker is located in the region of the medulla accessoria between the medulla and the lobula, the second and third optic neuropil (Fleissner 1982) (Fig. 4). Few neurons of this neuropil could be immunohistologically marked by an antibody raised against the clock-protein PER of the fruitfly *Drosophila* (Fleissner et al. 1993; Frisch et al. 1996). These cells may be homologous to the "lateral neurons" in the fruitfly which are shown to be essential for the circadian rhythm of locomotor activity of the fly (rev. Helfrich-Förster et al. 1998).

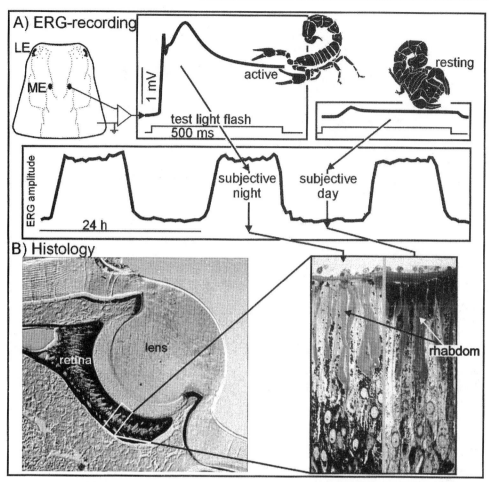

Fig. 3 Circadian ERG-rhythm in the median eyes of the scorpion *Androctonus australis*, demonstrated (A) by electrophysiological recordings of the electroretinogram (ERG) and (B) by histological methods. (A) upper row: Thin tungsten wires fixed to the cornea of the median eyes serve as surface electrodes (left side scheme: dorsal view of the prosoma); Identical light flashes evoke high ERG-amplitudes in the night, when the scorpion is in its active phase (middle), low ERG-amplitudes during the day, when the scorpion is resting (right side). Middle row: Continuous plots of the ERG-amplitudes in DD over several days reveal a clear circadian rhythm of sensitivity with high amplitudes during the subjective night and low amplitudes during the subjective day. (B) The median eye of the scorpion is of ocellar type (left side, The retina is kept in focus behind the lens by a vitreous body. Longitudinal section). (right side: Part of the retina at higher magnification in electro-physiologically determined circadian states in DD). The shielding pigment inside the photoreceptor cells is redrawn into the depth of the retina, thus exposing the rhabdoms to the incident light (top of figure), during night (left half) and pushed in front of the rhabdoms during day, then functioning like a sun-glass (right half), (after Fleissner and Fleissner 1998).

The circadian pacemaker in insects is bilaterally organised. The mutual interaction of both optic lobe pacemakers defines the subjective central circadian time of the animal (see below). The same holds true for scorpions, they also have a bilaterally organised circadian pacemaker. Though not yet

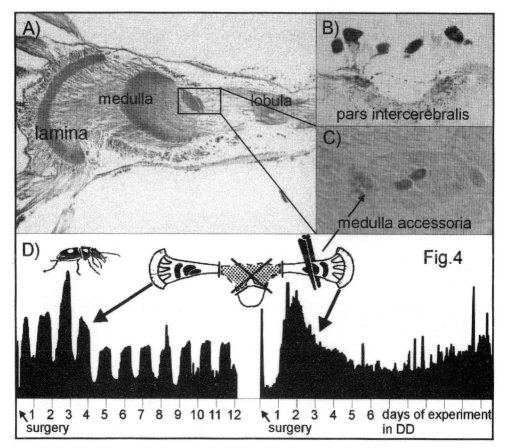

Fig. 4. Localisation of the circadian pacemaker in the beetle *Pachymorpha sexguttata* by immunohistological methods (A to C) and microsurgery (D). (A) In the optic lobe of the beetle, the optic neuropils are clearly discernible and due to their size easily accessible to microsurgery. (B and C) Immunocytology with an antibody raised against the *Drosophila* clock-protein PER marks few neurons, their axons and terminal regions in the brain (B) and the optic lobe (C). While the PER-immunoreactive neurons in the brain seem to be dispensable for the generation of the circadian rhythm, the neurons next to the medulla accessoria are essential components for rhythm expression. This can be derived from microsurgery and simultaneous ERG-recordings (D): In a beetle with excised brain, the optic lobes still preserve a complete circadian clock system. The ERG-rhythm persists (left side), but will be stopped by squeezing the optic lobe between medulla and lobula (right side). (D: changed from Fleissner 1982)

known in its structural identity, lesion experiments have shown that it is located near the optic ganglia "3 to 5" between the medulla and the central body (Fleissner and Fleissner 1985).

Efferent fibres signal the circadian time to the retina

But how is the pacemaker time transmitted to the retina and other effector systems? In scorpions we could trace this information processing pathway down to the synaptic contacts and here identify the information flow — due to a structural specificity of the arachnid optic system: Different from the insects, the arachnids, including the scorpions, have efferent terminals inside the retina (Fleissner and Schliwa 1977). They have been backfilled and thus traced along their entire central course (Fleissner and Heinrichs 1982): They stem from about 10 to 20 neurons bilaterally located near the esophagus. Each of these neurons innervates the eyes on both sides , the central body and the first optic neuropil. They receive synaptic input at the level of the optic ganglia 3-5 and in the central body only (Fig. 5) (Fleissner and Heinrichs 1982). It could be shown that these fibres are the only circadian pathways to the retina. The circadian night is given as a low frequency nervous discharge releasing octopamine inside the retina which drives the shielding pigment into the proximal ends of the receptor cells (see Fig. 3) (Fleissner and Fleissner 1985).

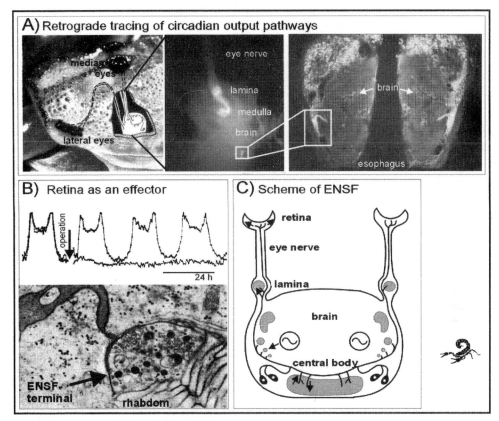

Fig. 5 **Tracing of the circadian signal in the median eyes and the brain of the scorpion *Androctonus australis*. (A) Lucifer yellow (LY) backfillings of the efferent neurosecretory fibres (ENSF) via the median eye nerve. (left side scheme: lateral view of the prosoma with the site of the brain; middle: whole mount of the brain under UV-light after LY applied to the eye nerve; right side: horizontal section through the brain at the height of the esophagus. The filled somata of the ENSF can be seen on each side of the brain. (B) The circadian ERG-rhythm of a median eye stops, after the ipsilateral optic nerve with the ENSF is cut (Operation) (upper row: ERG-recordings from both side median eyes). The circadian signal reaches the photoreceptor cells via efferent terminals next to their rhabdoms (lower inset: electronmicroscopic view of the median eye retina). (C) Schematic reconstruction of the central course and information flow in the ENSF (Shaded areas: optic neuropils and central body mass) (changed from Fleissner and Fleissner 1998).**

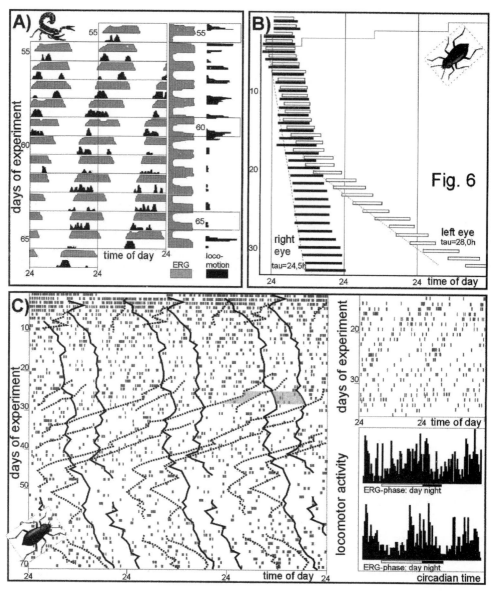

Fig. 6   The ERG-rhythm as monitor of subjective time. (A) Simultaneous recordings of ERG (grey) and
locomotor activity (black) of the scorpion *Androctonus australis*. The ERG-rhythm is free-running
with a period length (tau) shorter than 24h, locomotor activity occurs irregularly in a sloppy
phase relationship but obviously with the same tau (left side: double plot amplitudes normalised
to maximal values in each row; right side: continuous plot of absolute amplitudes; DD conditions).
(B) ERG-recordings from the left and right compound eye of the beetle *Blaps gigas*. The right and
left side circadian pacemaker internally desynchronise either spontaneously or when (this
experiment) one side eye receives a dim light (light bars: sensitive night states of the ERG of the
left eye), while the other one stays in darkness (black bars: sensitive night states of the ERG of the
right eye; triple plot). (C) Simultaneous recording of left and right eye ERG-rhythms and locomotor

The circadian day state is simply the result of lacking impulses in the efferent fibres, the pigment granules then migrate into the distal tips of the photoreceptors and stay there in front of the rhabdomeres. Here they are also trapped in de-efferented eyes (Fleissner and Fleissner 1978). In insect compound eyes there are also efferent fibres involved with serotonin as putative circadian signal, but these fibres do not reach into the retina but mostly end at the distal border of the lamina, first optic neuropil (Fleissner and Fleissner 1988). The pathways between the lamina and the retina are still unknown. Candidates might be scattered serotonergic fibres extending beyond the lamina and neurosecretory synapses directly behind the basal retinal membrane. In any case can the night state of the eye be completely simulated by increasing doses of serotonin (Fleissner and Fleissner 1998b). Similar efferent control systems of the circadian state of eyes have been shown in *Limulus* (Barlow et al. 1980; Calman and Battelle 1991) and spiders (Yamashita and Tateda 1981) (for review see: Fleissner and Fleissner 1988).

Retinal rhythms directly monitor the circadian "System time"

The retinal rhythms in arthropods are directly controlled by their ipsilateral circadian pacemaker which is one half of the "master clock" of the organism. In no case a separate retinal clock has been shown. This offers a great advantage when complex circadian periodicities at the system level are to be interpreted: The ERG rhythms may be used as direct monitors of pacemaker time with striking clarity compared to the animal's locomotor rhythms (Fig. 6A).

The circadian clock is a duplex system composed of two oscillators, one in both side optic lobes. Internal and external coupling of both pacemakers is essential: Spontaneously under constant conditions or due to experimental conditions, the loss or loosening of the coupling strength between right and left side may result in internal desynchronisation (beetles: Koehler and Fleissner 1978). In beetles, the two oscillators receive their own *zeitgeber* input, which may not affect the phase of the contra-lateral side, and they also lengthen or shorten their period lengths according to the ipsi-lateral illumination (Fig. 6 B) (Koehler 1987). In scorpions, both side eyes receive input from both oscillators, the relative contribution of the two input signals can be identified in split brain preparations or when one of the pacemakers or its coupling pathways has been destroyed (Michel 1989).

Experimental approaches to localise the pacemaker or to identify system variables on the molecular or cellular level are often cross-checked by the effects on the system level, e.g., on the locomotor rhythm. In general, a hurra-effect in this research is the loss of circadian rhythmicity and this is interpreted as hit to the rhythm generating processes. But simultaneous recordings of ERG and locomotor rhythms in the same animal show that "obvious" locomotor arrhythmicity can be recalculated from the persistent ERG-rhythms to a desynchronous bilateral clock system (Fig. 6 C) (Fleissner and Fleissner 1998a). In such a case, the experimental procedure may have hit bilateral coupling pathways, only.

(Fig. 6 *Contd.*)

activity of the beetle *Blaps gigas*. The circadian rhythm of locomotor activity is controlled by both sides pacemakers. Spontaneously the ERG-rhythm of one eye accelerates from about day 20 to day 40. (lines indicate the time course of ERG rhythms with their onset and end of night state sensitivity (shaded areas next day 30); locomotor activity is shown as grey stipples; triple plot). The insets (right side) show the locomotor activity pattern during this time of internal desynchronisation. Upper inset: original recordings; lower inset: Recalculation of the same set of data on the basis of modulo tau of the right and left ERG-rhythm. This calculation clearly indicates that both pacemakers have impact on the central locomotor control (A: data from Michel 1989; B: changed from Koehler and Fleissner 1978; C: data from Koehler 1987).

Conclusion

Retinal circadian rhythms in arthropods serve as a useful tool for the neurobiological analysis of the circadian clock at the organismic level. Retinal circadian rhythms in molluscs and vertebrates give insight into the cellular organisation of the clock system. The genetic analysis of rhythm generation unravels the underlying molecular mechanisms at the subcellular level. We now need research work which tries to link these different approaches before we can understand the complexity of circadian clock systems.

Acknowledgement

Supported by grants from the Deutsche Forschungsgemeinschaft. We thank our students and colleagues for their inspiring co-operation which helped to compile the data for this review.

References

Arechiga, H., Wiersma, C.A.G. (1969) Circadian rhythms of responsiveness in crayfish visual units. J. Neurobiol.**1**: 71–85.

Barlow, R.B. Jr., Bolanowski, S.J. Jr., Brachman, M.L. (1977) Efferent optic nerve fibers mediate circadian rhythms in the *Limulus* eye. Science **197**: 86–89.

Barlow, R.B. Jr., Chamberlain, S.C., Levinson, J.Z. (1980) *Limulus* brain modulates the structure and function of the lateral eyes. Science **210**: 1037–1039.

Besharse, J.C., Iuvone, P.M. (1983) Circadian clock in *Xenopus* eye controlling retinal serotonin N-acetyltransferase. Nature **305**: 133–135.

Block, G.D., Michel, S. (1997) Rhythms in Retinal Mechanisms. In: P.H. Redfern and B. Lemmer (eds) Physiology and Pharmacology of Biological Rhythms, Hb. Exp. Pharmacol. vol. 125. Springer, Berlin, pp 435–455.

Block, G.D., Roberts, M.H. (1981) Circadian pacemaker in the *Bursatella* eye: properties of the rhythm and its effect on locomotor behavior. J. Comp. Physiol. **142**: 403–410.

Block, G.D., Wallace, S. (1982) Localization of a circadian pacemaker in the eye of a mollusc, *Bulla*. Science **217**: 155–157.

Calman, B.G., Battelle, B.A. (1991) Central origin of the efferent neurons projecting to the eyes of *Limulus polyphemus*. Visual Neurosci. **6**: 481–495.

Chamberlain, S.C., Barlow, R.B. Jr. (1979) Light and efferent activity control rhabdom turnover in *Limulus* photoreceptors. Science **206**: 361–363.

Chen, D.M., Christianson, J.S., Sapp, R.J., Stark W.S. (1992) Visual receptor cycle in normal and period mutant *Drosophila*: microspectrophotometry, electrophysiology, and ultrastructural morphometry. Visual Neurosci. **9**: 125–135.

Day, M.F. (1941) Pigment migration in the eyes of the moth *Ephestia kuehniella* Zeller. Biol. Bull. **80**: 275–291.

Demoll, R. (1917) Die Sinnesorgane der Arthropoden, ihr Bau und ihre Funktion. Vieweg. Braunschweig.

Dube, C., Fleissner, G. (1986) Circadian rhythms in the compound eyes of the carabid beetle *Pachymorpha (Anthia) sexguttata*. II. Retinomotoric mechanism of exo- and endogenous sensitivity control. In: Den Boer P (ed) Adaptation, Dynamics and Evolution of Carabid Beetles. Gustav Fischer Verlag, Stuttgart, New York, pp 19-26.

Eguchi, E. (1999) Membrane turnover of rhabdom. In: Eguchi, E., Tominaga, Y. (eds.) Atlas of arthropod sensory receptors. Springer Verlag, Tokyo, pp 87–96.

Eskin, A., Harcombe, E. (1977) Eye of *Navanax*: Optic activity, circadian rhythm and morphology. J. Comp. Biochem. Physiol. A. **57**: 443–449.

Exner, S. (1981) Die Physiologie der facettirten Augen von Krebsen und Insecten. Franz Deuticke, Leipzig Wien.

Fahrenbach, W.H. (1981) The morphology of the horseshoe crab (*Limulus polyphemus*) visual system. VI. Innervation of photoreceptor neurons by neurosecretory efferents. Cell Tiss. Res. **216**: 655–659.

Fingerman, M., Lowe, M.E., Sundaraj, B.I. (1959) Dark-adapting and light-adapting hormones controlling the distal retinal pigment of the prawn *Palaemontes vulgaris*. Biol. Bull. **116**: 30–36.

Fleissner, G. (1971) Über die Sehphysiologie von Skorpionen. Belichtungspotentiale in den Medianaugen von *Androctonus australis L.* und ihre tagesrhythmischen Veränderungen. Inaug. Diss. J. W. Goethe-Universität, Frankfurt am Main.

Fleissner, G. (1972) Circadian sensitivity changes in the median eyes of the North African scorpion, *Androctonus australis*. In: Wehner R (ed) Information processing in the visual system of arthropods. Springer, Berlin Heidelberg New York, pp 133–139.

Fleissner, G. (1974) Circadian Adaptation und Schirmpigmentverlagerung in den Sehzellen der Medianaugen von *Androctonous australis* L. (Buthidae, Scorpiones). J. Comp. Physiol. **91**: 399–416.

Fleissner, G., (1982) Isolation of an insect circadian clock. J. Comp. Physiol. A. **149**: 311–316.

Fleissner, G., Fleissner, G. (1978) The optic nerve mediates the circadian pigment migration in the median eyes of the scorpion. Comp. Biochem. Physiol. A **61**: 69–71.

Fleissner, G., Fleissner, G. (1985) Neurobiology of a circadian clock in the visual system of scorpions. In: Barth, F. (ed.) Neurobiology of Arachnids. Springer Verlag, Berlin Heidelberg New York, pp 351–375.

Fleissner, G., Fleissner, G. (1986) Circadian rhythms in the compound eyes of the carabid beetle *Pachymorpha (Anthia) sexguttata*. I. Sensitivity-rhythms and the bilateral circadian oscillator system. In: Den Boer P. (ed.) Adaptation, Dynamics and Evolution of Carabid Beetles. Gustav Fischer Verlag, Stuttgart New York, pp 19–26.

Fleissner, G., Fleissner, G. (1988) Efferent control of visual sensitivity in arthropod eyes: with emphasis on circadian rhythms. Information processing in animals, vol. 5, Gustav Fischer Verlag, Stuttgart New York.

Fleissner, G., Fleissner, G. (1998a) The neurobiological background of the circadian system of beetles - a critical test of molecular approaches to clock function. Proc. SRBR. **6**: 112.

Fleissner, G., Fleissner, G. (1998b) Neuronale Grundlagen biologischer Uhren. Forschung Frankfurt. Jahrgang 98 (**2**): 56–67.

Fleissner, G., Fleissner, G. (1999) Simple eyes of arachnids. In: Eguchi, E., Tominaga, Y. (eds.) Atlas of arthropods sensory receptors. Springer, Tokyo, pp 55–70.

Fleissner, G., Heinrichs, S. (1982) Neurosecretory cells in the circadian-clock system of the scorpion, *Androctonus australis* Cell Tiss. Res. **224**: 233–238.

Fleissner, G., Schliwa, M. (1977) Neurosecretory fibres in the median eyes of the scorpion, *Androctonus australis* L. Cell Tiss. Res. **178**: 189–198.

Fleissner, G., Fleissner, G., Frisch, B. (1993) A new type of putative non-visual photoreceptors in the optic lobe of beetles. Cell Tiss. Res. **273**: 435–445.

Frisch, B., Fleissner, G., Brandes, C., Hall, J.C. (1996) Staining in the brain of *Pachymorpha sexguttata* mediated by an antibody against a *Drosophila* clock-gene product: labeling of cells with possible importance for the beetle's circadian rhythms. Cell Tiss. Res. **286**: 411–429.

Hamm, H.E., Menaker, M., (1980) Retinal rhythms in chicks: circadian variation in melatonin and serotonin N-acetyltransferase activity. Proc. Natl. Acad. Sci. USA **77**: 4998–5002.

Helfrich-Förster, C., Stengl, M., Homberg, U. (1998) Organization of the circadian system in insects. Chronobiol. Internat. **15**: 567–594.

Jacklet, J.W. (1969) Circadian rhythm of optic nerve impulses recorded in darkness from isolated eye of *Aplysia*. Science **164**: 562–563.

Jahn, T.J., Crescitelli, F. (1940) Diurnal changes in the compound eye. Biol. Bull. **78**: 42–52.

Kiesel, A. (1894) Untersuchungen zur Physiologie des facettirten Auges. S. B. Akad. Wiss. Wien, math.nat. Kl. **103**: 97–139.

Koehler, W. (1987) Die circadiane Uhr von Schwarzkäfern der Gattung *Blaps*, ein Multioszillatorsystem. Inaug. Diss, J.W. Goethe- Universität. Frankfurt am Main.

Koehler, W., Fleissner, G. (1978) Internal desynchronisation of bilaterally organised circadian oscillators in the visual system of insects. Nature **274**: 708–710.

Langer, H., Schmeincke, G., Anton-Erxleben, F. (1986) Identification and localization of visual pigments in the retina of the moth, *Antheraea polyphemus* (Insecta, Saturniidae). Cell. Tiss. Res. **245**: 81–89.

Leydig, F. (1864) Das Auge der Gliederthiere. H. Lauppsche Buchhandlung, Tübingen.

McMahon, D.G., Block, G.D. (1982) Organized photoreceptor layer is not required for light responses in three opisthobranch eyes. Neurosci. **8**: 33.

Meyer-Rochow, V.B. (1994) Light-induced damage to photoreceptors of spiny lobsters and other crustaceans. Crustaceana. **67**: 95–109.

Meyer-Rochow, V.B. (1999) Compound eye: Circadian rhythmicity, illumination, and obscurity. In: Eguchi, E., Tominaga, Y. (eds.) Atlas of Arthropod Sensory Receptors. Springer, Tokyo, pp 97– 124.

Michel, S. (1989) Laufaktivität und Augenempfindlichkeit von Skorpionen. Untersuchungen der Kopplung in einem circadianen Multioszillatorsystem. Inaug. Diss., J. W. Goethe-Universität. Frankfurt am Main.

Parker, G.H. (1932) The movement of the retinal pigment. Ergebn. Biol. **9**: 239–291.

Reme, C., Wirz-Justice A., Rhyner A., Hofmann S. (1986) Circadian rhythm in the light response of rat retinal disk-shedding and autophagy. Brain Res. **369**: 356–60.

Underwood, H., Siopes, T. (1985) Melatonin rhythms in quail: regulation by photoperiod and circadian pacemakers. J. Pineal Res. **2**: 133–143.

Welsh, H. (1930) Diurnal rhythm of the distal pigment cells in the eyes of certain crustaceans. Proc. Nat. Acad. Sci. USA **16**: 386–395.

Welsh, J.H. (1938) Diurnal rhythms. Quart. Rev. Biol. **13**: 123–139.

Wills, A.S., Page, T.L., Colwell, C.S. (1985) Circadian rhythms in the electroretinogram of the cockroach. J. Biol. Rhythms. **1**: 25–37.

Yamashita, S., Tateda, H. (1981) Efferent neural control in the eyes of orb weaving spiders. J. Comp. Physiol. A. **143**: 477–483.

Biological Rhythms
Edited by V. Kumar

8. Perception of Natural Zeitgeber Signals

G. Fleissner and G. Fleissner*

Zoologisches Institut, J.W. Goethe-Universität, Siesmayerstr. 70, D-60054 Frankfurt a. M., Germany

Biological clocks detect timing cues in their natural environment from the daily twilight transitions during dusk and dawn. Compared to lab conditions (lights on/off programs) these natural *zeitgeber* signals guarantee better external and internal synchronisation with respect to precision (onset of activity), threshold (of effective stimuli) and range of entrainment. This has been shown in humans, various rodent species, birds and arthropods. It is a so far unsolved problem, which sensory and neuronal mechanisms allow for twilight recognition, as image forming eyes extinguish timing cues via their light/dark adaptation. Neurobiological studies on the photoreceptor system of scorpions have lead to a preliminary network model which is mainly based on the re-afference principle. Scorpion eyes are controlled by the circadian clock and have a strong circadian sensitivity rhythm compensating for the external day/night changes. Furthermore, within these eyes there is a second afferent non-visual channel receiving a copy of the efferent circadian signal to correct the impact of the circadian adaptation on the perceived light program and to compare external and internal timing of dusk. This information flow may provide a clear signal to the pacemaker concerning its correct phase angle to the natural light dark program. Sensory models have been proposed in lower vertebrates and insects, where extraretinal and retinal photoreceptor systems may interact to recognise the spectral and intensity changes of twilight. The results and models discussed provide a basis for changing the experimental paradigms for the analysis of entrainment mechanisms in circadian systems.

Introduction

One of the basic problems of chronobiological research to understand the mechanisms that synchronise biological clocks with planetary cycles and related climatic periodicities, is far from being solved. There are formal descriptions of *zeitgeber* effects by phase response curves. They offer quantitative descriptions of phase shifting effects and, in the past, have helped in developing mathematical and functional models of biological clock systems (Pittendrigh 1965; Pittendrigh 1981; Pittendrigh and Daan 1976). Nevertheless PRCs could not explain the mechanisms of timing in nature which is the background of evolution of life on earth and the biological clocks in all living organisms (Daan 1981). Only in the last 50 years have biological clocks been tested under lab conditions by lights on/off programs, temperature and pharmacological pulses etc. However, animals have had millions of years they had to adopt to the gradual daily, lunar and annual changes in their environment.

Dawn and dusk deliver the most precise natural timing cues

Light, temperature and humidity are subjected to daily periodic changes and all are appropriate for

---

*E-mail: fleissner@zoology.uni-frankfurt.de

timing the circadian clocks, but these signals might be heavily distorted by climatic conditions, vegetation and the animals' behaviour – or simply not perceivable in the animals shelter[1]. The most reliable timing signal is the gradual change of light during dusk and dawn (DeCoursey 1989; Remmert 1978).

Aschoff and co-workers (Aschoff and Meyer-Lohmann 1954; Aschoff and Wever 1965) had already published convincing data on the significance of entrainment by simulated twilight conditions and had formulated a hypothesis that animals tend to entrain with the twilight at the beginning of their phase of activity: day active animals synchronise with dawn, night active animals with dusk (Aschoff 1969). In various organisms (human: Okudaira et al. 1983; Hebert et al. 1998; fish: Kavaliers and Ross 1981; monkeys, bats and hamsters: Erkert 1974; rodents: DeCoursey 1986; scorpions: Fleissner and Fleissner 1993; Lüttgen 1993) it could be shown that the organisms may recognise a certain moment during twilight (Fig. 1A) and follow this signal as an individually defined set point throughout changing seasonal conditions (Fig. 1B) (Kenagy 1976).

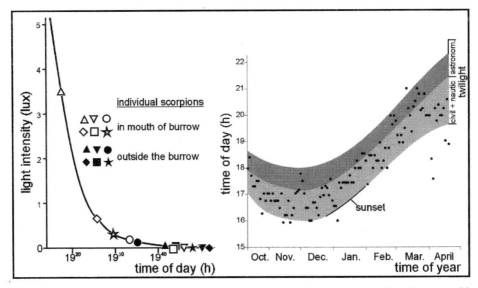

Fig. 1 **Synchronisation of locomotor activity of the scorpion *Androctonus australis* in its natural habitat. A: Individually marked scorpions as they appear in the mouth of their burrows (open symbols) and start walking around (closed symbols) during the lowest light levels of dusk (solid line). Single day observation. B: Civil, nautical and astronomical twilight (shaded areas) during dusk measured during October to April, the start of surface activity of several scorpions is indicated (dot). The scorpions clearly follow the seasonal changes of dusk. (A: Data from Dube 1989, B: Data from Lüttgen 1993).**

Simulated twilight results in better synchronisation than lights on/off

Simulations of twilight cannot only entrain the circadian clock system, as is shown for on/off light

---

[1]The Sahara scorpion *Androctonus australis* lives in self-digged caves about 40 to 50 cm underground. Daily temperature differences on the surface of the sand might exceed 60°C, they are damped down to less than 5°C inside the burrow. Interestingly the scorpion's surface activity is restricted to the short time span, when both, cave and surface temperature, match (Fleissner and Fleissner 1998).

programs, but they can do even better, as the internal synchronisation of the entire clock system is obviously profoundly improved. This can be seen in at least three parameters:

—The activity onset is more precisely locked to the light/dark changes (5 – 10 min versus about 2 hours).
—The threshold of synchronising light is lowered by at least 1 log unit.
—The range of entrainment is increased by about two hours.

Data for comparison of on/off light programs to light programs with simulated dawn or dusk are available from several animals and even humans (scorpions: Lüttgen 1993; rodents: Boulos et al. 1996a; Boulos et al. 1996c; humans: Wirz-Justice et al. 1998). This "dawn/dusk-effect" seems to be restricted to conditions very close to the physiological range of light intensity experienced by the animal under natural conditions (pilot studies with rodents: Boulos et al. 1996a and scorpions: Lüttgen 1993). If light is too bright or the slope of the transition between light and darkness is too steep, synchronisation is not different from the one achieved with on/off programs, if the slope is too small, the rhythm freeruns.

In lab studies with scorpions, the time pattern of locomotor activity resembles the observations under natural conditions (Fig. 2A) only under light/dark programs with simulated twilight transitions (Fig. 2D) (Fleissner and Fleissner 1998). Natural activity outside the burrow is restricted to the first few hours of the night. In DD (Fig. 2B) and under on/off entrainment (Fig. 2C) activity may occur during the day or late night, too, and usually is split into several ultradian bouts of activity. The precise synchronisation of activity onset with the simulated dusk persists even, when lunar illumination, similar to outside conditions, is added to the dawn/dusk program (Fig. 3).

Many fields of actual chronobiological research are aimed at analysing molecular and genetic mechanisms of timing by means of widely reduced model systems like tissue extracts, brain slices or cultured cells. Natural *zeitgeber* stimuli have been looked upon as far too complex and without special impact on the subcellular level of information. But this is simply not true as two sets of studies clearly demonstrated. (1) Cooper and colleagues (Cooper et al. 1998) investigated the dependence of c-fos expression in the SCN of rat brain slices on the L/D programs prior to the experiments. The time pattern of changing c-fos-immunoreactivity, indicative for the phase-shifting ability of light, was high at night and low during the day under DD, had a small evening peak and slightly higher morning peak in rats kept under L/D on/off programs, and surprisingly a remarkably higher evening peak in rats kept under natural light conditions. This finding matches the prediction of Aschoff (1969) that night active animals synchronise better with twilight at the beginning of their activity time. (2) Zucker and coworkers (Gorman and Zucker 1998) demonstrated that even the interpretation of photoperiodic responses in chronobiological and genetic experiments depends on the light programs used.

All these experiments lead us to assume that the *zeitgeber* receptors of biological clock systems can sense levels and the dynamics of entraining stimuli and compare them to pre-specified values via feedback loop systems. Model calculations for the human circadian system (Beersma et al. 1999) have theoretically demonstrated the relevance of natural *zeitgeber* stimuli for the accuracy of entrainment. First applications of this paradigm were successfully performed in psychiatry to improve light therapy for SAD-patients (Terman et al. 1989; Wirz-Justice et al. 1998).

Specialised photoreceptors and/or neuronal networks are necessary as *zeitgeber* receptors

All eyes evolved for image processing, have specialised to shift the steepest slope of their intensity

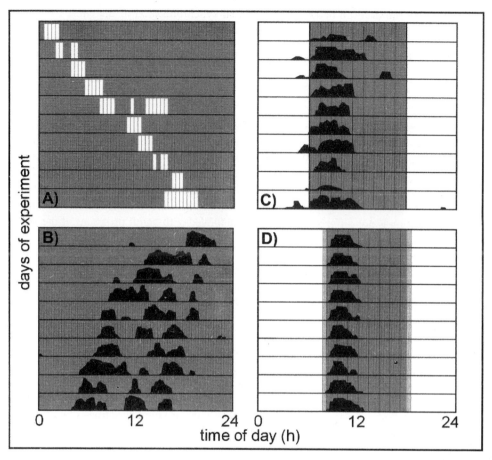

Fig. 2 **Time pattern of locomotor activity under various conditions. A: Locomotor activity in an arena which is partly shaded, lighted area is lit by star light intensitiy. During its activity time the scorpion always chooses the light part (light bars). B: In constant conditions (DD); C: In light/ dark programs with abrupt transitions (LD); D: In light/dark programs with simulated twilight transitions. Only when the scorpion can choose its hiding place and with simulated twilight, the time pattern of activity under lab conditions resembles that in the natural habitat.**

response function—where they can best discriminate visual differences in the environment—according to the actual lighting condition via light and dark adaptation. At least in night active animals light/ dark adaptation is regulated by a clock controlled change of sensitivity (see Chapter on retinal rhythms). These adaptational mechanisms effectively compensate for intensity changes during dusk and dawn and thus prevent the reception of timing cues from twilight transitions (Fleissner and Fleissner 1993). Therefore all animals investigated in this respect were shown to be equipped with specialised non-visual photoreceptors and/or neuronal networks detouring the visual information processing pathways. Best known is the pineal organ as eye of the circadian clock system in lower vertebrates and the retino-hypothalamic tract as non-visual pathway from the retina to the pacemaker in mammals. In invertebrates, lesion experiments and studies on a variety of mutants have lead to the hypothesis that extraretinal photoreceptors are the best candidates for photic *zeitgeber* receptors (rev: Helfrich-Förster et al. 1998; Page 1982; Truman 1976). Several brain neurons have been proposed as

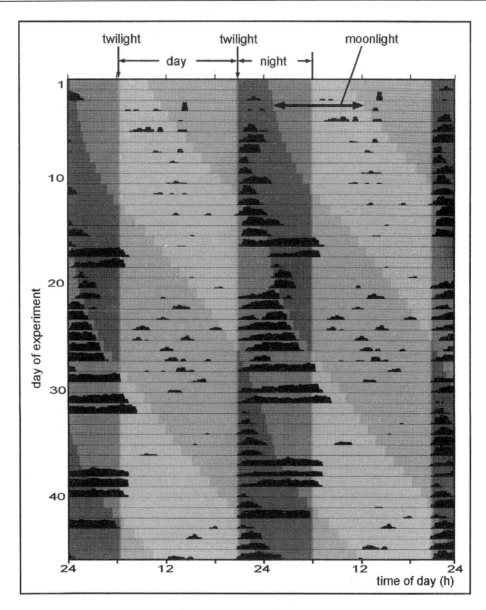

Fig. 3   **Locomotor activity of a scorpion in an arena under a light/low-light program with simulated twilight conditions (daylight intensity ca 100 lux; night/starlight intensity ca. 1mlux) and superimposed moonlight (following the natural lunar cycle, intensity ca. 30mlux). The scorpion has no shelter and cannot retreat from moonlight exposure, therefore the increased duration of activity time must be interpreted as flight from light. But still, activity onset keeps precisely entrained with dusk.**

additional candidates (Cymborowski and Korf 1995; Gao et al. 1999). But in no case have the eyes of the clock in arthropods been identified with certainty. Only in *Bulla*, a mollusc, it is known that the pacemaker neurons themselves can receive phase-shifting stimuli (McMahon and Block 1987)—

under on/off light programs, though they may have additionally indirect input from other photoreceptor cells in the eye-cup.

Summing up: The sensory and neuronal background of dawn/dusk detection is still totally unknown in all of the above mentioned animals.

The photoreceptor system of scorpions as a model of a dusk detector

So far, only in scorpions do we have a candidate system for a dusk detector. By field observations and various neurobiological methods we could identify a most promising neuronal network, which even in a mathematical modelling, seems to fulfil all prerequisites for this difficult task. It might function according to the "Reafferenzprinzip" as described by von Holst and Mittelstaedt (von Holst and Mittelstaedt 1950)[2]: The circadian signal in the scorpions' circadian system, controlling the pigment migration in the eyes, is released as night/non-night signal (Fleissner 1983). This means that the shielding pigment starts migrating towards the proximal ends of the photoreceptor cells during the circadian evening and stays there all night long thus increasing the light sensitivity of the eyes by 4 log units (Fleissner 1972; Fleissner 1977a). At the end of the circadian night, the circadian signal stops and the pigment migrates back towards the distal tips of the receptor cells, protecting the rhabdoms like a sunglass against incident light. When the circadian signal is blocked, e.g., by lesion of the optic nerve, the sensitivity rhythm is abolished and the eye permanently stays in its insensitive day state (Fleissner and Fleissner 1978).

This signal from the circadian pacemaker is delivered via efferent neurosecretory fibres (ENSF) (Fleissner and Schliwa 1977) to the peripheral ends of the photoreceptor cells. By tracing experiments it was shown that the afferent terminals and also interneurons in the lamina are innervated by these ENSF (Fig. 4) (Fleissner and Heinrichs 1982). This lamina network is astonishingly complex, for visual information is processed by a second neuronal pathway via arhabdomeric cells (Fleissner and Siegler 1978) that project from the retina directly to the second optic ganglion, the medulla, detouring the lamina (Fleissner 1985). But the lamina network may be used to enable the eyes to detect timing cues during natural dusk via a 3-step-feedback control system (Fleissner and Fleissner in press). In its natural environment the scorpion might not "see" light intensity changing during dusk as the sensitivity of the eyes increase with the same slope as environmental light intensity decreases. The circadian signal, however, controlling this pigment migration (Fig. 4A), presynaptically interferes with the photoreceptor afferents (Fig. 4B, C), thus perhaps delivering a signal to reconstitute the afferent information on external dusk to first interneurons in the lamina.

By a second step, this information on dusk might be compared to the circadian signal present in the interneurons. Here a comparison of the slope and timing of both signals, the decreasing intensity of twilight and the increasing impulse frequency of the circadian signal might yield information on the synchrony of both events. When they match, the clock is entrained with the natural day/night cycle, when the clock is early or late, then there will be unequivocal information on this "twilight" channel to advance or to delay the circadian phase, as already hypothesised by Fleissner (1977b).

---

[2]When we move our eyes the optic surrounding seems to stay constant, although the retinal image is also moving; when we passively move our eye by pressing a finger against it, the impression of our visual surrounding shifts accordingly. V. Holst interpreted this observation of a virtually stable environment as a result of a compensation of the afferent image distortion by the efferent pattern of muscle activity. The finger tip pushing the eye-ball induces a retinal displacement of the optic surroundings which has no neural correlate of efferent motor pattern and cannot be compensated for.

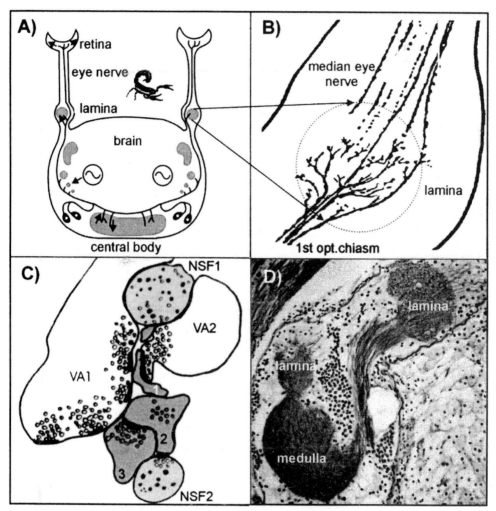

Fig. 4 Neuronal network in the scorpion's visual system as model for a "twilight receptor". A: When light intensity decreases during dusk, the visual sensitivity of the scorpion's median eyes—controlled via efferent neurosecretory fibres (ENSF) by the circadian clock—increases with the same slope and amplitude. This results in a subjectively constant range of environmental light conditions. (Scheme of the central course of the ENSF). B: In a second channel processing non-visual information in the first optic ganglion, the lamina, the information on "constant light levels" is recalculated by presynaptic information on the circadian signal, external dusk can be perceived here. (Camera-lucida drawing of terminals of the ENSF inside the lamina traced by Lucifer Yellow to the contralateral eye). C: Branches of the ENSF (NSF1, NSF2) terminate on the photoreceptor afferents (VA1, VA2) in the lamina. Interneurons (1 to 3) in the lamina also receive efferent signals from the ENSF. Here timing and slope of the external and internal dusk might be compared. If the clock signal is too fast or two early compared to the external dusk, unequivocal timing information for the pacemaker can be derived. (Scheme of a TEM section with traced ENSF). D: In order to prevent non-sense output from this proposed "twilight channel", there is a third control step in the medulla, where both information pathways, the visual and the non-visual one, and the lateral and median eye afferents meet. There is no *zeitgeber* input to the pacemaker, when both channels see the same signal. (Bodian stained section with the laminae of the lateral eyes (left) and the median eyes (right) and the common medulla neuropil.) (Data in A, B: after Fleissner and Heinrichs 1982).

A third step of control is necessary to avoid nonsense output from this channel. A neuronal filter comparing information processed on this non-visual twilight channel with the one processed on the visual channel could cut off photic input to the clock, whenever both deliver the same information (Fig. 4D). The morphological candidate for this third control site is the medulla, where both signalling pathways meet (Fleissner and Fleissner 1993). Several experimental data match this proposal.

This twilight channel is able to entrain the scorpion in a small circadian gate—at dusk–only (cf. Fig. 3), not with dawn, when the circadian signal breaks down and not at random circadian times. Non-natural entraining stimuli that are too steep (on/off), too high or step-functions cannot be interpreted as unequivocal timing cues and prevent precise synchronisation (see Fig. 2). Matching chronobiological data were obtained with hamsters and other rodent species (Boulos et al. 1996b), including a flat PRC (Boulos et al. 1996d). This may suggest that there exists a re-afferent feedback system in these animals as well.

When the clock has awakened the scorpion, the animal slowly approaches the mouth of its burrow–and the outside light conditions. Simultaneously dusk dims the light intensity and with the same slope the sensitivity in the eyes increases. It might well be that during that time span the scorpion moves on a subjectively "isophotic" level.

Eyes of the biological clocks in vertebrates and invertebrates

The photic Zeitgeber input to the circadian clock system in scorpions is very likely colour blind. The animals have only one spectral type of photoreceptors (Fleissner 1968). But there is evidence that other animals can be synchronized by the spectral shifts of environmental light during dusk and dawn, e.g., from the arctic fauna (Krüll et al. 1985). Spectral changes combined with intensity modulations can provide unequivocal timing cues. There is good evidence that the eyes of the clock make use of both parameters. It is striking that most of the proposed photic *zeitgeber* receptors show major differences in their spectral sensitivity for long and for short wavelengths where the most prominent changes occur during evening and morning twilight. In the pineal of lower vertebrates and birds there seem to be combined spectral and non-spectral responses to light (Meissl and Brandstätter 1992), the photic input to the SCN of mammals shows opposing spectral responses to scotopic and photopic stimuli (Aggelopoulos and Meissl 2000) via light to the retinal photoreceptors, which lead to modified hypotheses on dawn/dusk detectors for vertebrate systems (Meissl and Ekström 1993). The two extraretinal photoreceptor systems of beetles, the stemmata and the lamina and lobula organs (Fleissner et al. 1993), seem to have different photopigments, UV—and blue sensitive opsins in the stemmata and red/green opsins in the lamina and lobula organs (Waterkamp et al. 1998; Fleissner et al. submitted). In butterflies recordings from larval medulla neurons show similar opposing spectral responses (Ichikawa 1990). These system are good model candidates for *zeitgeber* receptors, but this has not yet been tested with adequate twilight programs. So far we do not know which of these receptor systems—alone or as specialised network—might be necessary to receive natural *zeitgeber* stimuli.

Conclusions

The hypothesis of the function of natural timing presented here is based on the pilot studies in few model organisms and suggests that:

— Twilight may deliver timing cues for an effective fine tuning of the exogenous (dawn or dusk) and endogenous day/night transitions.

— Photic input to the circadian pacemaker for natural timing cues might be restricted to a small circadian gate only.

— The *zeitgeber* receptors and information processing pathways are organised as neuronal networks that are different from the visual photoreceptors.

Therefore, analysing the functional mechanisms of natural timing might be altered to involve fundamentally changed paradigms for lab tests in various aspects:

— *zeitgeber* stimuli have to be adjusted to the physiological range of light/dark programs, i.e., species-specific needs and their possible changes according to season, age and physiological conditions must be tested prior setting up the experiments.

— We do not yet know which dynamics of the natural surroundings are decisive for transmitting timing cues to the clock systems—light intensity or spectral changes, alone or combined, or even in combination with other bioclimatic parameters. This must be individually tested.

— *zeitgeber* stimuli with simulated twilight conditions might be reasonably applied only near the subjective dusk or dawn, not randomised during subjective day and night.

— As critical tests for optimal internal and external synchronisation of the circadian clock system, it is necessary to at least evaluate precision of timing, (plus threshold of *zeitgeber* stimuli and range of entrainment), as this is different under abrupt versus graduated light/dark transitions.

— When testing the effectiveness of certain photoreceptor systems or underlying networks it may well be that a (sloppy) entrainment can be achieved in a lesioned or reduced system, but it must be checked whether this differs from the intact system.

The complexity of the natural surroundings and the unknown receptor and neuronal background of *zeitgeber* processes should not prevent us from restarting the analysis of *zeitgeber* mechanisms–this time not under maximally reduced lab conditions but oriented towards evolutionary successful natural timing cues.

Acknowledgements

Supported by grants from the Deutsche Forschungsgemeinschaft. We gratefully acknowledge the marvellous co-operation of our candidates, students and lab guests and their contributions to various aspects of this paper. Especially we thank C. Dube, M. Lüttgen and D. Neubacher who contributed data from their PhD-theses. W. Otto Friesen (Charlottesville, USA) helped us with valuable discussions and the English version of the manuscript.

References

Aggelopoulos, N.C., Meissl, H. (2000) Responses of neurones of the rat suprachiasmatic nucleus (SCN) to retinal illumination under photopic and scotopic conditions. J. Physiol. (in press).

Aschoff, J. (1969) Phasenlage der Tagesperiodik in Abhängigkeit von Jahreszeit und Breitengrad. Oecologia (Berlin) **3**: 125–165.

Aschoff, J., Meyer-Lohmann, J. (1954) Die Aktivität gekäfigter Grünfinken im 24-Stunden-Tag bei unterschiedlich langer Lichtzeit mit und ohne Dämmerung. Z. Tierpsychol. **12**: 254–265.

Aschoff, J., Wever, R. (1965) Circadian rhythms of finches in light-dark cycles with interposed twilights. Comp. Biochem. Physiol. **16**: 307–314.

Beersma, D.G., Spoelstra, K., Daan, S. (1999) Accuracy of human circadian entrainment under natural light conditions: Model simulations. J. Biol. Rhythms **14**: 524–531.

Boulos, Z., Macchi, M., Houpt, T.A., Terman, M. (1996a) Photic entrainment in hamsters: effects of simulated twilights and nest box availability. J. Biol. Rhythms **11**: 216–233.

Boulos, Z., Macchi, M., Terman, M. (1996b) Effects of twilights on circadian entrainment patterns and reentrainment rates in squirrel monkeys. J. Comp. Physiol. A **179**: 687–694.

Boulos, Z., Macchi, M., Terman, M. (1996c) Twilight transitions promote circadian entrainment to lengthening light-dark cycles. Am. J. Physiol. **271**: R813–818.

Boulos, Z., Terman, J.S., Terman, M. (1996d) Circadian phase-response curves for simulated dawn and dusk twilights in hamsters. Physiol. Behav. **60**: 1269–1275.

Cooper, H.M., Dkhissi, O., Sicard, B., Groscarret, H. (1998) Light evoked C-Fos expression in the SCN is different under on/off and twilight conditions. In: Touitou, Y. (ed.) Biological clocks. Mechanisms and Applications. Elsevier Science B.V., Paris Amsterdam. pp. 181–188.

Cymborowski, B., Korf, H.W. (1995) Immunocytochemical demonstration of S-antigen (arrestin) in the brain of the blowfly *Calliphora vicina*. Cell Tiss. Res. **279**: 109–114.

Daan, S. (1981) Adaptive daily strategies in behavior. In: Aschoff, J. (ed.) Biological Rhythms. Hb Behav. Neurbiol. Vol. 4. Plenum Press, New York. pp. 275–298.

DeCoursey, P.J. (1986) Light-sampling behaviour in photo-entrainment of a rodent circadian rhythm. J. Comp. Physiol. A **159**: 161–169.

DeCoursey, P.J. (1989) Photoentrainment of circadian rhythm. An ecologist's viewpoint. In: Hiroshige, T., Honma, K. (eds.) Circadian Clocks and Ecology. 3rd Sapporo Symposium on Biological Clocks. Hokkaido University Press, Sapporo. pp. 187–206.

Dubey, C. (1989). Die circadiane.

Erkert, H.G. (1974) Der Einfluß des Mondlichts auf die Aktivitätsperiodik nachtaktiver Säugetiere. Oecologia **14**: 269–287.

Fleissner, G. (1968) Untersuchungen zur Sehphysiologie der Skorpione. Verh Deutsche Zool. Ges. Innsbruck. pp. 375–380.

Fleissner, G. (1972) Circadian sensitivity changes in the median eyes of the North African scorpion, *Androctonus australis*. In: Wehner, R. (ed.) Information processing in the visual system of arthropods. Springer, Berlin Heidelberg New York. pp. 133–139.

Fleissner, G. (1977a) The absolute sensitivity of the median and lateral eyes of the scorpion, *Androctonus australis* L. (Buthidae, Scorpiones). J. Comp. Physiol. **108**: 109–120.

Fleissner, G. (1977b) Differences in the physiological properties of the median and the lateral eyes and their possible meaning for the entrainment of the scorpion's circadian rhythm. J. Interdiscipl. Cycle Res. **8**: 15–26.

Fleissner, G. (1983) Efferent neurosecretory fibres as pathways for the circadian clock signals in the scorpion. Naturwissenschaften **70**: 366.

Fleissner, G. (1985) Intracellular recordings of light responses from spiking and nonspiking cells in the median and lateral eyes of the scorpion. Naturwissenschaften **72**: 46–48.

Fleissner, G., Fleissner, G. (1978) The optic nerve mediates the circadian pigment migration in the median eyes of the scorpion. Comp. Biochem. Physiol. A **61**: 69–71.

Fleissner, G., Fleissner, G. (1993) Seeing time. In: Wiese, K., Gribakin, F., Popov, A.V., Renninger, G. (eds.) Sensory systems of invertebrates. Birkhäuser Verlag, Basel, Boston, Berlin. pp. 288–306.

Fleissner, G., Fleissner, G. (1998) Natural photic Zeitgeber signals and underlying neuronal mechanisms in scorpions. In: Touitou, Y. (ed.) Biological Clocks. Mechanisms and Applications. Elsevier, Paris. pp. 171–180.

Fleissner, G., Fleissner, G., Frisch, B. (1993) A new type of putative non-visual photoreceptors in the optic lobe of beetles. Cell Tiss. Res. **273**: 435–445.

Fleissner, G., Heinrichs, S. (1982) Neurosecretory cells in the circadian-clock system of the scorpion, *Androctonus australis*. Cell Tiss. Res. **224**: 233–238.

Fleissner, G., Schliwa, M. (1977) Neurosecretory fibres in the median eyes of the scorpion, *Androctonus australis* L. Cell Tiss. Res. **178**: 189–198.

Fleissner, G., Siegler, W. (1978) Arhabdomeric cells in the retina of the median eyes of the scorpion. Naturwissenschaften. **65**: 210–211.

Gao, N., Schantz, M.V., Foster, R.G., Hardie, J. (1999) The putative brain photoperiodic photoreceptors in the vetch aphid, *Megoura viciae*. J. Insect Physiol. **45**: 1011–1019.

Gorman, M.R., Zucker, I. (1998) Mammalian seasonal rhythms: New perspectives gained from the use of simulated natural photoperiods. In: Touitou, Y. (ed.) Biological Clocks. Mechanisms and Applications. Elsevier Science B.V. pp. 195–204.

Hebert, M., Dumont, M., Paquet, J. (1998) Seasonal and diurnal patterns of human illumination under natural conditions. Chronobiol. Int. **15**: 59–70.

Helfrich-Förster, C., Stengl, M., Homberg, U. (1998) Organization of the circadian system in insects. Chronobiol. Int. **15**: 567–594.

Ichikawa, T. (1990) Spectral sensitivities of elementary color-coded neurons in butterfly larva. J. Neurophysiol. **64**: 1861–1872.

Kavaliers, M., Ross, D.M. (1981) Twilight and day length affect the seasonality of entrainment and endogenous circadian rhythms in a fish, *Couesius plumbeus*. Can. J. Zool. **59**: 1326–1334.

Kenagy, G.J. (1976) The periodicity of daily activity and its seasonal changes in free-ranging and captive kangaroo rats. Oecologia **24**: 105–140.

Krüll, F., Demmelmeyer, H., Remmert, H. (1985) On the circadian rhythm of animals in high polar latitudes. Naturwissenschaften **72**: 197–203.

Lüttgen, M.A. (1993) Entrainment der circadianen Laufrhythmik durch Lichtzeitgeber: Untersuchung biologisch relevanter Lichtparameter am Beispiel der Lokomotionsrhythmik von *Androctonus australis* L. (Scorpiones, Buthidae). PhD-Thesis University Frankfurt am Main.

McMahon, D.G., Block, G.D. (1987) The *Bulla* ocular circadian pacemaker. II. Chronic changes in membrane potential lengthen free running period. J. Comp. Physiol. A **161**: 347–354.

Meissl, H., Brandstätter, R. (1992) Photoreceptive functions of the teleost pineal organ and their implications in biological rhythms. In: Ali, M.A. (ed.) Rhythms in Fishes. Plenum Press, New York. pp. 235–254.

Meissl, H., Ekström, P. (1993) Extraretinal photoreception by pineal systems: A tool for photoperiodic time measurements? Trends Comp. Biochem. Physiol. pp. 1223–1240.

Okudaira, N., Kripke, D.F., Webster, J.B. (1983) Naturalistic studies of human light exposure. Am. J. Physiol. **245**: R613–R615.

Page, T.L. (1982) Extraretinal photoreception in entrainment and photoperiodism in invertebrates. Experientia **38**: 1007–1013.

Pittendrigh, C.S. (1965) On the mechanism of entrainment of a circadian rhythm by light cycles. In: Aschoff, J. (ed.) Circadian clocks. North Holland Publ. Co. pp. 277–297.

Pittendrigh, C.S. (1981) Circadian systems: entrainment. In: Aschoff, J. (ed.) Biological rhythms. Hb Behav Neurobiol Vol. 4. Plenum Press, New York. pp. 95–124.

Pittendrigh, C.S., Daan, S. (1976) A functional analysis of circadian pacemakers in nocturnal rodents. IV. Entrainment: pacemaker as a clock. J. Comp. Physiol. A. **106**: 291–331.

Remmert, H. (1978) Ökologie. Springer-Verlag, Berlin-Heidelberg-New York.

Terman, M., Schlager, D., Fairhurst, S., Perlman, B. (1989) Dawn and dusk simulation as a therapeutic intervention. Biol. Psychiatry **25**: 966–970.

Truman, J.W. (1976) Extraretinal photoreception in insects. Photophysiology **23**: 215–225.

Von Holst, E., Mittelstaedt, H. (1950) Das Reafferenzprinzip. Naturwissenschaften **37**: 464–476.

Waterkamp, M., Fleissner, G., Fleissner, G. (1998) Information processing in the extraretinal photoreceptor systems of beetles. Proc. SRBR **6**: 135.

Wirz-Justice, A., Terman, M., Terman, J.S., Boulos, Z., Remé, C.E., Danilenko, K.V. (1998) Circadian functions and clinical applications of dawn simulation. In: Y., Touitou (eds.) Biological Clocks. Mechanisms and Applications. Elsevier Science B.V. 189–194.

Biological Rhythms
Edited by V. Kumar
Copyright © 2002, Narosa Publishing House, New Delhi, India

9. Photoreceptors for the Circadian Clock of the Fruitfly

C. Helfrich-Förster and W. Engelmann*

Department of Animal Physiology and Department of Botany, University of Tübingen, Germany

Light profoundly affects the circadian system of organisms. This chapter reviews the photoreceptors receiving the light and transducing it to the pacemakers controlling the eclosion and locomotor activity rhythms of *Drosophila*.

The compound eyes are not needed for entrainment of the eclosion rhythm. Nor are they needed for the locomotor activity rhythm, but without them the sensitivity is reduced and the action spectrum is narrower (blue region) as compared to the wild type. In contrast to the phase shifting of the eclosion rhythm, which is mediated by only one photoreceptor, several photoreceptors, including the Hofbauer-Buchner eyes, entrain the locomotor activity rhythm. Moreover, the pacemaker cells (the ventral lateral neurones) are also photoreceptive, being directly entrainable by light, as are other cells expressing rhythm generating genes.

We furthermore review the way in which the circadian rhythm is generated at the molecular level by at least one negative feedback loop in which clock genes and their products are involved. Entrainment, phase shifting and attenuation of the rhythm by light can be explained by a photochemical conformation change of the photopigment cryptochrome.

Cryptochrome seems to act in pacemaker neurones itself. It might be advantageous to use multiple photoreceptors for synchronising the circadian system. The signal to noise problem is reduced in this way, and the circadian system is enabled to utilise different qualities of light for entrainment.

Introduction

As in most organisms, circadian rhythms are widespread also in invertebrates. They govern the timing of much of their development, behaviour, physiology and biochemistry, as well as photoperiodic events.

The oscillators which produce these rhythms must be entrained to the 24 hour day of the earth in order to serve as reliable clocks. The daily light-dark cycle is the most important *zeitgeber* for synchronisation. In order to affect the clock and entrain it, light has to be received. Photopigments are used for this purpose. Since most cells are equipped with pigments, cells are principally able to absorb light and synchronise their circadian clocks. Some unicellular algae even use different photoreceptors in the same cell (Roenneberg and Foster 1997). A few multicellular organism are equipped with dermal photosensitivity or photosensitivity of nervous tissue. In others special neurones are light sensitive or photoreceptor cells are distributed diffusely over the body. However, most multicellular organisms developed special photoreceptor organs. They comprise normal eyes but also extraretinal eyes (for invertebrates see Ali 1984).

In order to clarify the mechanisms of photic entrainment several questions arise. What photoreceptors

---

*E-mail: engelmann@uni-tuebingen.de

and which pigments are involved? What are the elements of the transduction pathway? And how does the transduced light signal affect the circadian oscillator? These questions can be asked at the systemic level, at the cellular level and at the molecular level. We will review research on the fruitfly *Drosophila* that has focussed on these different questions. Space does not allow to report findings on light control of circadian rhythms in other animals. But see chapter of Fleissner and Fleissner in this volume.

Characteristics of circadian rhythms in *Drosophila*

The best studied circadian rhythms in *Drosophila* are eclosion and locomotor activity. Whereas the rhythm of locomotor activity can be recorded in individual flies, eclosion rhythm of the adult out of their pupal case can of course be observed only in a population of flies. At a certain time of development, a circadian clock opens a time gate for a restricted part of the day. If the particular fly misses this gate, it has to wait for the next gate to occur one circadian cycle later. The rhythm of eclosion was studied in detail in *Drosophila pseudoobscura* by Pittendrigh and co-workers, because the eclosion rhythm of this species is characterised by high precision (Pittendrigh 1967). Later studies focussed more on the activity rhythm of individual flies, and *Drosophila pseudoobscura* was replaced by *Drosophila melanogaster*, the model organism for geneticists. Here, we will not distinguish between both species since overall rhythmicity is similar in both flies, and since the same genes appear to be involved in the generation of rhythmicity (Peterson et al. 1988).

Eclosion and activity rhythms of *Drosophila* are very sensitive to light. They can be entrained to light-dark cycles of very low light-intensities (less than 1 lux for activity), and phase shifted by short light pulses (15 min of 0.1 lux for eclosion). Light pulses given at a special time are furthermore able to arrest the eclosion rhythm (Winfree 1970). But light pulses can also be used to initiate a rhythm in arrhythmic cultures (Honegger 1967). In both cases blue is the most effective light. Furthermore, continuous light lengthens the periods of both freerunning rhythms and suppresses rhythmicity when light intensity exceeds a certain limit (Chandrashekaran and Loher 1969, Konopka et al. 1989). For eclosion blue was again the most effective light (Winfree 1974). Phase response curves (PRCs) to pulses of white light have been established for both overt rhythms: they are characterised by a nonresponsive 'dead zone' in the subjective day followed by a phase-delay zone in the early subjective night and a phase-advance zone in the late subjective night (e.g. Pittendrigh and Minis 1964, Saunders et al. 1994; Myers et al. 1996; Figure 1).

Action spectra suggest the involvement of several photopigments in circadian photoreception

The phase delaying and advancing effects of monochromatic light have been compared to unravel the spectral sensitivity of the pacemaking system and to uncover the photopigments involved in circadian photoreception. An action spectrum for advances (CT20) and delays (CT17) for the eclosion rhythm showed a very broad sensitivity peak between 420 and 480 nm. Wavelengths longer than 540 nm were ineffective in shifting the eclosion rhythm (Frank and Zimmermann 1969). The spectra for advance and delay looked alike suggesting that the same photopigment mediates advance and delay phase shifts. However, advancing the rhythm phase shifts took several days of transients, whereas delay shifts were observed the day after the light pulse.

With weaker light, advancing the rhythm at CT19 was ten times more sensitive to 442 nm light as compared to delaying at CT18 (Chandrashekaran and Engelmann 1973). At higher—probably saturated—light intensities these differences disappeared, indicating two different photoreceptor pigments or different primary processes being responsible for light absorption during the advance and delay

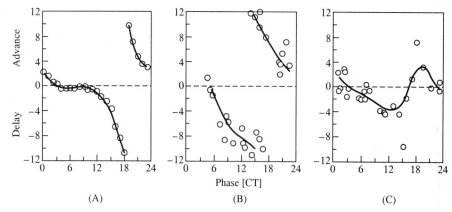

Fig. 1   Phase response curves of eclosion and activity rhythm of *Drosophila melanogaster* to light pulses. **A: 15 min pulses of white light (fluorescence tubes, 1000 lux) administered at different CTs leading to advances and delays of the eclosion rhythm. Note dead zone between CT4 and 11 with no phase shifting effect of the light, and the jump between delays and advances around CT19. After Pittendrigh and Minis 1964. B: 6 hour pulses of white light (fluorescence tubes, 1.13 W/m²) administered at different CTs leading to advances and delays of the locomotor activity rhythm. C: Same as B, but 1 hour pulses of white light (same intensity). B and C after Saunders et. al. 1994.**

portion of the PRC, respectively (for an alternative explanation see later). Two different photoreceptor pigments were also implicated in the PRC towards light pulses of the marine dinoflagellate *Gonyaulax polyedra* (for review: Roenneberg and Foster 1997). One receptor is mainly blue sensitive. It advances the system in the second half of the subjective night. The other is both red and blue sensitive, mostly delaying the rhythm throughout the subjective day and the first half of the subjective night.

A detailed action spectrum for delay phase shifts (at CT17) of the eclosion rhythm showed blue of 457 nm to be the most effective light with further maxima at 473, 435 and 375 nm (Klemm and Ninnemann 1976). The dose response curves of light intensity and magnitude of phase delays were parallel for wavelengths between 400 and 500 nm (Figure 2) indicating one single pigment as the photoreceptor for these wavelengths. The maxima of the action spectrum point to a flavoprotein and not to a carotenoid (rhodopsin) as the photoreceptor molecule for phase delays. This is supported by findings of Zimmerman and Goldsmith (1971): flies that were grown on a diet depleted of carotenoids showed that the sensitivity of the visual receptors (compound eyes) was decreased by three orders of magnitude whereas the photosensitivity of the circadian eclosion rhythm was not affected. Furthermore, flies lacking the compound eyes were still able to synchronise their eclosion rhythm after a LL/DD shift (Engelmann and Honegger 1966). Similarly, the activity rhythm of different blind mutants was entrained normally to LD-cycles (Helfrich and Engelmann 1983, Dushay et al. 1989, Wheeler et al. 1993). These results show that the compound eyes are not necessary for circadian light responses and that the circadian photoreceptor is different from that of vision.

Nevertheless, the compound eyes contribute to circadian photoreception at least in adult *Drosophila*: Action spectra for entrainment of locomotor activity are different in wildtype and carotenoid depleted, eyeless or otherwise blind flies (Blaschke et al. 1996, Ohata et al. 1998, Figure 3). Carotenoid depletion reduced the circadian light sensitivity by 3 log units (Ohata et al. 1998). This corresponded exactly to the decrease in sensitivity of the compound eyes. In eyeless flies (*sine oculis*) circadian light sensitivity was about 2 log units smaller as compared to wildtype flies and could not be reduced

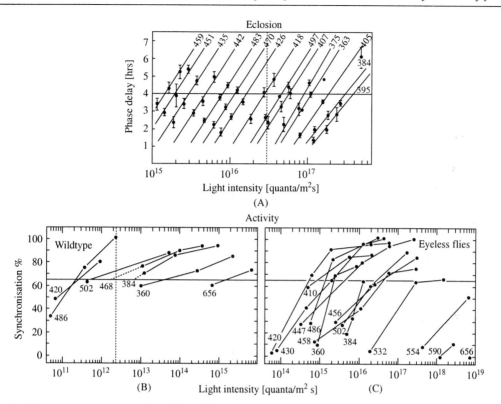

Fig. 2 Dose-response curves for delay shifts of eclosion in *Drosophila pseudoobscura* (top) and for synchronisation of activity in *Drosophila melanogaster* (bottom). Wavelengths are indicated in nm at the individual curves. Top: Phase delays provoked by 15 min coloured light administered at CT17. Note that slopes of curves are very similar to each other (after Klemm and Ninnemann 1976). The light intensities that provoked a phase delay of 4 hours (horizontal line) were used to plot the action spectrum shown in Fig. 3A. The phase delays provoked by coloured light of the same intensity (dotted vertical line) were plotted as a spectral response curve in Fig. 3C. Bottom: Percent synchronisation of the activity rhythm of wildtype (left) and eyeless flies (right) by 12 h coloured light: 12 h darkness of different intensity. The light intensities that synchronised 65% of the flies (horizontal line) were used for the action spectrum in Fig. 3A. The percentages of synchronised flies by coloured light of the same intensity (dotted vertical line) were plotted as spectral response curve in Fig. 3C. Note that the wildtype needs lower intensities for entrainment (x-axis from 10^{11} to 10^{16} quanta/m²s) as compared to the eyeless flies (x-axis from 10^{14} to 10^{19} quanta/m²s). Furthermore, the range of wavelengths successfully synchronising 65% of the flies is narrower in the eyeless flies (400 to 525 nm) as compared to the wildtype (350 to 650 nm). After Blaschke et al. (1996).

further by carotenoid depletion (Ohata et al. 1998; Figure 3B). Furthermore, eyeless flies showed a narrower action spectrum than wildtype flies. Whereas wildtype flies were entrainable by wavelengths from 350 to 650 nm with a broad maximum around 500 nm (which is close to the rhodopsin absorption peak of the peripheral retinal cells), eyeless flies entrained only from 400 to 525 nm and showed a maximum around 460 nm (Figure 3A, B). Thus, the action spectrum for eyeless flies shows some similarity to the action spectrum for phase delays of the eclosion rhythm. The main difference between the action spectra for phase delays in eclosion and for entrainment to LD-cycles of the

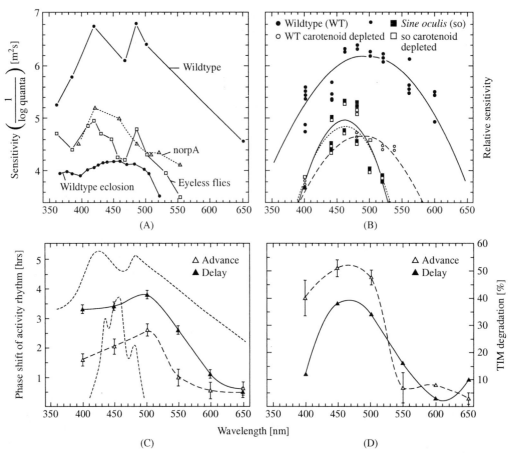

Fig. 3   Action spectra for entrainment of the activity rhythm of wildtype *Drosophila melanogaster* and several vision impaired mutants (A, adapted from Blaschke et al. 1996 and B, from Ohata 1998), for phase shifting wildtype eclosion (A, lowest curve, dots, adapted from Klemm and Ninnemann 1976) or wildtype activity (C, adapted from Suri et al. 1998) and for light dependent degradation of the clock protein TIMELESS (TIM) (D, adapted from Suri et al. 1998). The action spectra in A are derived from the dose response curves shown in Fig. 2. For activity the light intensities were determined that synchronised 65% of the flies, for eclosion those that provoked a phase delay of 4 hours (horizontal lines in Fig. 2). The data of the original action spectra for eclosion published by Klemm and Ninnemann were on a linear scale. Here these data were plotted logarithmically to allow comparison with the action spectra for entrainment. The spectral response curve for phase shifts of the activity rhythm (C) shows the phase shifts (advances or delays) provoked by a 10 minute light pulse at 5×10^{-3} W/cm^2 administered at CT15 and CT21, respectively. To compare these phase shifts with those of eclosion we calculated the phase shifts obtained at a special light intensity (15 minute light pulses at 7×10^{-7} W/cm^2) from the dose response curve of Fig. 2 (vertical dotted line). Similarly, we calculated the percentages of entrained flies at a light intensity of 5.38×10^{-11} W/cm^2 from the dose response curve of the wildtype shown in Fig. 2 (vertical dotted line). Both curves are shown as broken lines in C, that for phase shifts on eclosion on the bottom and that for entrainment of activity on the top. The scale is however different from that shown on the y-axis, which applies to advance and delay phase shifts of activity rhythm only. Maximal synchronisation at this light intensity was 100% and the maximal phase shifts for eclosion were 11 hours. For better clarity these scales were not indicated.

activity rhythm lies in the sensitivity to red light. Eclosion as well as the rhythm of activity of blind mutants are insensitive to red light (see also Helfrich-Förster 1997 for further examples of blind mutants). This suggests that rhodopsin contributes to the entrainment of the activity rhythm but not to phase shifts in eclosion. Entrainment involves parametric as well as non parametric effects of light, whereas phase shifts to short light pulses use only non parametric effects of light. The compound eyes and the larval photoreceptors which both utilise rhodopsins as photopigments may mainly be responsible for the parametric effects of light. Whether this is true and whether rhodopsins do also contribute to entrainment of the eclosion rhythm is unclear since no action spectrum for entrainment was conducted, so far. For the activity rhythm a spectral response curve for phase shifts is known (Suri et al. 1998; Figure 3C). This spectral response curve has a maximum at 500 nm with some sensitivity to red light. Thus it is more similar to the action spectrum for entrainment of activity than to the phase shifting action spectrum of eclosion (Figure 3C). This suggests that the compound eyes do also contribute to phase shifts of the activity rhythm (non parametric effects of light), and that differences in the composition of circadian photoreceptors in larvae and adults account for the different action spectra for eclosion and adult activity. Nevertheless, it is very likely that the larval eyes contribute to entrainment of the eclosion rhythm, because they project directly to the larval circadian pacemaker neurones and their terminals completely overlap with putative dendritic trees of the latter (Kaneko et al. 1997). Furthermore, functional larval eyes are necessary to entrain the molecular rhythms in the pacemaker cells in mutants that lack extraretinal photoreception (Kaneko et al. 2000, see below).

The contribution of different photoreceptors to entrainment is so far more evident in the adult activity rhythm. Blaschke et al. (1996) found two maxima in the action spectra for entrainment of activity of the wildtype and of eyeless flies. Furthermore, in dose response the lines representing the percentage of synchronised flies as a function of light intensity were not parallel for several of the tested wavelengths (Figure 2). This is again quite different from the dose response curves for delay shifts in the eclosion rhythm and indicates that different photoreceptors contribute to circadian photoreception mediating entrainment of locomotor activity. As already discussed, these could be the compound eyes and additional, still unknown extraretinal photoreceptors. Interestingly, these dose response curves were also not parallel in eyeless flies suggesting that even eyeless flies utilise different photoreceptors for entrainment (Figure 2). One of these different circadian photoreceptors could be a pair of extraretinal photoreceptors discovered by Hofbauer and Buchner (1989). Recent electron microscopic studies showed that each of these extraretinal eyes is composed of 4 photoreceptor cells with numerous microvilli arranged into coherent rhabdomeres (Yasuyama and Meinertzhagen 1999). The rhabdomeres exhibit rhodopsin (Rh6)-like immunoreactivity which provides evidence that the photoreceptors are functional. Furthermore, they are immunoreactive with antisera against arrestin—a molecule required in the phototransduction cascade—(van Swinderen and Hall 1995) and against the *period*-protein PER as are all other known photoreceptors (Hall 1998b). As neurotransmitters the extraretinal eyes use histamine (Pollock and Hofbauer 1991) and possibly acetylcholine (Yasuyama and Meinertzhagen 1999), and they project exactly into the brain region where *Drosophila's* pacemaker cells are located (for review see Helfrich-Förster 1996, Kaneko 1998, Hall 1998a). Therefore, the extraretinal eyes fulfil nicely the anatomical criteria for circadian photoreceptors. Analogous photoreceptive structures were also reported for other Diptera, for Mecoptera, Trichoptera and Coleoptera (for review: Yasuyama and Meinertzhagen 1999) but—like for *Drosophila's* extraretinal eyes—their physiological significance for circadian entrainment is not yet proven. As can be deduced from the dose response curves differing in slope (see above), the extraretinal eyes are not the only circadian photoreceptors in eyeless flies. Indeed, even *glass* mutants that lack all internal and external eye structures were still entrainable and phase shiftable (Helfrich-Förster 1997, Hall 1998b). This shows that other photoreceptors exist besides the compound eyes and the extraretinal eyes.

The best candidates for such extraretinal photoreceptors are *Drosophila's* pacemaker neurones themselves. Several experiments have indicated that all cells expressing the genes involved in rhythm generation (see below) are light sensitive by themselves and that the fruitfly's circadian system consists of a collection of autonomous photoreceptive oscillators (Plautz et al. 1997). Such photoreceptive autonomous clocks were explicitly found in the compound eyes (Cheng and Hardin 1998), in the chemosensory cells of the antennae (Plautz et al. 1997; Krishnan et al. 1999), in the malphigian tubules (Hege et al. 1997) and in the prothoracic glands (Emery et al. 1997). However, these cells and organs are not important for eclosion and activity rhythms. The pacemaker neurones controlling rhythmic eclosion and activity are located in the lateral brain and are therefore called Lateral Neurones (LN) (reviews: Hall 1998a; Kaneko 1998; Helfrich-Förster et al. 1998). The LN can be divided into a more dorsally located group (LN_d) and into a more ventral one (LN_v). The ventral group consists of neurones with large somata (large LN_v) and of neurones with small somata (small LN_v). The small LN_v send projections into the dorsal part of the central brain and several studies have indicated that the small LN_v are the most important cells for circadian behaviour. But the other cells may also contribute. To understand circadian photoreception in the LN we first have to describe how circadian rhythms are generated at the molecular level in these and other cells of *Drosophila*. For more elaborate reviews the reader is referred to Hall 1998a, b, Young 1998, Dunlap 1999, and Edery 1999.

Generation of circadian rhythmicity on the molecular level

Both rhythms—activity and eclosion—depend on a molecular feedback loop that is generated by interaction of several so called "clock genes" and their products. These are the *period* (*per*), *timeless* (*tim*), *Clock* (*Clk*), *cycle* (*cyc*) and *doubletime* (*dbt*) genes and the corresponding proteins PER, TIM, CLK, CYC, and DBT. Mutations in each of these genes have strong effects on both rhythms. They either render the flies arrhythmic or change the period of the freerunning rhythm. All clock genes and-proteins participate in a negative molecular feedback loop which will be characterised briefly in the following. CLK and CYC are basic helix-loop helix-PAS transcription factors. They form heterodimers and bind to an E-box element in the promoters of the *per* and *tim* genes thus activating transcription of both genes (Darlington et al. 1998; Rutila et al. 1998; Hao et al. 1997). Upon activation *per* and *tim* mRNA levels steadily increase reaching their peak levels early in the evening. However, PER and TIM levels do not peak until late evening. This delay appears to result from posttranscriptional regulation of *per* mRNA (Chen et al. 1998) and of PER (Dembinska et al. 1997; Stanewsky et al. 1997). One of the known regulatory mechanisms is the destabilisation of PER through DBT-dependent phosphorylation (Kloss et al. 1998; Price et al. 1998). Finally PER becomes stabilised via dimerization with TIM. After having reached a critical threshold PER-TIM dimers move into the nucleus where they repress their own transcription by inactivating the CLK-CYC transcriptional activators (Darlington et al. 1998). Due to the lag between mRNAs and proteins, this negative feedback results in a stable cycling in *per* and *tim* mRNA and protein levels. Recent data suggest that there are even two interlocked negative feedback loops: the just described *per/tim* loop, in which transcription is activated by CLK-CYC and repressed by PER-TIM, and a *Clk* loop, in which transcription is repressed by CLK-CYC and derepressed by PER-TIM (Glossop et al. 1999). Although the principle of circadian negative feedback loop(s) is now well established, not all participating components seem to be discovered yet. Recently, Belvin et al. (1999) reported that the cAMP response element binding protein (CREB) also participates in the circadian feedback loop promoting oscillations of PER and TIM. More such clock factors will be found in the future.

Entrainment of the molecular feed back oscillator

Similar to the overt rhythms the molecular feedback loops can be entrained by light-dark cycles, phase shifted by light pulses and attenuated by continuous light (for review Edery 1999). These effects of light on the molecular rhythm are achieved through light dependent degradation of TIM; other clock proteins are not degraded by light (Hunter-Ensor et al. 1996; Lee et al. 1996; Zeng et al. 1996; Myers et al. 1996). Under continuous light the TIM level is permanently very low. Because PER needs dimerization with TIM to be protected from degradation, the PER level is low, too (Price et al. 1995). *per* and *tim* mRNA remain at median constitutive levels which are similar to those of the per^0 and tim^0 mutants (Qiu and Hardin 1996). As a consequence no cycling in the clock genes and proteins can be observed under LL. This fits well to the arrhythmicity of the overt rhythms under these conditions (see above).

The light induced degradation of TIM involves an ubiquitin dependent process in the proteasomes (Naidoo et al. 1999). Interestingly, TIM degradation measured in the LN after short light pulses of different energy correlates well with the magnitude of phase shifts in the activity rhythm elicited by such light pulses (Yang et al. 1998). Furthermore, the spectral response curves for both TIM degradation and behavioural phase shifts (advances and delays) show maximal responses at wavelengths of 400 to 450 nm (Suri et al. 1998; Figure 3C, D) indicating that both events are causally related. A mutation in the *tim* gene turned out to be hypersensitive to light, and TIM degradation occurred also in the absence of PER and thus in the absence of a functioning circadian clock (Suri et al. 1998). TIM degradation was also not dependent on functional compound eyes (Yang et al. 1998). This indicates that circadian resetting can be mediated via TIM through an extraretinal pathway. Its existence was predicted by the above mentioned behavioural studies. All these results indicate that TIM degradation is the crucial step in circadian light perception. Nevertheless, TIM degradation does not always occur after lights-on. During the larval stages, TIM cycles in antiphase to the LN in two cells of the so called Dorsal Neurones (DN), thus being high during the day and low during the night (Kaneko et al. 1997). The biological significance of this antiphase cycling is unclear but it shows that TIM can be regulated differentially and is not a light-sensitive molecule by itself. Obviously, the light signal has to be transduced biochemically through blue-light absorbing photopigments to TIM. This transduction pathway is largely unknown but there is growing evidence that the blue light absorbing photopigment *Drosophila*-cryptochrome (DCRY) is involved in this process (Emery et al. 1998; Stanewsky et al. 1998).

Cryptochromes are flavoproteins that have similarities with photolyases, their presumptive evolutionary ancestors (review: Cashmore et al. 1999). The cryptochrome absorption spectrum directly overlaps with the behavioural action spectra of *Drosophila* (Selby and Sancar 1999). The DCRY level in head extracts of *Drosophila* is dramatically affected by light exposure (Emery et al. 1998): Under a 12:12 LD-cycle DCRY level decreases upon illumination reaching very low levels during the second half of the day; after lights-off it increases again and reaches its peak level at the end of the night. This cycling is also found in the clock mutants per^0, tim^0, *Clk* and cyc^0 and is thus independent of the clock genes. The *DCry* mRNA level did not show any immediate response to lights-on, indicating that *DCry* gene expression is not light-inducible (Egan et al. 1999). Nevertheless *DCry* mRNA level did show a prominent and circadian cycling which was absent in the clock mutants (Emery et al. 1998; Ishikawa et al. 1999). This indicates that *DCry* gene transcription is clock-controlled whereas the DCRY protein level is merely controlled gene dose are correlated by light and independent of the clock molecules. A lower *DCry* gene dose reduces the magnitude of phase shifts (Egan et al. 1999) whereas overexpression of *DCry* leads to larger phase shifts (Emery et al. 1998).

Both occurred at lower light intensities only, indicating that the system is saturated when DCRY exceeds a certain level. Interestingly, the overexpression effect was less striking and more variable in the advance zone of the PRC (at ZT21) than in the delay zone (at ZT15). The accumulation profile of DCRY might be the reason for this: at ZT15 the DCRY level is still rather low and could probably be raised considerably by overexpression. Thus the effects of light on DCRY could be amplified; at ZT21, DCRY is already close to its peak level and possibly saturated. Therefore, only mild effects on behavioural responses are expected after DCRY overexpression (see Emery et al. 1998 for further discussion). This also implies that the circadian system of normal flies should respond less well to weak light in the delay zone than it does in the advance zone due to the different levels of available DCRY. Indeed, that is exactly what was found by Chandrashekaran and Engelmann years ago (see above). Their interpretation that different photopigments or transduction pathways are involved in advance or delay zone of the PRC shift could still be valid, but the explanation with the different DCRY levels appears more simple.

In summary, the results point to a crucial role of DCRY in circadian photoreception and suggest that TIM might be a direct target of DCRY. Indeed, an in vitro cell based assay has shown recently that DCRY undergoes a photochemical conformation change that allows it to interact directly with TIM (Ceriani et al. 1999). Furthermore, upon illumination of the cells DCRY is able to move into the nucleus and to interact with the PER/TIM complex. This interaction renders the PER/TIM complex inactive and unable to participate in the negative feedback loop. According to these results degradation of TIM is a consequence of the PER/TIM blockage by DCRY and it is not the first step in circadian phototransduction. This is interesting because it may explain the hypersensitivity to light of the per^s mutant (for discussion see Hall 1998b). If light sensitivity depends primarily on TIM degradation it seems clear that mutations in the *tim* gene can eventually cause light hypersensitivity (Suri et al. 1998). But it is hard to understand why mutations in the *per* gene can do the same. However, if the critical first step of light is deactivation of the PER/TIM complex by CRY, it may be that the mutant PER^s/TIM complex is inactivated more effectively by CRY than the wildtype PER/TIM complex. Although TIM is the direct partner of CRY and CRY interacts with TIM leading to its degradation even in the absence of PER (Suri et al. 1998, see above), it is conceivable that an altered conformation of the PER^s/TIM dimer may lead to faster deactivation of this complex and thus to a hypersensitivity to light.

Further evidence that DCRY is involved in circadian photoreception was gained by isolation of a mutation in the *DCry* gene that was called cry^{baby} (cry^b) (Stanewsky et al. 1998). The cry^b mutation affects a highly conserved amino-acid probably involved in binding FAD, one of the two cofactors necessary for cryptochrome functioning. cry^b mutants are unable to reset their clock to short light pulses, and TIM and PER levels on Western blots of head extracts were not anymore entrainable by LD-cycles although they oscillate normally in temperature cycles (Stanewsky et al. 1998). Furthermore, cry^b mutants do not become arrhythmic under continuous light conditions, indicating that not only the parametric but also the non parametric effects of light on the circadian system are affected by this mutation (Emery et al. 2000a).

Surprisingly, despite all these alterations in circadian photoresponses, cry^b mutants can still entrain their activity rhythm to LD-cycles, and PER and TIM still cycle in the small LN_v (Stanewsky et al. 1998). This shows that other light pathways involving the compound eyes or the extraretinal eyes participate in circadian photoentrainment as already implied by the experiments with eyeless mutants and carotenoid depleted flies mentioned above. Indeed, cry^b exhibits poor entrainment in a genetic background in which the phototransduction cascade of the compound eyes is interrupted. This is the case in *no receptor potential* (*norpA*); cry^b double mutants (Stanewsky et al. 1998). After a phase shift of the LD-cycle, the *norpA*; *cry^b* double mutants take many cycles to adjust to the new phase and some flies do not adjust at all. But they are still not completely circadianly blind and the TIM-cycling in the small LN$_v$ is still entrainable (Helfrich-Förster et al. 2001). Their residual capability to entrain might be due to the extraretinal eyes from which it is unknown whether they use the same phototransduction cascade as the compound eyes. Recent experiments with glass *cry^b* double mutants that lack all external and internal photoreceptive structures plus cryptochrome have shown that such mutants are totally blind to circadian time cues (Helfrich-Förster et al 2001). This indicates that all these photosensitive structures contribute to entrainment.

At present it is not completely clear where the action site of DCRY is since a DCRY-antiserum that works on brain sections is not yet available. To exert the proposed interaction with TIM, DCRY must be coexpressed with the clock genes. Indeed, *in situ* hybridisations on head sections have shown that *DCry* is expressed in cell bodies that could be identical with the LN$_v$ (Egan et al. 1999). Furthermore, the entrainment deficiency of *norpA*; *cry^b* double mutants was rescued by expressing the *DCry* gene exclusively in the LN$_v$ (Emery et al. 2000b). This indicates strongly that DCRY acts in the pacemaker neurones itself. The same appears to apply for *Drosophila* larvae. The small LN$_v$ are present from the first larval instar onward and show a cycling in the level of PER and TIM (Kaneko et al. 1997). This cycling is still entrainable in *cry^b* mutants as well as in *glass* and *norpA* mutants that impair the function of the larval photoreceptors (Kaneko et al. 2000). All these mutants show also an entrainable eclosion rhythm (Stanewsky, personal communication; Helfrich-Förster, unpublished results). Interestingly, TIM-cycling is abolished in *norpA*; *cry^b* double mutants (Kaneko et al. 2000). Whether entrainment of the eclosion rhythm is also impaired in such double mutants is not yet known, but this result strongly suggests that both the larval photoreceptors and DCRY contribute to entrainment. Thus, the larval small LN$_v$ appear to be entrained by cryptochrome plus the larval eyes and fail to be entrained when both pathways are impaired. Future experiments must show whether the same is true for entrainment of the eclosion rhythm.

Conclusions

In summary, *Drosophila* uses multiple photoreceptors for entrainment. Larvae and adults appear to utilise the blue-light photopigment cryptochrome which most probably exerts its action in the pacemaker neurones itself. Additionally, they seem to use the retinal eyes at least for entrainment. In adults the extraretinal eyes may play a further role in adjusting activity to the environmental changes in light intensity. Future studies are necessary to analyse the specific roles of these different photoreceptors in *Drosophila's* circadian system.

Using multiple photoreceptors may enable circadian systems to profit from the different qualities of external light stimuli. The signal-to noise problem is reduced by combining several inputs. Natural light-dark cycles do not simply consist of white lights-on and lights-off—the conditions that are commonly used in the laboratory for entraining circadian rhythms. Already Bünning pointed out that organisms most probably choose the very low predawn and postdusk light intensities for measuring daylength: These provide the most reliable reference points that are rather independent of seasonal and daily weather conditions. During natural twilight at dusk and dawn the quality of light changes in three important features: There are large changes in the amount of light, its spectral composition and in the sun's position relative to the horizon. All three features could be used by organisms depending on their ecology and strategy for sampling light (see Roenneberg and Foster 1997 and chapter by Fleissner and Fleissner for further discussion). For instance, entraining by dawn and dusk has proven to be more effective than lights-on /off programs in all animals tested so far including man (Fleissner and Fleissner this volume).

Acknowledgements

We thank Ralf Stanewsky for discussion and for communication of unpublished results.

References

Ali, M.A. (1984) Photoreception and vision in invertebrates. NATO ASI series A: Life Sciences Vol. 92. ISBN 0–306–41626–3.

Belvin, M.P., Zhou, H., Yin, J.C.P. (1999) The *Drosophila* dCREB2 gene affects the circadian clock. Neuron **22**: 777–787.

Blaschke, I., Lang, P., Hofbauer, A., Engelmann, W., Helfrich-Förster, C. (1996) Preliminary action spectra suggest that the clock cells of *Drosophila* are synchronized to the external LD-cycle by the compound eyes plus extraretinal photoreceptors. In: Elsner, N., Schnitzler, H-U. (eds.) Brain and Evolution. Proceedings of the 24th Göttingen Neurobiology Conference, Volume I. p: 30.

Cashmore, A.R., Jarillo, J.A., Wu, Y-J., Liu, D. (1999) Cryptochromes: Blue light receptors for plants and animals. Science **284**: 760–765.

Ceriani, M.F., Darlington, T.K., Staknis, D., Mas, P., Petti, A.A., Weitz, C.J., Kay, S.A. (1999) Light-dependent sequestration of TIMELESS by CRYPTOCHROME. Science **285**: 553–568.

Chandrashekaran, M.K., Engelmann, W. (1973) Early and late subjective night phases of the *Drosophila pseudoobscura* circadian rhythm require different energies of blue light for phase shifting. Z. Naturforsch. **28c**: 750–753.

Chandrashekaran, M.K., Loher, W. (1969) The effect of light intensity on the circadian rhythms of eclosion in *Drosophila pseudoobscura*. Z. Verg. Physiol. **62**: 337–347.

Chen, Y., Hunter-Ensor, M., Schotland, P., Seghal, A. (1998) Alterations of *per* RNA in non-coding regions affect periodicity of circadian behavioral rhythms. J: Biol. Rhythms **13**: 364–379.

Cheng, Y., Hardin, P.E. (1998) *Drosophila* photoreceptors contain an autonomous circadian pacemaker that can function without *period* mRNA cycling. J. Neurosci. **18**: 741–750.

Darlington, T.K., Wagner-Smith, K., Ceriani, M.F., Staknis, D., Gekakis, N., Steeves, T.D.L., Weitz, C.J., Takahashi, J.S., Kay, S.A. (1998) Closing the circadian loop: CLOCK-induced transcription of its own inhibitors *per* and *tim*. Science **280**: 1599–1603.

Dembinska, M.E., Stanewsky, R., Hall, J.C., Rosbash, M. (1997) Circadian cycling of a PERIOD-beta-galactosidase fusion protein in *Drosophila*: evidence for cyclical degradation. J. Biol. Rhythms **12**: 157–172.

Dunlap, (1999) Molecular bases for circadian clocks. Cell **96**: 271–290.

Dushay, M.S., Rosbash, M., Hall, J.C. (1989) The disconnected visual system mutations in *Drosophila* drastically disrupt circadian rhythms. J. Biol. Rhythms **4**: 1–27.

Edery, I. (1999) Role of postransciptional regulation in circadian clocks: lessons from *Drosophila*. Chronobiol. Internat. **16**: 377–414.

Egan, E.S., Franklin, T.M., Hilderbrand-Chae, M.J., McNeil, G.P., Roberts, M.A. Schroeder, A.J., Zhang, X., Jackson, F.R. (1999) An extraretinally expressed insect cryptochrome with similarity to the blue light photoreceptors of mammals and plants. J. Neurosci. **19(10)**: 3665–3673.

Emery, I.F., Noveral, J.M., Jamison, C.F., Siwicky, K.K. (1997) Rhythms of *Drosophila period* gene expression in culture. Proc. Natl. Acad. Sci. USA **94**: 4092–4096.

Emery, P., So, W.V., Kaneko, M., Hall, J.C., Rosbash, M. (1998) CRY, a *Drosophila* clock and light-regulated cryptochrome, is a major contributor to circadian rhythm resetting and photosensitivity. Cell **95(5)**: 669–679.

Emery, P., Stanewsky, R., Hall, J.C. and Rosbash, M. (2000a) A unique circadian-rhythm photoreceptor. Nature **404**: 456–457.

Emery, P., Stanewsky, R., Helfrich-Förster, C., Emery-Le, M., Hall, J.C. and Rosbash, M. (2000b) *Drosophila* CRY confers circadian light-sensitivity to behavioral pacemaker neurons. Neuron **26**: 493–504.

Engelmann, W., Honegger, H.W. (1966) Tagesperiodische Schlüpfrhythmik einer augenlosen *Drosophila melanogaster* Mutante. Naturwissenschaften **53**: 588.

Frank, K.D., Zimmermann, W.F. (1969) Action spectra for phase shifts of a circadian rhythm in *Drosophila*. Science **163**: 688–689.

Glossop, N.R.J., Lyons, L.C., Hardin, P.E. (1999) Interlocked feedback loops within the *Drosophila* circadian oscillator. Science **286**: 766–768.

Hall, J.C. (1998a) Molecular neurogenetics of biological rhythms. J. Neurogenet. **12**: 115–181.

Hall, J.C. (1998b) Genetics of biological rhythms in *Drosophila*. Adv. Genet. **38**: 135–184.

Hao, H., Allen, D.L., Hardin, P.E. (1997) A circadian enhancer mediates PER-dependent mRNA cycling in *Drosophila melanogaster*. Mol. Cell. Biol. **17**: 3687–3693.

Hege, D.M., Stanewsky, S., Hall, J.C., Giebultowicz, J.M. (1997) Rhythmic expression of a PER-reporter in the malphighian tubules of decapitated *Drosophila*: evidence for a brain-independent circadian clock. J. Biol. Rhythms **12**: 300–308.

Helfrich, C., Engelmann, W. (1983) Circadian rhythms of the locomotor activity rhythm in *Drosophila melanogaster* and its mutant 'sine oculis' and 'small optic lobes'. Physiol. Entomol. **8**: 257–272.

Helfrich-Förster, C. (1996) *Drosophila* rhythms: From brain to behaviour. Semin. Cell Dev. Biol. **7**: 791–802.

Helfrich-Förster, C. (1997) Photic entrainment of *Drosophila* activity rhythm occurs via retinal and extraretinal pathways. ESC-Workshop "Photic and Non-Photic Entrainment of Biological Rhythms". Biol. Rhythm Res. **28S**: 119.

Helfrich-Förster, C. (1998) Robust circadian rhythmicity of *Drosophila melanogaster* requires the presence of Lateral Neurons: A brain-behavioral study of *disconnected* mutants, J. Comp. Physiol. A **182**: 435–453.

Helfrich-Förster, C., Stengl, M., Homberg, U. (1998) Organization of the circadian system in insects. Chronobiol. Internat. **15(6)**: 567–594.

Helfrich-Förster, C., Winter, C., Hofbauer, A., Hall, J. C., Stanewsky, R. (2001) This circadian clock of fruit is blind after elimination of all known photoreceptors. Neuron **30**: 249–261.

Hofbauer, A., Buchner, E. (1989) Does *Drosophila* have seven eyes? Naturwissenschaften **76**: 335–336.

Honegger, (1967) Zur Analyse der Wirkung von Lichtpulsen auf das Schlüpfen von *Drosophila pseudoobscura*. Z. Vergl. Physiol. **57**: 244–262.

Hunter-Ensor, M., Ousley, A., Seghal, A. (1996) Regulation of the *Drosophila* protein *timeless* suggests a mechanism for resetting the circadian clock by light. Cell **84**: 677–685.

Ishikawa, T., Matsumoto, A., Kato, T. Jr., Togashi, S., Ryo, H., Ikenaga, M., Todo, T., Ueda, R., Tanimura, T. (1999) DCRY is a *Drosophila* photoreceptor protein implicated in light entrainment of circadian rhythm. Genes to Cells **4(1)**: 57–65.

Kaneko, M. (1998) Neural substrates of *Drosophila* rhythms revealed by mutants and molecular manipulations. Curr. Opin. Neurobiol. **8**: 652–658.

Kaneko, M., Helfrich-Förster, C., Hall, J.C. (1997) Spatial and temporal expression of the *period* and the *timeless* genes in the developing nervous system of *Drosophila*: Newly identified pacemaker candidates and novel features of clock-gene product cyclings. J. Neurosci. **17**: 6745–6760.

Kaneko, M., Hamblen, M., Hall, J.C. (2000) Involvement of the *period* gene in developmental time-memory: Effect of the *per^short* mutation. J. Biol. Rhythms **15**: 13–30.

Klemm, E., Ninnemann, H. (1976) Detailed action spectrum for the delay shift in pupae emergence of *Drosophila pseudoobscura*. Photochem. Photobiol. **24**: 369–371.

Kloss, B., Price, J.L., Saez, L., Blau, J., Rothenfluh, A., Wesley, C.S., Young, M.W. (1998) The *Drosophila* clock gene *double-time* encodes a protein closely related to human casein kinase I. Cell **94**: 97–107.

Konopka, R.J., Pittendrigh, C., Orr, D. (1989) Reciprocal behavior associated with altered homeostasis and photosensivity of *Drosophila* clock mutants. J. Neurogenet. **6**: 1–10.

Krishnan, B., Dryer, S.E., Hardin, P.E. (1999) Circadian rhythms in olfactory responses of *Drosophila*. Nature **400**: 375–378.

Lee, C., Parikh, V., Itsukaichi, T., Bae, K., Edery, I. (1996) Resetting the *Drosophila* clock by photic regulation of PER and a PER-TIM complex. Science **271**: 1740–1744.

Myers, M.P., Wagner-Smith, K.K., Rothenfluh-Hilfiker, A., Young, M.W. (1996) Light-induced degradation of TIMELESS and entrainment of the *Drosophila* circadian clock. Science **271**: 1736–1740.

Naidoo, N., Song, W., Hunter-Ensor, M., Seghal, A. (1999) A role for the proteasome in the light response of the *timeless* clock protein. Science **285**: 1737–1741.

Ohata, K., Nishiyama, H., Tsukahara, Y. (1998) Action spectrum of the circadian clock photoreceptor in *Drosophila melanogaster*. In: Touitou, Y. (ed.) Biological clocks: Mechanisms and Applications. Amsterdam, Elsevier. pp. 167–171.

Peterson, G., Hall, J.C., Rosbash, M. (1988) The *period* gene of *Drosophila* carries species-specific behavioral instructions. EMBO J. **7**: 3939–3947.

Pittendrigh, C.S. (1967) The driving oscillation and its assay in *Drosophila pseudoobscura*. Proc. Natl. Acad. Sci. USA **58**: 1762–1767.

Pittendrigh, C.S., Minis, D.H. (1964) The entrainment of circadian oscillations by light and their role as photoperiodic clocks. Am. Naturalist **98**: 261–294.

Plautz, J.D., Kaneko, M., Hall, J.C., Kay, S.A. (1997) Independent photoreceptive circadian clocks throughout *Drosophila*. Science **278**: 1632–1635.

Pollock, I., Hofbauer, A. (1991) Histamine-like immunoreactivity in the visual system and brain of *Drosophila melanogaster*. Cell Tiss. Res. **266**: 391–398.

Price, J.L., Dembinska, M.E., Young, M.W., Rosbash, M. (1995). Supression of PERIOD protein abundance and circadian cycling by the *Drosophila* clock mutation *timeless*. EMBO J. **14**: 4044–4049.

Price, J.L., Blau, J., Rothenfluh, A., Abodeely, M., Kloss, B., Young, M.W. (1998) *double-time* is a novel *Drosophila* clock gene that regulates PERIOD protein accumulation. Cell **94**: 83–95.

Qiu, J., Hardin, P. (1996) *per* mRNA cycling is locked to lights-off under photoperiodic conditions that support circadian feedback loop function. Mol. Cell. Biol. **16**: 4182–4188.

Roenneberg, T., Foster, R.G. (1997) Twilight times: light and the circadian system. Photochem. Photobiol. **66**: 549–561.

Rutila, J.E., Suri, V., Le, M., So, W.V., Rosbash, M., Hall, J.C. (1998) CYCLE is a second bHLH-PAS clock protein essential for circadian rhythmicity and transcription of *Drosophila period* and *timeless*. Cell **93**: 805–814.

Saunders, D.S., Gillanders, S.W., Lewis, R.D. (1994) Light-pulse phase response curves for the locomotor activity rhythm in *period* mutants of *Drosophila melanogaster*. J. Insect Physiol. **40**: 957–968.

Selby, C.P., Sancar, A. (1999) A third member of the photolyase/blue-light photoreceptor family in *Drosophila*: A putative circadian photoreceptor. Photochem. Photobiol. **69**: 105–107.

Stanewsky, R., Frisch, B., Brandes, C., Hamblen-Coyle, M.J., Rosbash, M., Hall, J.C. (1997) Temporal and spatial expression patterns of transgenes containing increasing amounts of the *Drosophila* clock gene *period* and a lacZ reporter: mapping elements of the PER protein involved in circadian cycling. J. Neurosci. **17**: 676–696.

Stanewsky, R., Kaneko, M., Emery, P., Beretta, B., Wager-Smith, K., Kay, S.A., Rosbash, M., Hall, J.C. (1998). The *cryb* mutation identifies cryptochrome as a circadian photoreceptor in *Drosophila*. Cell **95**: 681–692.

Suri, V., Qian, Z., Hall, J.C., Rosbash, M. (1998) Evidence that the TIM light response is relevant to light-induced phase shifts in *Drosophila melanogaster*. Neuron **21**: 225–234.

Swinderen B., van, Hall, J.C. (1995) Analysis of conditioned courtship in *dusky-Andante* rhythm mutants of *Drosophila*. Learn. Mem. **1**: 49–61.

Wheeler, D.A., Hamblen-Coyle, M.J., Dushay, M.S., Hall, J.C. (1993) Behavior in light-dark cycles of *Drosophila* mutants that are arrhythmic, blind or both. J. Biol. Rhythms **8**: 67–94.

Winfree, A.T. (1970) Integrated view of resetting a circadian clock. J. Theoret. Biol. **28**: 327–374.

Winfree, A.T. (1974) Suppressing *Drosophila* rhythm with dim light. Science **183**: 970–972.

Yang, Z., Emerson, M., Su, H.S., Seghal, A. (1998) Response of the *timeless* protein to light correlates with behavioral entrainment and suggests a non-visual pathway for circadian photoreception. Neuron **21**: 215–223.

Yasuyama, K., Meinertzhagen, I.A. (1999). Extraretinal photoreceptors at the compound eye's posterior margin in *Drosophila melanogaster*. J. Comp. Neurol. **412**: 193–202.

Young, M.W. (1998) The molecular control of circadian behavioral rhythms and their entrainment in *Drosophila*. Annu. Rev. Biochem. **67**: 135–152.

Zeng, H., Qian, Z., Myers, M.P., Rosbash, M. (1996) A light-entrainment mechanism for the *Drosophila* circadian clock. Nature **380**: 129–135.

Zimmerman, W.F., Goldsmith, T.H. (1971) Photosensitivity of the circadian rhythm and of visual receptors in carotenoid-depleted *Drosophila*. Science **171**: 1167–1169.

10. Photoentrainment of Vertebrate Circadian Rhythms

R.G. Foster*

Department of Integrative and Molecular Neuroscience,
Division of Neuroscience and Psychological Medicine,
Imperial College School of Medicine, Charing Cross Hospital,
Fulham Palace Road, London, W6 8RF, UK

A circadian clock is of no use unless biological time is adjusted to local environmental time, and most organisms use the changes in the quantity and quality of light at twilight as their primary zeitgeber to effect photoentrainment. The sensory demands of photoentrainment have imposed a unique set of selection pressures, which has led to the evolution of specialised photoreceptor systems. In the non-mammalian vertebrates, pineal and deep brain photoreceptors play an important, but poorly defined, role in circadian organisation. By contrast, photoentrainment in the mammals relies exclusively upon ocular photoreceptors. Although superficially very different, the entraining photoreceptor inputs of mammals and non-mammals appear to be both specialised (employing novel photopigments), and complex (utilising multiple photopigments). Why there should be this multiplicity of photic inputs to the circadian system remains unclear, but must surly be related to the sensory task of twilight detection. During twilight, the quality of light changes in three important respects: (1) the amount of light; (2) the spectral composition of light; (3) and the position of the sun. In theory all of these parameters could be used by the circadian system to detect the phase of twilight, but each would be subject to considerable variation or "noise". Furthermore, the impact of this noise will depend upon the organism and the environment in which it inhabits. Thus the task of twilight detection is likely to be very complex and show considerable variation between species. If we are to place the molecular dissection of the circadian system into a functional context, then the ecology of photoentrainment must be given serious consideration.

Introduction

The function of the circadian system is to co-ordinate the phase of a biological event to a specific feature of the 24-h environmental cycle, and to ensure that the phases of multiple rhythmic events within the organism are appropriately coupled. To achieve this timing, the circadian system must remain synchronised with the solar day. The entrainment of a biological clock requires an input pathway consisting of a receptor and transduction elements for the detection of specific environmental signals (*zeitgebers*). Furthermore, elements of the circadian clock must be capable of transforming the incoming signals to appropriate changes of the rhythm's phase. When entrained, the circadian clock adopts a distinct phase relationship with the astronomical day, and each of the expressed rhythms adopts its own phase relationships with the clock. Depending upon the species, biological clocks respond to a variety of different *zeitgeber*. For example, many microorganisms, plants, and heterothermic animals can be entrained by rhythmic changes in environmental temperature, whilst

---

*E-mail: r.foster@ic.ac.uk

social signals such as feeding schedules in humans can act as *zeitgeber*. However, the stable daily change in the light environment at dawn or dusk provides the most reliable indicator of the time of the day. As a result, most organisms use the changes in the quantity and quality of light at twilight as their primary *zeitgeber* to effect photoentrainment (Roenneberg and Foster 1997).

Photoentrainment in non-mammals

Non-mammalian vertebrates possess a diverse complement of photoreceptors, with several different types of photoreceptor organ developing from the embryonic forebrain (telencephalon). These photoreceptor organs can be classified as: (1) an intracranial pineal organ or pineal body (*epiphysis cerebri*) which is photoreceptive in all non-mammalian vertebrates; (2) an intracranial parapineal organ, found in many teleost fish and some agnatha; (3) an extracranial 'third eye', variously called a frontal organ (anura) or parietal eye/body (lacertids); (4) deep brain photoreceptors, which are located in several sites of the brain and are found in all non-mammalian vertebrates; and (5) lateral eyes, which contain photoreceptors in all the vertebrate classes. Of these extraretinal photoreceptors, the deep brain and pineal photoreceptors have been shown to play a critical role in the regulation of temporal physiology in the non-mammalian vertebrates. Why the mammals have lost extraocular photoreceptors remains speculative, but has been correlated with the early evolutionary history of eutherian mammals and their passage through a nocturnal bottleneck (Foster and Menaker 1993). Very recent work in zebra fish has suggested that circadian rhythms of gene expression can be entrained by light in isolated organs (heart and kidney) (Whitmore et al. 2000). However, critical experiments using light as a physiological stimulus need to be undertaken to distinguish between a truly photic effect and an energetic/thermal effect of light.

Extraretinal photoreceptors

Although highly variable in form, the vertebrate pineal has been shown to act as a photoreceptor in all non-mammalian species examined (Meissl and Yanez 1994). Illumination of the non-mammalian pineal *in vitro* can modify melatonin synthesis in several different ways: (a) by entraining a circadian rhythm of melatonin synthesis (for example the lizard *Anolis*, see Menaker and Wisner 1983); (b) by regulating melatonin synthesis acutely (that is, melatonin synthesis is driven by a light:dark cycle and there is no endogenous clock driven rhythm in melatonin synthesis under constant conditions, for example the trout, see Max and Menaker 1992); (c) by regulating melatonin synthesis both acutely and by entraining a circadian rhythm of melatonin synthesis (for example the chicken, see Takahashi et al. 1989). Whether clock-driven, light-driven, or both, melatonin synthesis and release is confined to the dark portion of a light:dark cycle. The extent to which the melatonin signal is use by the circadian system varies greatly, both within and between the vertebrate classes. For example, in some birds (such as sparrows) the rhythmic release of pineal melatonin is essential for sustained rhythms in circadian behaviour under constant conditions (Gaston and Menaker 1968), and infusion of melatonin can entrain circadian behaviour (for example in pigeons, Chabot and Menaker 1992). By contrast, pinealectomy has no obvious effects upon the circadian behaviour of other birds (for example Chicken and quail, Menaker et al. 1981). Why the pineal should have such differing roles in such closely related species remains a fascinating issue (Menaker and Tosini 1996). Gwinner has proposed that the role of the pineal in avian circadian organisation is related to whether the species shows a long-distance or local migration each year (Gwinner et al. 1997).

Deep brain photoreceptors were first linked to circadian entrainment as a result of the experiments undertaken by Menaker and colleagues during the 1960s and early 1970s. Studies on the house sparrow (*Passer domesticus*) showed that the removal of both eyes and pineal did not block photoentrainment (Menaker and Underwood 1976). This observation was subsequently duplicated in many species of bird, fish, amphibia and reptiles (for review see Foster et al. 1993). It is important to note that neural tissue is remarkably permeable to light, and although light is scattered and filtered, photons penetrate deep into the brain (Foster and Follett 1985). Although this light cannot be used to generate an image of the world, it can be used to calculate the overall amount of environmental light, and hence time of day. Deep brain photoreceptors have also been shown to play an important role in the regulation of the photoperiodic response of non-mammalian vertebrates. In the 1930s Benoit showed that blinded ducks exposed to spring-like photoperiods would be stimulated to breed (Benoit 1935a; Benoit 1935b). More recent studies, involving blinding and shielding light from entering the brain, have confirmed these original findings in a range of bird species (for review see Foster and Follett 1985). The pineal organ was originally assumed to regulate the avian reproductive responses, but it is now known that the pineal plays little or no role in the photoperiodic response. In quail, for example, the specific long-day illumination of the pineal does not stimulate gonadal growth (Homma et al. 1980), and pinealectomy of either blind or eye intact quail leaves the photoperiodic response unaffected (Simpson et al. 1983). By contrast, local illumination of the basal brain using fine fibre optics in blinded and pinealectomised quail, causes gonadal growth at normal rates (Oliver et al. 1979; Yokoyama et al. 1978).

The characterisation of the extraretinal photoreceptors

Surgical lesions, directed illumination using fine optic fibres, and *in vitro* analysis have demonstrated that the pineal (and other regions of the brain) contain photoreceptors, but characterisation of these photoreceptors has proved difficult. Describing the spectral sensitivity profile (action spectrum) of a light-dependent response is a critical step in identifying the regulatory photopigment of any system. The absorption spectrum of a photopigment gives a relative probability of photons being absorbed as a function of wavelength. As a result, the absorption spectrum of the photopigment dictates the spectral response of the photoreceptor. Accounting for any confounding factors such as screening pigments, the action spectrum of a light-response must match the absorbance spectrum of the photopigment/s upon which it relies. For example, the rod and cone photopigments of the vertebrates consist of an opsin protein coupled to a chromophore derived from an 11-*cis* form of vitamin A retinaldehyde. Despite great variation in their spectral maxima, rod and cone photopigments have spectra that match the standard template for a vitamin-A based photopigment ("Dartnall nomogram"). Significantly. action spectra for the light suppression of melatonin in many non-mammals (Deguchi 1981; Meissl and Yanez 1994), and an action spectrum for the deep brain photoreceptors mediating the photoperiodic response of quail (Foster et al. 1985), also fit a Dartnall nomogram. Thus opsin-retinaldehyde based photopigments were implicated in mediating these responses, and as a result, opsin and 11-*cis* retinaldehyde were sought within the CNS.

Rod-and cone-opsin specific antibodies have been found to label pinealocytes in a great variety of vertebrates, suggesting that both rod-like and cone-like photopigments are located within the pineal (Vigh and Vigh-Teichmann 1988). However, in several studies, sub-sets of pinealocytes remained unlabelled by any opsin antibodies (Foster et al. 1987). This failure was largely attributed to problems associated with either tissue fixation or antigenic sensitivities. The alternative possibility, that the pineal contained photopigments different from the rod and cone opsins of the eye, was given some consideration, but not demonstrated until the mid 1990s (see Novel extraretinal photopigments below). A functional photopigment requires a chromophore and, in the limited number of studies undertaken, 11-*cis* retinoid has always been identified within the pineal of non-mammals (Foster et al. 1989; Tabata et al. 1985).

In contrast to the pineal, the localisation of opsins within the basal brain has proved much more of a problem. Many studies over a period of fifteen years, using a range of different anti-opsin antibodies, failed to give any clear localisation (Foster et al. 1987; Vigh et al. 1980; Vigh and Vigh-Teichmann 1988). At the time it seemed inexplicable that anti-opsin antibodies that labelled pinealocytes within the avian pineal gland would fail to label any cells within the brain. This failure caused many researchers to dismiss the whole notion of encephalic photoreceptors. However, in 1988 Rae Silver's laboratory demonstrated that cerebrospinal fluid (CSF)-contacting neurones within the septal and tuberal areas of the brain of the ring dove, quail and duck could be labelled with an anti-rod opsin antibody. The impact of this finding was blunted because Western blots of these brain regions failed to validate the antibodies used (Silver et al. 1988). Several years later, three new anti-cone opsin antibodies produced an intense immunostaining of CSF-contacting cells within the septal area of the brain of a lizard (*Anolis carolinensis*). Significantly, Western blots recognised a single 40 kD protein in ocular, anterior brain and pineal extracts, suggesting that the immunostaining observed was specific (Foster et al. 1993). Furthermore, the isolation of 11-*cis* retinoid from the *Anolis* forebrain, suggested that these opsins were functional (Foster et al. 1993). Since 1993, several papers have considered opsin localisation within the central nervous system of a range of different vertebrates. For example, opsin-like labelling was found within the neurosecretory cells of the NMPO (nucleus magnocellularis preopticus) of the hypothalamus of several fish (Foster et al. 1994) and amphibia (Yoshikawa et al. 1994), and cells within the subhabenular of fish showed opsin labelling (Ekström et al. 1987). Studies on the adult lamprey identified multiple populations of opsin immunoreactive CSF-contacting neurones throughout the hypothalamus, and non-CSF-contacting neurones were labelled in the epithalamic and caudal diencephalon (Garcia-Fernandez et al. 1997).

Collectively, these results have implicated a number of different encephalic populations of photoreceptors: the pinealocytes of the pineal, CSF-contacting neurones adjacent to the third and lateral ventricles, non-CSF contacting cells of the NMPO, and other regions of the forebrain. Furthermore, these receptors utilise rod- and cone-like opsins and have additional photopigments that differ from both. The immunocytochemical studies suggested that extraretinal photoreception might be complex, and recent molecular findings have confirmed and extended this view.

Novel extraretinal photopigments

A key breakthrough in our understanding of the extraretinal photoreceptors has come with the application of molecular approaches to isolate novel extraocular opsins. The first of these opsins was isolated in 1994 by Fukada and colleagues who isolated a cDNA from the chicken pineal that encodes a photopigment with an absorption maximum near 470 nm (Okano et al. 1994). This opsin was called 'pinopsin', and orthologues of pinopsin have subsequently been isolated from the pineal of several different birds (Kawamura and Yokoyama 1996) and lizards (Kawamura and Yokoyama 1997). Pinopsin expression appears restricted to the pineal in birds, but may be expressed in both the pineal and retina of some reptiles. In 1997, the sequence of a novel fish opsin was described, VA (*vertebrate ancient*) opsin. VA opsin forms a photopigment with an absorption maxima at 451 nm (with vitamin A_1) and is expressed in a sub-set of horizontal and amacrine cells of the retina, the pineal, and subhabenular region of the brain (Kojima et al. 2000; Moutsaki et al. 2000; Philp et al. 2000b; Soni et al. 1998). The role of VA opsin photopigment in both retinal and extraretinal photoreception remains unknown. Clearly, these results have important implications to both retinal and extraretinal photoreception. Another novel fish opsin, parapineal opsin, was isolated from catfish. This opsin is expressed primarily within the parapineal organ, but is also weakly expressed in the pineal (Blackshaw and Snyder 1997). The functional properties of this opsin have yet to be determined. The most recent extraretinal opsin to be discovered is melanopsin (Provencio et al. 1998b). Melanopsin was originally isolated from photosensory melanophores of *Xenopus laevis* but is also expressed in the hypothalamus (NMPO region), iris (photosensory in amphibia), and the horizontal cell layer of the retina (similar to VA opsin). Furthermore, an ortholog of *Xenopus* melanopsin has been isolated from chickens (Provencio et al. 1998c) and fish (Rollag and Provencio—personal communication). Unfortunately expression studies are lacking, and we do not know whether melanopsin is capable of forming a functional photopigment in any of these vertebrates.

In addition to the discovery of novel opsins, recent molecular studies have suggested that there may be forms of rod and cone opsin that are unique to the extraretinal photoreceptors. For example, in several species of teleost fish, opsin cDNAs were isolated which share approximately 75% nucleotide and amino acid identity with their corresponding rod-opsins from the retina (Mano et al. 1999; Philp et al. 2000a). The basis for the sequence differences between the "extraretinal rod-like" (ERrod)-like opsins and retinal rod-opsins remains unclear, but may be related to the differing photosensory roles of the retinal and extra-retinal photoreceptors, and/or as a result of genetic drift of these opsins after a gene duplication event. Functional analysis of the ERrod-like photopigments, and the isolation of additional extra-retinal opsins should help resolve these alternatives. An additional example of pineal specialisation is the recent discovery of retinal and pineal specific arylalkylamine N-acetyltransferase (AANAT) genes isolated from two teleost fish-trout and pike. The evolution of two AANAT genes may represent a strategy for tissue optimisation of the photic regulation of melatonin synthesis (Coon et al. 1999). This suggests that a number of the photosensory elements of the vertebrate pineal might be specialised for encephalic light detection.

Multiple extraretinal photopigments

Evidence for "encephalic photoreceptors" was first demonstrated in fish by Karl von Frisch in 1911. He showed that light-induced colour changes in the skin of the European minnow (*Phoxinus laevis*) would still occur in the absence of the eyes and pineal complex, and that lesions within the basal brain would block this response to light. von Frisch concluded that there must be photoreceptive elements within the diencephalon of fish (Frisch 1911). Ninety years later we have evidence for not one, but multiple populations of photoreceptors within the diencephalon of fish and other vertebrates. Furthermore, the pineal complex of non-mammals appears to possess both novel and classical photopigments (Okano et al. 1994; Philp et al. 2000b). The future challenge will be to place these different photoreceptors into a physiological and behavioural context, and link the various opsins to a circadian or photoperiodic response in the whole animal.

Photoentrainment in mammals

Until recently, all of the experimental evidence suggested that the circadian system of mammals is entrained by photoreceptors within the eye. However, in a fairly recent report in Science that was widely reported in the press, Campbell and Murphy suggested that bright light of 13,000 lux (lx) applied to the popliteal region (skin behind the knee) can shift circadian rhythms of body temperature and melatonin (Campbell and Murphy 1998). The conclusions drawn by these researchers are controversial, and there has been much reluctance to accept these results by most circadian and vision biologists (Foster 1998). This hesitance arises because enucleation in humans (Czeisler et al. 1995; Lockley et al. 1997) and other mammals (Nelson and Zucker 1981), always blocks photoentrainment. In addition, popliteal illumination produces no effect whatsoever on the suppression of nocturnal levels of pineal melatonin (Lockley et al. 1998). Furthermore, a very recent study examined the phase-shifting effect of exposing the chest and abdomen to 13,000 lx broad spectrum light on phase shifting melatonin, cortisol and thyrotropin rhythms. These treatments had no effect (Lindblom et al. 2000). The lack of independent support for extraocular photoentrainment of the human circadian system suggests that the original observations by Campbell and Murphy are not robust, and may result from of some unrecognised artefact of their experimental procedures (Yamazaki et al. 1999). Experiments that duplicate their methodology precisely should resolve this issue.

Although the entraining photoreceptors appear to be exclusively ocular, the response characteristics of the circadian and visual systems to light are markedly different. In those mammals studied, the threshold for phase shifting is significantly higher than that required to elicit a visual response. For example, hamsters can recognise optical gratings at a luminance level 200 times less than the level necessary to elicit phase shifts in locomotor rhythms (Emerson 1980). In addition, the circadian system appears markedly insensitive to light stimuli of a short duration. Indeed, the circadian system of the hamster is relatively insensitive to a stimulus duration of less than 30 seconds (Nelson and Takahashi 1991). These features of the circadian system have been associated with filtering out those light stimuli, such as moon light and lightning, that would not provide information about the time of day (Roenneberg and Foster 1997).

In mammals light information from the eye reaches the primary circadian oscillator in the suprachiasmatic nuclei (SCN) via a specific neural projection called the retino-hypothalamic tract (RHT). This tract arises from a small sub-set of retinal ganglion cells and forms a relatively small number of the fibres of the optic nerve. For example, in the mouse retina, only 0.1% of the RGCs form the RHT projection to the SCN (Provencio et al. 1998a). Although the RGCs that form the RHT have been identified, the photoreceptors that are connected to these cells have not.

Novel ocular photoreceptors

Disentangling which retinal cells mediate photoentrainment from the mass of neurones dedicated to image detection has been a major problem. The retina of all mammals contains two types of known photoreceptors: rods which are typically associated with dim light vision, and cones which are associated with colour vision under bright light conditions. Trying to establish whether these photoreceptors mediate the effects of light on the biological clock has been a complex issue. However, the use of mice with naturally occurring genetic disorders of the eye has provided a partial solution to this issue. Mice homozygous for *retinal degeneration* (*rd/rd*) were the first animals to be used for such studies. Despite the massive loss of rods and cones, *rd/rd* mice show circadian responses to light that are unattenuated when compared with non-degenerate control mice (*rd/+, +/+*). Thus the sensitivity of the circadian system to light does not parallel loss of either rod or cone photoreceptors (Provencio et al. 1994). However, removal of the eyes abolishes all circadian responses to light in *rd/rd* mice (Foster et al. 1991). The remarkable findings in *rd/rd* mice showed that rods are not required for photoentrainment, but the eyes must contain these light sensing cells. In more recent studies in humans, a substantial portion of patients who have eyes but had lost their vision due to retinal disease, also retained their ability to suppress melatonin (Czeisler et al. 1995) as well as to shift their circadian phase (Lockley et al. 1997).

The most recent experiments to address the impact of rod and cone photoreceptor loss on the circadian system employed mouse models that lacked all rods and cones. In rodless+coneless mice, circadian responses to light (photoentrainment and pineal melatonin suppression) were intact (Freedman et al. 1999; Lucas et al. 1999). These results lead to the striking conclusion that mammals must use some unidentified photoreceptor outside the rod and cone receptors. Again, removal of the eyes abolished all circadian responses to light demonstrating that the eyes must house these unrecognised photoreceptors (Freedman et al. 1999).

A novel opsin photopigment in the mammalian retina

Although the requirement of rod and cone opsins in circadian light reception has been excluded, there is evidence for involvement of an opsin-based pigment in this response. In rodents two detailed action spectra for photoentrainment have been undertaken. In both the golden hamster (Takahashi et al. 1984) and mouse (Provencio and Foster 1995), the data showed a very close fit to a standard template for a vitamin-A based photopigment (see The characterisation of extraretinal photoreceptors). These findings, taken together with the results from rodless+coneless mice, suggest that the mammalian eye contains unrecognised non-rod, non-cone opsin-based photoreceptors within the inner retina which regulate circadian rhythms. The conclusion that the vertebrate inner retina contains novel photoreceptors is given direct support by the discovery of VA photopigments within the inner retina of fish (Soni et al. 1998) (see Novel extraretinal photopigments). To date, no VA opsin homologs have been identified in mammals, but a number of other candidates exist. These include RGR (retinal-binding G protein-coupled receptor), peropsin, and encephalopsin, and melanopsin. With the exception of encephalopsin (Blackshaw and Snyder 1999), which is expressed in a variety of neural and non-neural tissues, and therefore falls outside of the scope of this review, they are all ocular. RGR, differs from rod and cone opsins in that its preferred chromophore is all-*trans*-retinaldehyde rather than 11-*cis*-retinal. Upon exposure to light, RGR photoisomerises the all-*trans* chromophore to the 11-*cis* confirguration. Its likely function, therefore, is not that of a signalling photopigment, but rather that of a photoisomerase (Hao and Fong 1999). Peropsin, is also localised to the RPE. The role of peropsin is unknown, but several lines of evidence suggest that it may function like RGR (Sun et al. 1997a). The most recent, and arguably the best candidate to date, is a mammalian homolog of *Xenopus* melanopsin (see Novel extraretinal photopigments) which differs from all of the other novel mammalian opsins in that it is expressed within the neural retina (Provencio et al. 2000). Specifically, melanopsin is expressed in a small number of cells within the ganglion and amacrine cell layers in the inner retina. This distribution is strikingly similar to the distribution of murine retinal ganglion and amacrine-like cells known to form the retinohypothalamic tract (Provencio et al. 1998a). Once again, functional expression studies are lacking, and we do not know whether this mammalian form of melanopsin is capable of forming a functional photopigment.

It is worth noting that in addition to circadian physiology, many other aspects of mammalian biology are influenced by gross changes in environmental light, including pupil size, blood pressure, mood and attention (Wetterberg 1993). It is possible, therefore, that uncharacterised ocular photoreceptors might form the basis of a general "irradiance detection" pathway mediating many, if not all, non-visual responses to light. Support for this hypothesis comes from our recent work on rodless + conless mice. In addition to circadian responses to light, these animals also show a pupillary light reflex. The action spectrum for this response demonstrates the involvement of an opsin/vitamin A-based photopigment with a wavelength of maximum sensitivity at 479 nm. The known mouse photopigments have maximal absorbances at wavelengths of 360 nm (UV cones, Jacobs et al. 1991), 498 nm (rods, Bridges 1959) and 508 nm (Green cones, Sun et al. 1997b), and so cannot account for the pupillary responses observed in rodless + coneless mice (Lucas et al. 2001). Until, we have matched the action spectra for pupillary and circadian responses to light it remains possible that these aspects of physiology are driven by different novel photoreceptors. However, the principle of parsimony would argue against this, and directs the search for circadian photopigment to those candidates that conform to the basic opsin:11-cis retinaldehyde composition and, in mice, have spectral maxima near 479nm.

In addition to opsin-based photopigments, recent studies in the plant *Arabidopsis*, the fruit-fly *Drosophila* and mice suggested that a group of proteins, called the cryptochromes (CRY), might be responsible for detecting light and mediating photoentrainment in phylogenetically diverse organisms

(Devlin and Kay 1999). In mammals at least, the evidence for this hypothesis was always weak (Lucas and Foster 1999a; Lucas and Foster 1999b), and has recently collapsed as a result of detailed studies by Griffin et al (Griffin et al. 1999) who failed to uncover any effect of light on the activity of these proteins in cell culture. Furthermore, the disruption of the cryptochrome genes (CRY1 and CRY2) does not block the light-induced expression of two "clock genes" in the SCN of mice (Okamura et al. 1999).

A role for rods and cones

There is overwhelming evidence that unidentified (non-rod, non-cone) photoreceptors within the mammalian eye mediate photoentrainment. However, this does not mean that the classical rod and cone photoreceptors play no role in this process. The experiments on rodless+coneless mice outlined above merely suggest that these receptors are not required. Indeed, indirect evidence for the involvement of cone photoreceptors in photoentrainment comes from studies on an extraordinary animal—the blind mole rat (*Spalax ehrenbergi*). *Spalax* is a subterranean fossorial rodent with subcutaneous atrophied eyes and shows a massive reduction (87–97%) of those regions of the brain associated with the image-forming visual system (Cooper et al. 1993). Although visually blind (Haim et al. 1983; Rado et al. 1992), the minute eyes (little more than 0.5 mm in diameter) can perceive light and are used to entrain circadian rhythms (David-Gray et al. 1998; Goldman et al. 1997; Rado et al. 1991). Photoentrainment is thought to occur in the wild when *Spalax* removes debris from the tunnel complex and is exposed to brief periods of natural light (Rado and Terkel 1989). Over the past 30 million years, evolutionary processes appear to have disentangled and eliminated the image-forming visual system of this animal whilst retaining those components of the eye that regulate the biological clock. Remarkably, a cone opsin has been isolated from the eye of *Spalax*, and this opsin has been shown to form a fully functional photopigment (David-Gray et al. 1998). These results provide strong, although indirect, evidence that cone photopigments contribute to photoentrainment in *Spalax*, and by implication, other mammals (David-Gray et al. 1998, 1999). The second line of evidence that implicates a role for the rods and cones in irradiance measures comes from electrophysiological studies. Extracellular electrodes placed in the SCN of anaesthetised rats suggest that the SCN receive inputs from both rod and cone photoreceptors (Aggelopolus and Meissl 2000). This input from the rods and cones presumably augment the photic information received by inner retinal photoreceptors in some manner (Roenneberg and Foster 1997)

Multiple photopigments and twilight detection

The conclusion, based upon studies in *Spalax*, that cone photopigments might contribute to photoentrainment in all mammals (see A role for rods and cones) would appear to contradict the findings that the loss of both rod and cone photoreceptors has no effect on rodent photoentrainment, and that the retina contains novel circadian photoreceptors (Freedman et al. 1999; Lucas et al. 1999). These results appear less contradictory, however, when the pineal organs of non-mammals are considered. As discussed above, in non-mammalian vertebrates the pineal organ may be both an important part of the circadian timing system (in some ways analogous to the mammalian SCN) and is itself directly light sensitive. This photosensitivity is attained using multiple photopigments, with rod-and cone-like opsins, as well as novel opsins, co-existing within the same organ (Shand and Foster 1999). Thus both novel and classical photopigments seem to mediate the effects of light on temporal physiology in all the vertebrate classes. And the question is why? If this question is to be addressed, then we must first define those features of the light environment, which are important for the regulation of temporal physiology.

During twilight, the quality of light changes in three important respects: (1) the amount of light, (2) spectral composition of light, and (3) the source of light (i.e. the position of the sun). These photic parameters all change in a systematic way, and in theory all could be used by the circadian system to detect the phase of twilight (Roenneberg and Foster 1997). However, each of these parameters is subject to considerable sensory "noise" (Table 1), and the impact of this noise will depend upon the organism and the environment in which it inhabits. One can also make the general point that in all sensory systems, much of the complexity observed is associated with noise reduction. A classic example of this in the visual system in colour vision. Colour vision is a mechanism for increasing the signal to noise of an object against its background by exploiting the fact that different objects do not reflect the same wavelengths of light equally.

Table 1 **The major sources of "noise" associated with the photic regulation of temporal physiology. Like other sensory systems (Dusenbery 1992), the two main sources of noise for twilight detection are associated with variation in the light stimulus and variation in exposure to the light stimulus. In each case the impact of this noise will depend upon the organism and the environment in which it inhabits. Some examples of the possible types of noise that might be expected to complicate photoentrainment are listed.**

(1) Variation in the stimulus	
Channel/Signal Noise	Fluctuations in the light signal, e.g. cloud cover or day length.
Environmental Noise	Extraneous signals from other sources of light, e.g. starlight, moonlight, and lightning.
Receptor Noise	Molecular noise of the receptor pathway, e.g. variation in external temperature.
(2) Variation in stimulus exposure	
Sensory Adaptation	Changing receptor thresholds, e.g. receptor habituation, changes in pupil size.
Behavioural Noise	Behavioural state, e.g. emergence from burrow, feeding, courting etc., etc.
Developmental Noise	Stage of development, e.g. feeding niche, body pigmentation, neural connections, developmental niche (egg or in uterus) etc.

Until recently circadian biologists have tended to use light merely as a means of shifting the clock, but of course twilight detection is not a straightforward stimulus. It is highly dynamic and subject to considerable noise. On the basis of what we know about other sensory systems we would predict that the photic inputs regulating temporal physiology will be both specialised and complex. Indeed, we have evidence for both novel (specialised) and multiple (complex) inputs regulating temporal physiology in all the vertebrate classes. Furthermore, multiple photopigments appear to contribute to photoentrainment in organisms as diverse as *Gonyaulax* (Roenneberg and Deng 1997) and *Drosophila* (Stanewsky et al. 1998). In view of the considerable progress that has been made in associating different photoreceptors to the circadian system, it is now time for circadian biologists to stop asking "what is the circadian photopigment" and ask the more sophisticated question "how do multiple photic channels interact to reduce the noise problem inherent in twilight detection".

Acknowledgements

Our research is sponsored by the Biotechnology and Biological Sciences Research Council (BBSRC), UK and EU BioMed2.

References

Aggelopolus, N., Meissl, H. (2000) Responses of neurones of the rat suprachiasmatic nucleus to retinal illumination under photopic and scotopic conditions. J. Physiol. **523**: 211–222.

Benoit, J. (1935a) Le role des yeux dans l'action stimulante de la lumiere sur le developpement testiculaire chez le canard. C.R. Seances. Soc. Biol. Fil. **118**: 669–671.

Benoit, J. (1935b) Stimulation par la lumiere artificielle du developpement testiculaire chez des canards aveugles par section du nerf optique, C.R. Seances. Soc. Biol. Fil. **120**: 133–136.

Blackshaw, S., Snyder, S. (1997) Parapineal opsin, a novel catfish opsin localized to the parapineal organ, defines a new gene family. J. Neurosci. **17**: 8083–8092.

Blackshaw, S., Snyder, S. (1999) Encephalopsin: A novel mammalian extraretinal opsin discretely localized in the brain. J. Neurosci. **19**: 3681–3690.

Bridges, C. (1959) The visual pigments of some common laboratory animals. Nature **184**: 727–728.

Campbell, S.S., Murphy, P. (1998) Extraocular circadian phototransduction in humans. Science **279**: 396–399.

Chabot, C.C., Menaker, M. (1992) Effects of physiological cycles of infused melatonin on circadian rhythmicity in pigeons. J. Comp. Physiol. A **170**: 615–622.

Coon, S.L., Begay, V., Deurloo, D., Falcon, J., Klein, D.C. (1999) Two arylalkylamine N-acetyltransferase genes mediate melatonin synthesis in fish. J. Biol. Chem. **274**: 9076–9082.

Cooper, H.M., Herbin, M., Nevo, E. (1993) Visual system of a naturally microphthalmic mammal: The blind mole rat, *Spalax ehrenbergi*. J. Comp. Neurol. **328**: 313–350.

Czeisler, C.A., Shanahan, T.L., Klerman, E.B., Martens, H., Brotman, D.J., Emens, J.S., Klein, T., Rizzo III J.F. (1995) Suppression of melatonin secretion in some blind patients by exposure to bright light. The New England Journal of Medicine **332**: 6–11.

David-Gray, Z.K., Janssen, J.W., DeGrip, W,J., Nevo, E., Foster, R.G. (1998) Light detection in a 'blind' mammal. Nat. Neurosci. **1**: 655–656.

David-Gray, Z., Cooper, H.M., Janssen, J.W.H., Nevo, E., Foster, R.G. (1999) Spectral tuning of a circadian photopigment in a subterranean "blind" mammal (*Spalax ehrenbergi*) FEBS Lett. **461**: 343 347.

Deguchi, T. (1981) Rhodopsin-like photosensitivity of isolated chicken pineal gland. Nature **290**: 702–704.

Devlin, P.F., Kay, S.A. (1999) Cryptochromes—bringing the blues to circadian rhythms. Trends in Cell Biol. **9**: 295–299.

Dusenbery, D.B. (1992) Sensory Ecology: How organisms acquire and respond to information. W.H. Freeman and Company, New York.

Ekström, P., Foster, R.G., Korf H-G, Schalken, J.J. (1987) Antibodies against retinal photoreceptor-specific proteins reveal axonal projections from the photosensory pineal organ of teleosts. J. Comp. Neurol. **265**: 25–33.

Emerson, V.F. (1980) Grating acuity in the golden hamster: The effects of stimulus orientation and luminance. Exp. Brain Res. **38**: 43–52.

Foster, R.G. (1998) Shedding light on the biological clock. Neuron. **20**: 829–832.

Foster, R.G., Follett, B.K. (1985) The involvement of a rhodopsin-like photopigment in the photoperiodic response of the Japanese quail. J. Comp. Physiol. A **157**: 519–528.

Foster, R.G., Follett, B.K., Lythgoe, J.N. (1985) Rhodopsin-like sensitivity of extra-retinal photoreceptors mediating the photoperiodic response in quail. Nature **313**: 50–52.

Foster, R.G., Garcia-Fernandez, J.M., Provencio, I., DeGrip, W.J. (1993) Opsin localization and chromophore retinoids identified within the basal brain of the lizard *Anolis carolinensis*. J. Comp. Physiol. A **172**: 33-45.

Foster, R.G., Grace, M.S., Provencio, I., Degrip, W.J., Garcia-Fernandez, J.M. (1994) Identification of vertebrate deep brain photoreceptors. Neurosci. Biobehav. Rev. **18**: 541–546.

Foster, R.G., Korf, H.G., Schalken, J.J. (1987) Immunocytochemical markers revealing retinal and pineal but not hypothalamic photoreceptor systems in the Japanese quail. Cell Tiss. Res. **248**: 161–167.

Foster, R.G., Menaker, M. (1993) Circadian photoreception in mammals and other vertebrates. In: Wetterberg, L. (ed) Light and biological rhythms in man. Pergamon, pp. 73–91.

Foster, R.G., Provencio, I., Hudson, D., Fiske, S., De Grip, W., Menaker, M. (1991) Circadian photoreception in the retinally degenerate mouse (rd/rd). J. Comp. Physiol. A **169**: 39–50.

Foster, R.G., Schalken, J.J., Timmers, A.M., De Grip, W.J. (1989) A comparison of some photoreceptor characteristics in the pineal and retina: I. The Japanese quail (*Coturnix coturnix*). J. Comp. Physiol. A **165**: 553–563.

Freedman, M.S., Lucas, R.J., Soni, B., von Schantz, M., Munoz, M., David-Gray, Z.K., Foster, R.G. (1999) Regulation of mammalian circadian behavior by non-rod, non-cone, ocular photoreceptors. Science **284**: 502–504.

Frisch K.v. (1911) Beitrage zur Physiologie der Pigmentzellen in der Fischhaut. Pfluger's Archv. Gesamte Physiol. Menschen Tiere **138**: 319–387.

Garcia-Fernandez, J.M., Jimenez, A.J., Gonzalez, B., Pombal, M.A., Foster, R.G. (1997) An immunocytochemical study of encephalic photoreceptors in three species of lamprey. Cell Tiss. Res. **288**: 267–278.

Gaston, S., Menaker, M. (1968) Pineal Function: The biological clock in the sparrow? Science **160**: 1125–1127.

Goldman, B.D., Goldman, S.L., Riccio, S.L., Terkel, J. (1997) Circadian patterns of locomotor activity and body temperature in blind mole-rats, *Spalax ehrenbergi*. J. Biol.Rhythms **12**: 348–361

Griffin, E., Staknis, D., Weitz, C. (1999) Light-ndependent role of CRY1 and CRY2 in the mammalian circadian clock. Science **286**: 768–771.

Gwinner, E., Hau, M., Heigl, S. (1997) Melatonin: generation and modulation of avian circadian rhythms. Brain Res. Bull. **44**: 439-444.

Haim, A.G., Heth, H., Pratt, H., Nevo, E. (1983) Photoperiodic effects on thermoregulation in a "blind" subterranean mammal. J. Exp. Biol. **107**: 59–64.

Hao, W., Fong, H.K. (1999) The endogenous chromophore of retinal, G. protein-coupled receptor opsin from the pigment epithelium. J. Biol. Chem. **274**: 6085–6090.

Homma, K., Ohta, M., Sakakibara, Y. (1980) Surface and deep photoreceptors in photoperiodism in birds. In: Tanake, Y., Tanaka, K., Ookawa, T. (ed) Biological Rhythms in Birds. Springer-Verlag Berlin, pp. 149–156.

Jacobs, G.H., Neitz, J., Deegan, J.F. (1991) Retinal receptors in rodents maximally sensitive to ultraviolet light. Nature **353**: 655–656.

Kawamura, S., Yokoyama, S. (1996) Molecular characterization of the pigeon P-opsin gene. Gene **182**: 213–214.

Kawamura, S., Yokoyama, S. (1997) Expression of visual and nonvisual opsins in American chameleon. Vis. Res. **37**: 1867–1871.

Kojima, D., Mano, H., Fukada, Y. (2000) Vertebrate ancient-long opsin: A green-sensitive photoreceptive molecule present in zebrafish deep brain and retinal horizontal cells. J. Neurosci. **20**: 2845–2851.

Lindblom, N., Hejskala, H., Hatonen, T., Mustanoja, S., Alfthan, H., Alila-Johansson A, Laasko M-L (2000) No evidence for extraocular light induced phase shifting of human melatonin, cortisol and thyrotropin rhythms. NeuroReport **11**: 713–717.

Lockley, A.U., Skene, S.W., Arendt, D.J., Tabandeh, H., Bird, A.C., Defrance, R. (1997) Relationship between melatonin rhythms and visual loss in the blind. J. Clin. Endocrinol. Metab. **82**: 3763–3770.

Lockley, S.W., Skene, D.J., Thapan, K., English, J., Ribeiro, D., Haimov, I., Hampton, S., Middleton, B., von Schantz, M., Arendt, J. (1998) Extraocular light exposure does not suppress plasma melatonin in humans. J. Clin. Endocrinol. Metab. **83**: 3369–3372.

Lucas, R.J., Douglas, R.H., Mrosovsky, N., Foster, R.G. (2001) Inner retinal photoreceptors regulate diverse responses to light: Characterisation of a novel ocular photopigment in mice. Neuron (under review).

Lucas, R.J., Foster, R.G. (1999a) Circadian Clocks: A cry in the dark? Curr. Biol. **9**: 825–828.

Lucas, R.J., Foster, R.G. (1999b) Circadian Rhythms: Something to cry about? Curr. Biol. **9**: 214–217.

Lucas, R.J., Freedman, M.S., Munoz, M., Garcia-Fernandez J.M., Foster, R.G. (1999) Regulation of the mammalian pineal by non-rod, non-cone, ocular photoreceptors. Science **284**: 505–507.

Mano, H., Kojima, D., Fukada, Y. (1999) Exo-rhodopsin: a novel rhodopsin expressed in the zebrafish pineal gland. Mol. Brain Res. **73**: 110–118.

Max, M., Menaker, M. (1992) Regulation of melatonin production by light, darkness, and temperature in the trout pineal. J. Comp. Physiol. A **170**: 479–489.

Meissl, H., Yanez, J. (1994) Pineal photosensitivity. A comparison with retinal photoreception. Acta. Neurobiol. Exp. **54**: 19–29.

Menaker, M., Tosini, G. (1996) The evolution of vertebrate circadian systems. In: Honma, K., Honma, S. (eds) Circadian organization and oscillatory coupling: Proceedings of the sixth Sapporo Symposium on Biological Rhythms. Hokkaido University Press, Sapporo, Japan, pp. 39–52.

Menaker, M., Underwood, H. (1976) Extraretinal photoreception in birds. Photochem. Photobiol. **23**: 299–306.

Menaker, M., Wisner, S. (1983) Temperature-compensated circadian clock in the pineal of Anolis. Proc. Natl. Acad. Sci. USA **80**: 6119–6121.

Menaker, M., Hudson, D.J., Takahashi, J.S. (1981). Neural and endocrine components of circadian clocks in birds. In: Follett, B.K., Follett, D.E. (eds.) Biological clocks in seasonal reproductive cycles. John Wright & Sons Ltd., Bristol vol. 32, pp. 171–183.

Moutsaki, P., Bellingham, J., Soni, B.G., David-Gray, Z.K. Foster, R.G. (2000) Sequence, genomic structure, and tissue expression of carp (*Cyprinus carpio* L.) vertebrate ancient (VA) opsin. FEBS Lett. **473**: 316–322.

Nelson, D.E., Takahashi, J.S. (1991) Sensitivity and integration in a visual pathway for circadian entrainment in the hamster (*Mesocricetus auratus*). J. Physiol. **439**: 115–145.

Nelson, R.J., Zucker, I. (1981) Absence of extraocular photoreception in diurnal and nocturnal rodents exposed to direct sunlight. Comp. Biochem. Physiol. A **69**: 145–148.

Okamura, H., Miyake, S., Sumi, Y., Yamaguchi, S., Yasui A, Muijtjens, M., Hoeijmakers J.H.J., van der Horst, G.T.J. (1999) Photic induction of mPer1 and mPer2 in Cry-deficient mice, lacking a biological clock. Science **286**: 2531–2534.

Okano, T., Yoshizawa, T., Fukada, Y. (1994) Pinopsin is a chicken pineal photoreceptive molecule. Nature **372**: 94–97.

Oliver, J., Jallageas, M., Bayle, J.D. (1979) Plasma testosternone and L.H. levels in male quail bearing hypothalamic lesions or radioluminous implants. Neuroendocrinol. **28**: 114–122.

Philp, A.R., Bellingham, J., Garcia-Fernandez J-M, Foster, R.G. (2000a) A novel rod-like opsin isolated from the extra-retinal photoreceptors of teleost fish. FEBS Lett. **468**: 181–188.

Philp, A.R., Garcia-Fernandez J-M, Soni, B.G., Lucas, R.J., Bellingham, J., Foster, R.G. (2000b) Vertebrate ancient (VA) opsin and extraretinal photoreception in the Atlantic salmon (*Salmo salar*). J. Exp. Biol. **203**: 1925–1936.

Provencio, I., Foster, R.G. (1995) Circadian rhythms in mice can be regulated by photoreceptors with cone-like characteristics. Brain Res. **694**: 183–190.

Provencio, I., Cooper, H.M., Foster, R.G. (1998a) Retinal projections in mice with inherited retinal degeneration: implications for circadian photoentrainment. J. Comp. Neurol. **395**: 417–439.

Provencio, I., Jiang, G., DeGrip, W.J., Hayes, W.P., Rollag, M.D. (1998b) Melanopsin: An opsin in melanophores, brain and eye. Proc.Nat.Acad. Sci. USA **95**: 340–345.

Provencio, I., Jiang, G., Hayes, W.P., Zatz, M., Rollag, M.D. (1998c) Novel skin and brain opsin, melanopsin, is found in the chicken. IOVS **39**: S236 (abstract 1075).

Provencio, I., Rodriguez, I.R., Jiang, G., Hayes, W.P., Moreira, E.F., Rollag, M.D. (2000) A novel human opsin in the inner retina. J. Neurosci. **20**: 600–605.

Provencio, I., Wong, S., Lederman, A., Argamaso, S.M. Foster, R.G. (1994) Visual and circadian responses to light in aged retinally degenerate mice. Vis. Res. **34**: 1799–1806.

Rado, R., Terkel, J. (1989) Circadian activity of the blind mole rat, *Spalax ehrenbergi*, monitored by radio telemetry, in seminatural and natural conditions. In: Spanier, E., Steinberger, Y., Luria, M. (eds) Environmental quality and ecosystem stability. vol. IV-B ISEEQS, Jerusalem, Israel, pp. 391–400.

Rado, R., Bronchti, G., Wollberg, Z., Terkel, J. (1992) Sensitivity to light of the blind molerat: Behavioral and neuroanatomical study. Israel J. Zool. **38**: 323–331.

Rado, R., Gev, H., Goldman, B.D., Terkel, J. (1991) Light and Circadian activity in the blind mole rat. In: Riklis, E. (ed) Photobiology. Plenum Press, New York, pp. 581–587.

Roenneberg, T., Deng, T-S (1997) Photobiology of the Gonyaulax circadian system I: Different phase response curves for red and blue light. Planta. **202**: 494–501.

Roenneberg, T., Foster, R.G. (1997) Twilight Times: Light and the circadian system. Photochem. Photobiol. **66**: 549–561.

Shand, J., Foster, R.G. (1999) The extraretinal photoreceptors of non-mammalian vertebrates. In: Archer, S., Djamgoz, M., Loew, E. (eds.) Adaptive mechanisms in the ecology of vision. Kluwer Academic Publishers,Dordrecht/Boston/London, pp. 197–222.

Silver, R., Witkovsky, P., Horvath, P., Alones, V., Barnstable, C.J., Lehman, M.N. (1988) Coexpression of opsin- and VIP-like-immunoreactivity in CSF-contacting neurons of the avian brain. Cell Tiss. Res. **253**: 189–198.

Simpson, S.M., Urbanski, H.F., Robinson, J.E. (1983) The pineal gland and the photoperiodic control of luteinizing hormone secretion in intact and castrated Japanese quail. J. Endocrinol. **99**: 281–287.

Soni, B.G., Philp, A., Knox, B.E., Foster, R.G. (1998) Novel Retinal Photoreceptors. Nature **394**: 27–28.

Stanewsky, R., Kaneko, M., Emery, P., Beretta, B., Wager-Smith, K., Kay, S., Rosbash, M., Hall, J.. (1998) The cry[b] mutation identifies cryptochrome as a circadian photoreceptor in *Drosophila*. Cell **95**: 681–692.

Sun, H., Gilbert, D.J., Copeland, N.G., Jenkins, N.A., Nathans, J. (1997a) Peropsin, a novel visual pigment-like protein located in the apical microvilli of the retinal pigment epithelium. Proc. Natl. Acad. Sci. USA **94**: 9893–9898.

Sun, H., Macke, J.P., Nathans, J. (1997b) Mechanisms of spectral tuning in the mouse green cone pigment. Proc. Natl. Acad. Sci. USA **94**: 8860–8865.

Tabata, M., Suzuki, T., Niwa, H. (1985) Chromophores in the extraretinal photoreceptor (pineal organ) of teleosts. Brain Res. **338**: 173–176.

Takahashi, J., DeCoursey, P., Bauman, L., Menaker, M. (1984) Spectral sensitivity of a novel photoreceptive system mediating entrainment of mammalian circadian rhythms. Nature **308**: 186–188.

Takahashi, J.S., Murakami, N., Nikaido, S.S., Pratt, B.L., Robertson, L.M. (1989) The avian pineal, a vertebrate model system of the circadian oscillator: Cellular regulation of circadian rhythms by light, second messengers, and macromolecular synthesis. Recent Prog. Horm. Res. **45**: 279–352.

Vigh, B., Röhlich, P., Vigh-Teichmann, I., Aros, B. (1980) Comparison of the pineal complex, retina and cerebrospinal fluid-contacting neurons by immunocytochemical antirhodopsin reaction. Z. Mikrosk. Anat. Forsch. **94**: 623–640.

Vigh, B., Vigh-Teichmann, I. (1988) Comparative neurohistology and immunocytochemistry of the pineal complex with special reference to CSF-contacting neuronal structures. Pineal Res. Rev. **6**: 1–65.

Wetterberg, L., ed. (1993) Light and biological rhythms in man. Wenner-Gren International Series, Pergamon Press.

Whitmore, D., Foulkes, N.S., Sassone-Corsi, P. (2000) Light acts directly on organs and cells in culture to set the vertebrate circadian clock. Nature **404**: 87–91.

Yamazaki, S., Goto, M., Menaker, M. (1999) No evidence for extraocular photoreceptors in the circadian system of the Syrian hamster. J. Biol. Rhythms **14**: 197–201.

Yokoyama, K., Oksche, A., Darden, T.R., Farner, D.S. (1978) The sites of encephalic photoreception in the photoperiodic induction of growth of the testes in the white crowned sparrow, *Zonotrichia leucophrys gambelii*. Cell. Tiss. Res. **189**: 441–467.

Yoshikawa, T., Yashiro, Y., Oishi, T., Kokame, K., Fukada, Y. (1994) Immunoreactivities to rhodopsin and rod/cone transducin antisera in the retina, pineal complex and deep brain of the Bullfrog, *Rana catesbeiana*. Zool. Sci. **11**: 675–680.

Biological Rhythms
Edited by V. Kumar
Copyright © 2002, Narosa Publishing House, New Delhi, India

11. Circadian Organization in Fish and Amphibians

G.M. Cahill*

Department of Biology and Biochemistry, University of Houston, Houston, TX, USA

Circadian systems in fish and amphibians share several characteristics with other non-mammalian vertebrates. The most salient of these characteristics at the physiological level is the presence in multiple tissues of independently photosensitive and self-sustaining circadian oscillators. The circadian oscillators that regulate melatonin synthesis in the pineal and retina have been the most extensively investigated. In particular, studies of teleost pineals and of the retina of *Xenopus laevis* have contributed to our understanding of the cellular and molecular bases of rhythm generation, entrainment and output pathways in these organs. However, our understanding of how these and other oscillatory structures interact to drive rhythmicity in intact fish and amphibians has lagged behind progress in other vertebrates, primarily because convenient and reliable measures of behavioral rhythmicity were lacking. Recent technical advances have made it possible to record robust swimming activity rhythms from larval zebrafish. These methods may also be applicable to measurement of behavioral rhythms in other fish and amphibians. Recent studies of the zebrafish and *Xenopus* homologs of mammalian clock-related genes indicates that molecular clock mechanisms in fish and amphibians are similar, but not identical to those in other vertebrates. In particular, these studies have revealed new complexities in molecular mechanisms of vertebrate circadian rhythmicity, and they have also contributed to our understanding of system organization in these animals.

Physiology of fish circadian systems

In vertebrates, the pineal complex, the retinas and the hypothalamic suprachiasmatic nuclei (SCN) have long been viewed as key components of a centralized circadian timing system. In one or another species, each of these structures has been shown to contain a circadian oscillator, and each has been shown to act as a pacemaker for regulation of some aspect of whole-animal rhythmicity. In addition, non-mammalian vertebrates have pineal and deep brain photoreceptors that may mediate entrainment of rhythmicity. The pineal and retinas produce the hormone melatonin rhythmically, and also may transmit rhythmicity through neuronal efferents, while the primary output of the SCN is presumably neural. Research on fish circadian systems has focused primarily on the function of the pineal and retina, with the greatest attention paid to melatonin rhythms. Although their roles in overall circadian system function in fish are poorly understood, the evidence suggests that the pineal, retinas and melatonin may play modulatory roles. Relatively little is known about the contribution of oscillators in the hypothalamus or other parts of the fish brain. Recent evidence indicates that, in addition to the three central circadian oscillatory structures, some peripheral tissues in fish contain light-sensitive circadian oscillators (Whitmore et al. 2000), suggesting that the pacemaking system *in vivo* may be more complex than was assumed previously.

---

*E-mail: gcahill@uh.edu

The pineals of all fish that have been studied are directly photosensitive; they contain photoreceptor cells that resemble those of the lateral eyes, with well-developed inner and outer segments and presynaptic processes (Korf et al. 1998). The pineals of lampreys and of most, but not all, teleosts contain independent circadian oscillators that can drive rhythms of melatonin synthesis when the pineal is removed and cultured in constant darkness (Falcón et al. 1989; Kezuka et al. 1989; Iigo et al. 1991; Zachmann et al. 1992; Bolliet et al. 1996; Cahill 1996; Samejima et al. 1997). In pike, there is strong evidence that individual pineal photoreceptor cells contain self-sustaining, cellular circadian oscillators (Bolliet et al. 1997). The best-studied exception to the generalization that teleost pineals contain circadian oscillators is the trout pineal, which produces melatonin at constant high levels when cultured in constant darkness (Gern and Greenhouse 1988; Max and Menaker 1992; Coon et al. 1998). If the trout pineal contains a circadian oscillator at all, it is not coupled to melatonin synthesis. In all teleost species, including trout, exposure of the pineal to light *in vitro* acutely suppresses melatonin synthesis. Furthermore, light:dark cycles *in vitro* can reset the circadian oscillators in pineals that are intrinsically rhythmic. These data indicate teleost pineals have the capacity to produce daily melatonin rhythms in the absence of any physiological input from the rest of the circadian system. In trout these rhythms result from light acting directly on the pineal to suppress melatonin during the day, and in other species, this same mechanism is combined with regulation by light-entrained intrinsic pineal rhythmicity. In fact, there is only limited evidence for any efferent control of teleost pineal melatonin production. There is some evidence that trout and pike pineals are responsive to some neurotransmitters and hormones (Falcón et al. 1991). However, there is no anatomical evidence in teleosts for the sympathetic innervation that is critical in the efferent control of pineal function in mammals and birds (Korf et al. 1998), and attempts to demonstrate any effects of catecholamines on zebrafish pineal rhythms failed (Cahill 1997). Central neuronal projections from the pineal gland have been described in lamprey (Yáñez et al. 1993), salmon (Holmqvist et al. 1992) and goldfish (Jimenez et al.1995), and the targets of pineal projections were found to overlap retinal targets. It is unknown whether these pineal neuronal projections play any role in the circadian system.

Many aspects of teleost retinal function are controlled by circadian oscillators *in vivo*. Well-known rhythmic retinal phenomena include the metabolism of neurotransmitters and neuromodulators, such as serotonin, melatonin and dopamine, and rhythmic changes in the morphology of photoreceptors, including retinomotor movement of outer segments and changes in synaptic structure (reviewed by Cahill and Besharse 1995). More recently, it was demonstrated that the circadian system also modulates the relative dominance of rod and cone input to horizontal cells in goldfish (Wang and Mangel 1996), as well as visual sensitivity in zebrafish (Li and Dowling 1998). The only *in vitro* evidence that a retinal oscillator controls rhythmicity is from zebrafish retina, which produces damped rhythms of melatonin release in culture (Cahill 1996).

Given the dominant role of the SCN as a pacemaker in other vertebrate circadian systems, it seems likely that hypothalamic oscillators also play important roles in teleost circadian systems. Currently however, there is no direct evidence for a circadian oscillator in the hypothalamus of any teleost. Ablation studies have shown that the preoptic nucleus is necessary for circadian behavior in a cyclostome (Ooka-Souda et al. 1993). A candidate homolog of the SCN has been identified in lampreys based on its retinal innervation and neuropeptide content (Weigle et al. 1996). However, it has been difficult to identify the functional homolog of the SCN in teleosts by these criteria. Retinohypothalamic (RHT) projections have been described in some teleosts (reviewed by Holmqvist et al. 1992), including embryonic and larval (but not adult) zebrafish (Burrill and Easter 1991). The teleost RHT innervates several areas of the hypothalamus, including the preoptic nucleus, the anterior periventricular nucleus, and the periventricular preoptic nucleus (as defined by Holmqvist et al.

1992). In addition to the widespread retinal projections, neuropeptides that are localized within the SCN in mammals are found in neurons distributed throughout the periventricular hypothalamus in fish (Holmqvist et al. 1992). It is possible that some subset of these retinorecipient hypothalamic areas act as circadian pacemaker in teleosts. Alternatively, pacemaker function may be distributed among several of these areas. Some direct measure of rhythmicity will be required to determine whether pacemaker function is localized discretely.

There have been numerous studies over the years of various types of behavioral rhythmicity in fish (reviewed in Ali 1992). However, most of these have not addressed circadian control directly because the animals were maintained under cyclic lighting conditions. Most studies of free running circadian rhythmicity have focused on general locomotor (swimming) activity. Long-term recording of locomotor activity in constant conditions presents special difficulties in aquatic animals. For example, it is difficult to provide food ad libitum to fish without fouling the water, and bout feeding can alter activity and can set a food-entrainable circadian oscillator (Sanchez-Vazquez et al. 1996). Furthermore, if fish are fed they must be kept either in a relatively large volume of water or in a filtered, recirculating system to prevent buildup of toxic metabolites.

A number of approaches have been used to sample fish locomotor activity. The most common approach has been to measure the frequency with which swimming movements interrupt one or more infrared beams projected across the aquarium (e.g. Sanchez-Vazquez et al. 1996; Hurd et al. 1998). This approach has the advantages of economy and adaptability to many types of aquarium. However, it can only detect that portion of the activity that results in interruption of the beam, and it provides little information about the nature of that activity. We found considerable variability in the activity patterns of adult zebrafish using such a system (Hurd et al. 1998). Others have also reported both intra-and inter-individual variability in swimming rhythms using such a system, including apparent switches between nocturnal and diurnal activity (Iigo and Tabata 1996; Sanchez-Vazquez et al. 1996). These switches clearly reflect some change in behavior, but it is unclear whether it is the amount or the kind of activity that is changing. Another system that produced impressive records of behavioral rhythmicity from individual and groups of fish was an ultrasonic detector devised by Kavaliers (1978, 1979, 1980), but there have been no reports of studies that employ this system in the past 20 years.

We have used an on-line computerized image analysis system to measure circadian rhythms in the locomotor activity of larval zebrafish (Cahill et al. 1998). These fish can survive and remain rhythmic when maintained for more than a week in less than 1 ml of water, without food or water changes. Example activity records are shown in Fig. 1. The imaging system automatically measures the

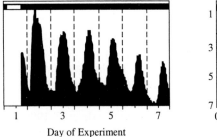
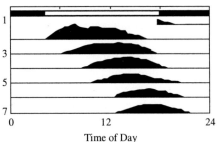

Day of Experiment   Time of Day

Fig. 1   Behavioral rhythm recorded by automated video image analysis of a larval zebrafish, 10–16 days of age, under constant infrared illumination. Left, linear plot of activity; Right, actogram plot of activity peaks from the same record. In both records, the activity data were smoothed by a 4 h moving average.

distance that each fish moves during a 30–60 second sampling period every 4-5 minutes. Advances in computer processing speed and image analysis software now make it possible to measure activity simultaneously from up to 150 individuals with one of these systems. The activity records of over 95% of larval zebrafish display statistically significant circadian rhythmicity with this system, and satisfactory phase and period estimates can typically be determined from 85–90% of the animals. We are now using this system as an assay for the effects of physiological and pharmacological manipulations on circadian rhythmicity, as well as in a screen for genetic mutations that alter the period of the rhythm. It seems likely that variants of this technique could be useful for measuring behavioral rhythmicity from other fish and amphibians. A major limitation of this technique is that the conditions are unnatural and restrictive, and this might disrupt rhythmicity in more sensitive species. However, in cases where rhythmicity does persist, this system enables easy measurement of rhythmicity from many animals simultaneously.

There have been limited attempts to localize components of the circadian pacemaking and photoreceptive systems that regulate fish behavioral rhythms. The pineal is essential for behavioral rhythmicity in lampreys, but the eyes are not (Morita et al. 1992). Pinealectomy does not abolish circadian rhythmicity in any teleost species tested so far, but it does alter the period, stability and/or amplitude of behavioral circadian rhythms in lake chub, burbot and white sucker (Kavaliers 1979, 1980; Kavaliers and Ralph 1980). Data from catfish suggest that the eyes and the pineal contribute to synchronization of behavioral rhythms by a light:dark cycle, but that other photoreceptors and oscillators also contribute (Tabata et al. 1988). However, in preliminary experiments with larval zebrafish, we have been unable to detect any effect of pinealectomy or ocular enucleation on the period or amplitude of freerunning rhythmicity, or on entrainment (Hurd and Cahill, unpublished). Taken together, the data from teleosts indicate that other oscillators (possibly in the hypothalamus) are primarily responsible for regulation of behavioral rhythmicity, and that the pineal and retina play a modulatory role.

Molecular biology of teleost circadian systems

Studies of clock-regulated genes, and of gene products that are likely to be part of the oscillator mechanism itself in teleosts have begun to accumulate. These studies have revealed that the molecular mechanisms underlying teleost circadian rhythmicity are in many ways similar to those found in other vertebrates, but they have also revealed some interesting differences.

In the pineals of pike and zebrafish, mRNA for arylalkylamine-N-acetyltransferase (AA-NAT), the penultimate enzyme in the melatonin synthetic pathway, is regulated by the clock (Bégay et al. 1998). In pike pineal, mRNA for tryptophan hydroxylase (TPH), another enzyme in this pathway, is also regulated by the clock. This suggests that at least part of the circadian control of melatonin synthesis result from rhythms in transcript levels. Supporting this interpretation, in trout pineal, where melatonin production is not rhythmic in constant conditions, neither are these mRNAs. Interestingly, light does not acutely affect AA-NAT or tryptophan hydroxylase mRNA in any of these species, suggesting that the acute suppression of melatonin is due to post-transcriptional effects of light.

Recently, it was discovered that pike and trout have two separate genes for AA-NAT (Coon et al. 1999). As described below, this is also true of genes that code for components of the central oscillator in zebrafish. These findings are consistent with other data indicating that for many single-copy genes in mammals, there are two paralogous genes in teleosts. This suggests that a whole-genome duplication occurred in the teleost lineage, after it split from the tetrapod lineage (Postlethwait et al. 1998). An intriguing aspect of the two teleost AA-NAT genes is that they are not simply redundant. The enzymes have different biochemical characteristics and substrate specificities, and interestingly, one is expressed almost exclusively in the pineal, and the other almost exclusively in the retina. In this case, it appears that the two paralogs have evolved to support complementary subsets of the ancestral gene's functions, and this may be useful in defining molecular mechanisms of retinal and pineal rhythmicity.

CLOCK and BMAL1 are central components of the molecular circadian oscillator in animals (reviewed by Young 1999). These transcription factors dimerize, bind to the promoters of negatively acting clock genes, and activate transcription. One ortholog of *clock* and two paralogs of *bmal1* from zebrafish have been cloned, and their expression patterns have been described (Whitmore et al. 1998, 2000; Cermakian et al. 2000). The mRNA levels for all three of these genes are rhythmic, both in central clock structures and in peripheral tissues. This is intriguing because in other animal systems rhythmicity has been observed in either *clock* or *bmal1* expression, but not both. The abundance of all three transcripts peaks in the late day or early night in zebrafish, although there are subtle, tissue-dependent differences in timing. The most interesting result of these studies to date is that rhythmicity in *clock* expression persists in cultured peripheral tissues, including the heart and kidney. Furthermore, the oscillators that control these rhythms in cultured peripheral tissues can be reset in vitro by light:dark cycles. This indicates that in fish, many tissues have fully functional and light-entrainable molecular clocks. The physiological roles of these molecular oscillators can only be guessed at now, but these findings certainly broaden the view how a circadian pacemaking system might be constructed in fish.

The importance of zebrafish as a genetic model system for developmental biology has led to genomic projects to develop the infrastructure required for genetic mapping and cloning of mutated zebrafish genes. These projects will aid in the study of circadian rhythms as well. For example, an expressed sequence tag (EST) project is planned to provide partial sequence information for 100,000 zebrafish genes (http://zfish.wustl.edu/). This project has so far identified clones containing at least ten homologs of candidate circadian oscillator genes. A genetic map based on simple sequence length polymorphisms has been developed to enable linkage mapping of mutants (Shimoda et al. 1999), and radiation hybrid maps have been developed that enable rapid physical mapping of cloned zebrafish genes and EST's (Kwok et al. 1998). Together, these resources will facilitate cloning of clock genes identified by mutations in zebrafish. With this in mind, we and others have initiated screens for mutations that alter circadian rhythmicity in zebrafish. The behavioral assay described above makes it possible to easily screen large numbers of animals for mutants.

Physiology of amphibian circadian systems

The majority of research on amphibian circadian systems has focused on the cell biology of the retinal circadian oscillator of *Xenopus laevis* and relatively little is known about the circadian function of other candidate pacemaker structures. Amphibian pineal complexes, which in anurans include the frontal organ as well as the pineal proper, contain true photoreceptors cells and synthesize melatonin rhythmically under cyclic lighting conditions (Korf et al. 1998). Pineals from *Xenopus* tadpoles are also capable of expressing circadian rhythms of melatonin synthesis *in vitro*, although these rhythms damp relatively quickly in constant darkness (Green et al. 1999).

Two reports have addressed the roles of candidate pacemaker structures in circadian regulation of behavior in *Xenopus*. Anderson (1987) investigated the roles of the pineal, frontal organ and eyes in circadian regulation of behavior. He reported that ablation of these organs altered the period of freerunning activity rhythms, but that light-entrainable circadian oscillators remained after removal of all three of these structures. Harada et al. (1998) also found that ocular enucleation shortened the periods of *Xenopus* activity rhythms, but did not abolish rhythmicity or entrainment. These investigators also measured the effects of hypothalamic lesions directed at the suprachiasmatic area. They found that freerunning behavioral rhythmicity was disrupted by these lesions, and that arrhythmicity was correlated with the extent of the lesion. These findings suggest that hypothalamic oscillators play a central role in regulation of amphibian activity rhythms, and that the pineal and retinas modulate these rhythms.

The presence of a circadian oscillator in vertebrate retina was first demonstrated by the persistence of AA-NAT rhythms in cultured eyecups from *Xenopus* (Besharse and Iuvone 1983), and cultured *Xenopus* preparations are still among the most robust systems available for analysis of retinal rhythmicity. Tissue reduction studies showed that the circadian oscillator, entrainment mechanisms, and output pathways for regulation of melatonin synthesis are all preserved in isolated photoreceptor cell layers from the *Xenopus* retina (Cahill and Besharse 1993). This photoreceptor layer preparation, which contains only rods and cones, has simplified studies of cellular mechanisms of circadian entrainment in the retina. The removal of other cell types eliminates multineuronal synaptic pathways as well as the contributions of other cell types to biochemical measurements.

The circadian oscillator in retinal photoreceptors can be entrained by cycles of light or of agents that activate D2-like dopamine receptors (Cahill and Besharse 1991, 1993). Dopamine is released by amacrine and interplexiform cells in the retina, and its release is stimulated by light (reviewed by Cahill and Besharse 1995). Light and dopamine receptor agonists have nearly identical effects on the phase of the photoreceptor cell oscillator. Both treatments reset the oscillator toward a daytime phase. However, the signaling pathways that mediate these effects differ at the level of second messengers. The effects of dopamine agonists are completely blocked by treatments that elevate cyclic AMP, but light can still reset the oscillator in the presence of high cyclic AMP (Hasegawa and Cahill 1999a). Cyclic AMP itself mimics darkness in its effects on the oscillator, resetting it toward a nighttime phase (Hasegawa and Cahill 1999b). This second messenger is suppressed by dopamine, and it appears that this suppression is required for dopamine-induced phase resetting.

Molecular biology of amphibian circadian systems

In *Xenopus* retina, TPH is the rate-limiting enzyme in melatonin synthesis, and TPH mRNA levels are rhythmic in cultured retinas (Green et al. 1995). The majority of TPH mRNA is located in the photoreceptors, and its rhythm peaks during the nighttime. The activity of this enzyme, however, is highest in the inner retina, and the rhythm of this inner retinal TPH activity peaks during the daytime (Valenciano et al. 1999). These findings underscore the importance of separating cell types for interpretation of molecular and biochemical rhythms. The expression of AA-NAT mRNA in *Xenopus* retina has not been studied yet, but the enzyme activity is rhythmic. Given the results from pineal, it would be surprising if this is not due at least in part to rhythmicity in its mRNA.

In a screen for clock-regulated mRNAs in *Xenopus* retina, Green and Besharse (1996a) used differential display to identify randomly amplified cDNAs from cultured retina. To avoid the selection of false positives that has plagued other investigators who have used this technique, they pursued only candidates that met the stringent criterion of consistent rhythmicity through two cycles in constant darkness. They found that surprisingly few transcripts (4/2000) are rhythmically expressed in retina. One of these, dubbed nocturnin, is a novel gene of unknown function that expressed exclusively in the retina, with a circadian rhythm that peaks at night (Green and Besharse 1996b). Nocturnin has sequence motifs similar to a yeast transcription factor, CCR4, so it may be involved in rhythmic regulation of other clock-controlled genes.

Of the known molecular oscillator components, the only one so far reported for *Xenopus* retina is *clock*, which is constitutively expressed in photoreceptor cells (Zhu et al. 2000). It is expected that, as with the recent studies of zebrafish clock-related genes, analysis of molecular mechanisms in *Xenopus* will reveal interesting variations in the nature of the molecular feedback loop. New technology that enables rapid generation of transgenic *Xenopus*, in which transgene expression can be targeted to photoreceptors (Knox et al. 1998), will be particularly useful in these studies.

References

Ali, M.A. (1992) Rhythms in Fishes. Plenum Press, New York.

Anderson, K.D. (1987) Role of the eyes, frontal organ and pineal organ in the generation of the circadian activity rhythm and its entrainment by light in the South African clawed frog, *Xenopus laevis*. Ph.D. Dissertation, Northwestern University.

Bégay, V., Falcón, J., Cahill, G.M., Klein, D.C., Coon, S.L. (1998) Transcripts encoding two melatonin synthesis enzymes in the teleost pineal organ: circadian regulation in pike and zebrafish, but not in trout. Endocrinol. **139**: 905–912.

Besharse, J.C., Iuvone, P.M. (1983) Circadian clock in *Xenopus* eye controlling retinal serotonin N-acetyltransferase. Nature **305**: 133–135.

Bolliet, V., Ali, M.A., Lapointe, F.J., Falcón, J. (1996) Rhythmic melatonin secretion in different teleost species: an *in vitro* study. J. Comp. Physiol. B **165**: 677–683.

Bolliet, V., Bégay, V., Taragnat, C., Ravault, J.P., Collin, J.P., Falcón, J. (1997) Photoreceptor cells of the pike pineal organ as cellular circadian oscillators. Eur. J. Neurosci **9**: 643–653.

Burrill, J.D., Easter, S.S. Jr. (1991) Development of the retinofugal projections in the embryonic and larval zebrafish (*Brachydanio rerio*), J. Comp. Neurol. **346**: 583–600.

Cahill, G.M. (1996) Circadian regulation of melatonin production in cultured zebrafish pineal and retina. Brain Res. **708**: 177–181.

Cahill, G.M. (1997) Circadian melatonin rhythms in cultured zebrafish pineals are not affected by catecholamine agonists. Gen. Comp. Endocrinol. **105**: 270–275.

Cahill, G.M., Besharse, J.C. (1991) Resetting the circadian clock in cultured *Xenopus* eyecups: regulation of retinal melatonin rhythms by light and D2 dopamine receptors. J. Neurosci. **11**: 2959–71.

Cahill, G.M., Besharse, J.C. (1993) Circadian clock functions localized in *Xenopus* retinal photoreceptors. Neuron **10**: 573–577.

Cahill, G.M., Besharse, J.C. (1995) Circadian rhythmicity in vertebrate retinas: Regulation by a photoreceptor oscillator. Prog. Retinal Eye Res. **14**: 267–291.

Cahill, G.M., Hurd, M.W., Batchelor, M.M. (1998) Circadian rhythmicity in the locomotor activity of larval zebrafish. Neuroreport **9**: 3445–3449.

Cermakian, N., Whitmore, D., Foulkes, N.S., Sassone-Corsi, P. (2000) Asynchronous oscillations of two zebrafish CLOCK partners reveal differential clock control and function. Proc. Natl. Acad. Sci. USA **97**: 4339–4344.

Coon, S.L., Bégay, V., Falcón, J., Klein, D.C. (1998) Expression of melatonin synthesis genes is controlled by a circadian clock in the pike pineal organ but not in the trout. Biol. Cell. **90**: 399–405.

Coon, S.L., Bégay, V., Deurloo, D., Falcón, J., Klein, D.C. (1999) Two arylalkylamine N-acetyltransferase genes mediate melatonin synthesis in fish. J. Biol. Chem. **274**: 9076–9082.

Falcón, J., Marmillon, J.B., Claustrat, B., Collin, J.P. (1989) Regulation of melatonin secretion in a photoreceptive pineal organ: an in vitro study in the pike. J. Neurosci. **9**: 1943–1950.

Falcón, J., Thibault, C., Martin, C., Brun-Marmillon, J., Claustrat, B., Collin, J.P. (1991) Regulation of melatonin production by catecholamines and adenosine in a photoreceptive pineal organ. An *in vitro* study in the pike and the trout. J. Pineal Res. **11**: 123–134.

Gern, W.A., Greenhouse, S.S. (1988) Examination of *in vitro* melatonin secretion from superfused trout (*Salmo gairdneri*) pineal organs maintained under diel illumination of continuous darkness, Gen. Comp. Endocrinol. **71**: 163–174.

Green, C.B., Besharse, J.C. (1996a) Use of a high stringency differential display screen for identification of retinal mRNAs that are regulated by a circadian clock. Mol. Brain Res. **37**: 157–165.

Green, C.B., Besharse, J.C. (1996b) Identification of a novel vertebrate circadian clock-regulated gene encoding the protein nocturnin. Proc. Natl. Acad. Sci. USA **93**: 14884–1488.

Green, C.B., Cahill, G.M., Besharse, J.C. (1995) Regulation of tryptophan hydroxylase expression by a retinal circadian oscillator *in vitro*. Brain Res. **677**: 283–290.

Green, C.B., Liang, M.Y., Steenhard, B.M., Besharse, J.C. (1999) Ontogeny of circadian and light regulation of melatonin release in *Xenopus laevis* embryos. Dev. Brain Res. **117**: 109–116.

Harada, Y., Goto, M., Ebihara, S., Fujisawa, H., Kasegawa, k, Oishi, T. (1998) Circadian locomotor activity rhythms in the African clawed frog, *Xenopus laevis*: The role of the eye and the hypothalamus. Biol. Rhythm Res. **29**:30–48.

Hasegawa, M., Cahill, G.M. (1999a) A role for cyclic AMP in entrainment of the circadian oscillator in *Xenopus* retinal photoreceptors by dopamine but not by light. J. Neurochem. **72**: 1812–1820.

Hasegawa, M., Cahill, G.M. (1999b) Modulation of rhythmic melatonin synthesis in *Xenopus* retinal photoreceptors by cyclic AMP. Brain Res. **824**: 161–167.

Hurd, M.W., Debruyne, J., Straume, M., Cahill, G.M. (1998) Circadian rhythms of locomotor activity in zebrafish. Physiol. Behav. **65**: 465–472.

Holmqvist, B.I., Östholm, T., Ekström, P. (1992) Retinohypothalamic projections and the suprachiasmatic nucleus of the teleost brain. In Ali, M.A. (ed.) Rhythms in Fishes, Plenum Press, New York, pp. 293–318.

Iigo, M., Tabata, M. (1996) Circadian rhythms of locomotor activity in the goldfish *Carassius auratus*. Physiol. Behav. **60**: 775–781.

Iigo, M., Kezuka, H., Aida, K., Hanyu, I. (1991) Circadian rhythms of melatonin secretion from superfused goldfish (*Carassius auratus*) pineal glands in vitro. Gen. Comp. Endocrinol. **83**: 152–158.

Jimenez, A.J., Fernandez-Llebrez, P., Perez-Figares, J.M. (1995) Central projections from the goldfish pineal organ traced by HRP-immunocytochemistry. Histol Histopathol. **10**: 847–852.

Kavaliers, M. (1978) Seasonal changes in the circadian period of the lake chub, *Couesius plumbeus*. Can. J. Zool. **56**: 2591–2596.

Kavaliers, M. (1979) Pineal involvement in the control of circadian rhythmicity in the lake chub, *Couesius plumbeus*. J. Exp. Zool. **209**: 33–40.

Kavaliers, M. (1980) Circadian locomotor activity rhythms of the burbot, *Lota lota*: Seasonal differences in period length and the effect of pinealectomy. J. Comp. Physiol. **136**: 215–218.

Kavaliers, M., Ralph, C.L. (1980) Pineal involvement in the control of behavioral thermoregulation of the white sucker, *Catostomus commersoni*. J. Exp. Zool. **212**: 301–303.

Kezuka, H., Aida, K., Hanyu, I. (1989) Melatonin secretion from goldfish pineal gland in organ culture. Gen. Comp. Endocrinol. **75**: 217–221.

Knox, B.E., Schlueter, C., Sanger, B.M., Green, C.B., Besharse, J.C. (1998) Transgene expression in *Xenopus* rods. FEBS Lett. **423**: 117–121.

Korf, H.W., Schomerus, C., Stehle, J.H. (1998) The pineal organ, its hormone melatonin, and the photoneuroendocrine system. Adv. Anat. Embryol. Cell Biol. **146**: 1–100.

Kwok, C., Korn, R.M., Davis, M.E., Burt, D.W., Critcher, R., McCarthy L, Paw, B.H., Zon, L.I., Goodfellow, P.N., Schmitt, K. (1998) Characterization of whole genome radiation hybrid mapping resources for non-mammalian vertebrates. Nucleic Acids Res . **26**: 3562–3566.

Li, L., Dowling, J.E. (1998) Zebrafish visual sensitivity is regulated by a circadian clock. Visual Neurosci. **15**: 851–857.

Max, M., Menaker, M. (1992) Regulation of melatonin production by light, darkness, and temperature in the trout pineal. J. Comp. Physiol. A **170**: 479–489.

Morita, Y., Tabata, M., Uchida, K., Samejima, M. (1992) Pineal-dependent locomotor activity of lamprey, *Lampetra japonica*, measured in relation to LD cycle and circadian rhythmicity. J. Comp. Physiol. A **171**: 555–562.

Ooka-Souda, S., Kadota, T., Kabasawa, H. (1993) The preoptic nucleus: the probable location of the circadian pacemaker of the hagfish, *Eptatretus burgeri*. Neurosci. Lett. **164**: 33–36.

Postlethwait, J.H., Yan Y-L, Gates, M.A., Horne, S., Amores, A., Brownlie, A., Donovan, A., Egan, E.S., Force, A., Gong, Z., Goutel, C., Fritz, A., Kelsh, R., Knapik, E., Liao, E., Paw, B., Ransom, D., Singer, A., Thomson, M., Abduljabbar, T.S., Yelick, P., Beier, D., Joly J-S, Larhammar, D., Rosa, F., Westerfield, M., Zon, L.I., Johnson, S.L., Talbot, W.S. (1998) Vertebrate genome evolution and the zebrafish gene map. Nat. Genet. **18**: 345–349.

Samejima, M., Tamotsu, S., Uchida, K., Moriguchi, Y., Morita, Y. (1997) Melatonin excretion rhythms in the cultured pineal organ of the lamprey, *Lampetra japonica*. Biol. Signals **6**: 241–246.

Sanchez-Vazquez, F.J., Madrid, J.A., Zamora, S., Iigo, M., Tabata, M. (1996) Demand feeding and locomotor circadian rhythms in the goldfish, *Carassius auratus*: dual and independent phasing. Physiol. Behav. **60**: 665–74.

Shimoda, N., Knapik, E.W., Ziniti, J., Sim, C., Yamada, E., Kaplan, S., Jackson, D., de Sauvage, F., Jacob, H., Fishman, M.C. (1999) Zebrafish genetic map with 2000 microsatellite markers. Genomics **58**: 219–232.

Tabata, M., Minh-Nyo, M., Oguri, M. (1988) Involvement of retinal and extraretinal photoreceptors in the mediation of nocturnal locomotor activity rhythms in the catfish, *Silurus asotus*. Exp. Biol. **47**: 219–225.

Valenciano, A.I., Alonso-Gomez, A.L., Iuvone, P.M. (1999) Diurnal rhythms of tryptophan hydroxylase activity in *Xenopus laevis* retina: opposing phases in photoreceptors and inner retinal neurons. Neuroreport. **10**: 2131–5.

Wang, Y., Mangel, S.C. (1996) A circadian clock regulates rod and cone input to fish retinal cone horizontal cells. Proc. Natl. Acad. Sci. USA **93**: 4655–4660.

Weigle, C., Wicht, H., Korf, H.W. (1996) A possible homologue of the suprachiasmatic nucleus in the hypothalamus of lampreys (*Lampetra fluviatilis L.*). Neurosci. Lett. **217**: 173–176.

Whitmore, D., Foulkes, N.S., Sassone-Corsi, P. (2000) Light acts directly on organs and cells in culture to set the vertebrate circadian clock. Nature **404**: 87–91.

Whitmore, D., Foulkes, N.S., Strahle, U., Sassone-Corsi, P. (1998) Zebrafish Clock rhythmic expression reveals independent peripheral circadian oscillators. Nature Neurosci. **1**: 701–707.

Yáñez, J., Anadon, R., Holmqvist, B.I., Ekström, P. (1993) Neural projections of the pineal organ in the larval lamprey (*Petromyzon marinus L.*) revealed by indocarbocyanine dye tracing, Neurosci. Lett. **164**: 213–216.

Young, M.W. (1999) Molecular control of circadian behavioral rhythms. Recent Prog. Horm. Res. **54**: 87–94.

Zachmann, A., Falcón, J., Knijff, S.C.M., Bolliet, V., Ali, M,A. (1992) Effects of photoperiod and temperature on rhythmic melatonin secretion from the pineal organ of the white sucker (*Catostomus commersoni*) in vitro. Gen. Comp. Endocrinol. **86**: 26–33.

Zhu, H., LaRue, S., Whiteley, A., Steeves, T.D., Takahashi, J.S., Green, C.B. (2000) The *Xenopus* clock gene is constitutively expressed in retinal photoreceptors. Mol. Brain Res. **75**: 303–308.

Biological Rhythms
Edited by V. Kumar
Copyright © 2002, Narosa Publishing House, New Delhi, India

12. The Circadian Organization of Reptiles

C. Bertolucci, A. Foà and G. Tosini[#][*]

Department of Biology, University of Ferrara, Ferrara, Italy

[#]Neuroscience Institute, Morehouse School of Medicine, 720 Westview dr., Atlanta, 30310, Georgia, USA

The present review summarizes the current knowledge of the circadian organization of Reptiles. This taxonomic group has provided (and continue to provide) a very useful experimental model for the understanding of the vertebrate circadian system. The circadian organization of reptiles is multioscillatory in nature. The retinas, the pineal and the parietal eye (and, possibly, the SCN) contain circadian clocks. Of particular interest is the observation that the role these structures play in the circadian organization varies considerably among species and within the same species in different seasons. Another remarkable feature of this class is the redundancy of circadian photoreceptors: retinas of the lateral eyes, pineal, parietal eye and the brain all contain photoreceptors.

Introduction

It is well known that many of biochemical, physiological and behavioral parameters exhibited by organisms show daily fluctuations, and most of these fluctuations persist when the organisms are maintained in constant environmental conditions, thus demonstrating that they are driven by an endogenous oscillator. Rhythms that persist in constant conditions with a period close to 24 hours are called circadian. Circadian rhythms have been now described in almost all groups of organisms from bacteria to humans and a considerable amount of information about the physiological, the cellular, and the molecular mechanisms responsible for the generation of circadian rhythmicity are now available.

In vertebrates there are few structures, the removal of which have significant effects on the behavioral circadian rhythmicity, and therefore they can be considered as part of the circadian system. These structures are the suprachiamastic nuclei of the hypothalamus (SCN), the lateral eyes and the pineal complex. This set of organs constitutes what is now called the "Vertebrates Circadian Axis" (Menaker and Tosini 1996). Although these structures are present in all the vertebrates, their contribution to the circadian system may vary considerably among classes and even within the same class. The SCN, for example, are the central circadian pacemaker in mammals, and their lesion abolishes almost all circadian rhythms (the only known exceptions are some circadian rhythms within the retinas), while in non-mammalian vertebrates the evidence for SCN involvement in circadian rhythmicity is far less extensive. The lateral eyes (retinas) contain self-sustained circadian oscillators in all vertebrates' classes and their removal may affect physiological and/or behavioral rhythms in amphibians, reptiles, birds and also in mammals. In addition, in mammals the eyes are the only structures capable of perceiving light and thus necessary for circadian entrainment (Yamazaki et al. 1999).

---

*E-mail: tosinig@msm.edu

The pineal gland is a central component in the regulation of circadian rhythmicity of reptiles and other non-mammalian vertebrates (Underwood 1990), whereas the role the gland plays in mammals is marginal, since its removal has little or no effects on overt rhythms (Cassone 1990).

Reptiles because of their phylogenetic position and ecology have provided—and still provide—the circadian field with some of the most interesting model to understand circadian organization, its evolution and its variability. The present review summarizes the current knowledge about the circadian organization of this fascinating taxonomic group.

Pineal complex in the regulation of circadian rhythms

The pineal complex (pineal gland and parietal eye) is a morphologically and functionally related set of organs that arises as an evagination of the roof of the diencephalon. The pineal organ is present in almost all vertebrates (alligator and owl have only a very rudimentary pineal organ) whereas the parietal eye is present only in some lizards species and in the tuatara (*Sphenodon punctatus*).

In Reptiles the pineal gland contains photosensory cells with secretory activity. The major product of these secretory cells is the hormone melatonin (however other indoles are also synthesized in the pineal gland), and this hormone is believed to play an important role in the circadian system of reptiles (see below). Melatonin is synthesized from the amino acid tryptophan throughout a well known biosynthetic pathway. Because of its capability to respond to change in illumination and temperature the pineal gland is considered to be the photothermoendocrine transducer (via the action of the hormone melatonin) of changes in photoperiod and environmental temperature (Underwood 1990).

The parietal eye has a lens, cornea and retina; the parietal eye retina is very simple (i.e. it is made of photoreceptors and ganglion cells only) and the photoreceptors synapse directly onto the ganglion cells, the axons of which form the parietal nerve. The parietal eye nerve innervates several areas of the brain (but does not project to the visual part). The parietal eye seems to be involved in many physiological functions of lizards (thermoregulation, reproduction, and orientation), but, in general, its role seems marginal or redundant. Almost unknown is the relationships between the parietal eye and the pineal gland. The parietal eye synthesizes melatonin, but in much lower quantity with respect to the pineal gland. It is likely that melatonin may simply fulfill a local function within the parietal eye.

Recently the parietal eye has become an interesting model to study the evolution of phototransduction mechanisms in vertebrate photoreceptors (Xiong et al. 1998).

Circadian oscillation *in vitro*

As it occurs in all vertebrate classes, in reptiles melatonin levels (pineal and blood) show a clear daily rhythmicity. For example in *Testudo hermanni*—during the period of activity—melatonin is high at night and low during the day (Vivien- Roels et al. 1979). Clear daily rhythms are present in the snake *Nerodia rhombifera* (Tilden and Hutchison 1993) and in *S. punctatus* (Firth et al. 1989), *Anolis carolinensis* (Underwood 1985) *Dipsosaurus dorsalis* (Janik and Menaker 1990), *Trachydosaurus rugosus* (Firth et al. 1979; Firth and Kennaway 1980, 1987), *Tiliqua rugosa* (Firth et al. 1999) and *Iguana iguana* (Tosini and Menaker 1996, 1998). In *T. rugosus* (Firth et al. 1979), *A. carolinensis* (Underwood 1985), *D. dorsalis* (Janik and Menaker 1990), *Podarcis sicula* (Foa et al. 1992a) and *I. iguana* (Tosini and Menaker 1998) the melatonin rhythm persisted also when the animals were held in constant darkness and temperature demonstrating, therefore, its true circadian nature.

However, the presence of a circadian melatonin rhythm *per se* cannot be used as a reliable indicator of the presence of a self-sustained oscillator, since the rhythmic melatonin synthesis/release could be driven by circadian oscillators located outside the pineal as it occurs in mammals.

An easy way to demonstrate the presence or absence of a circadian clock in the pineal is that of preparing an *in vitro* culture of the gland for few days (at least 3), while simultaneously measuring melatonin release at fixed intervals of time. This approach was firstly pioneered in chicken pineal (Kasal et al. 1979) and since then has been applied to the pineal (but also to the retina) of many other animals. In the lizards *A. carolinensis*, *Sceloporus occidentalis*, *D. dorsalis*, *I. iguana* (Iguanidae) and *Christinus marmoratus* (Gekkonidae) it has been demonstrated that in isolated cultured pineal melatonin synthesis and release persisted for several days, and the synthesis/release was rhythmic under light:dark cycles (Menaker and Wisner 1983; Menaker 1985; Janik and Menaker 1990; Pickard and Tang 1993; Moyer et al. 1995; Tosini and Menaker 1998).

However, only the pineal of *A. carolinensis*, *S. occidentalis* and *I. iguana* showed a persistent rhythm in melatonin release when cultured in constant darkness and temperature, thus demonstrating the presence of a circadian oscillator within the pineal itself (Figure 1). On the other hand, cultured *D. dorsalis* pineals secreted melatonin in large–but not rhythmic–quantities. Exposing the cultured

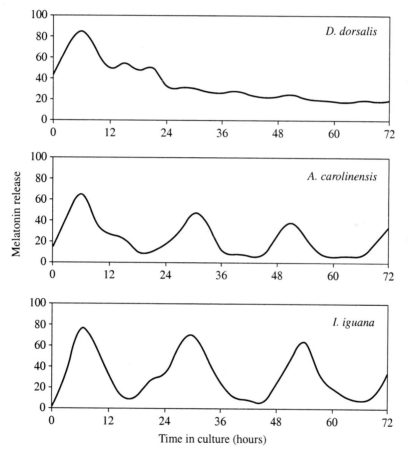

Fig. 1   *In vitro* **pattern of melatonin release (as % of the mean) from the pineal glands of three different species of iguanid lizard. A circadian rhythm of melatonin release is present in *A. carolinensis* and *I. iguana*, but not in *D. dorsalis*. The figures are redrawn from the following sources: *A. carolinensis* (Menaker and Wisner 1983); *D. dorsalis* (Janik and Menaker 1990); *I. iguana* (Tosini and Menaker 1998).**

pineal to bright illumination greatly reduced melatonin synthesis and/or release and abolished rhythmicity (Moyer et al. 1995; Tosini 1997).

In *A. carolinensis* and *I. iguana* the circadian rhythm of melatonin synthesis/release from cultured pineal has been shown to be temperature compensated (Menaker and Wisner 1983; Tosini et al. 1998) and this rhythms could also be entrained by temperature cycles. Recent studies have also shown that the parietal eye may contain a circadian clock controlling the synthesis/release of melatonin (Tosini 1997; Tosini and Menaker 1998).

Role of pineal and melatonin in regulation of circadian rhythms

As we have mentioned before, the pineal gland is considered as a neuroendocrine transducer of variation in environmental (light and temperature) conditions, and such action is likely to be mediated by the hormone melatonin. An easy way to address the role that the pineal plays in the circadian organization is to remove this gland and observe the effect that this removal has on the circadian rhythms.

Circadian rhythms in locomotor activity have been reported for several species of reptiles (see Underwood 1990). Removal of the pineal gland abolishes the circadian rhythms in locomotor activity in *A. carolinensis* (Underwood 1983), *Gallotia galloti* (Molina-Borja 1996), affects the period of the rhythm in *S. olivaceous* (Underwood 1977), *S. occidentalis* (Underwood 1981), *P. sicula* (Foa 1991; Innocenti et al. 1996) and has no effect in *D. dorsalis* (Janik and Menaker 1990) or in *I. iguana* (Tosini and Menaker 1998). Pineal transplantation in previously pinealectomized lizards induced significant changes in the free-running period of locomotor activity rhythms (Foa et al. 1997). In *I. iguana* a circadian rhythm in body temperature has been also demonstrated (Tosini and Menaker 1995), and the pineal organ is centrally involved in the generation and control of this rhythm, since rhythmicity disappears after its removal (Tosini and Menaker 1998).

Circadian rhythms in temperature selection have been also reported for several reptiles (Cowgell and Underwood 1979; Jarling et al. 1989; Refinetti and Susalka 1997). Pinealectomy temporarily abolished the circadian rhythms in behavioral thermoregulation in *P. sicula* (Innocenti et al. 1993), and reduced the amplitude of rhythms in behavioral thermoregulation in the *I. iguana* (Tosini and Menaker 1996).

Finally, also a circadian rhythm in electroretinogram has been reported in several lizards (review in: Shaw et al. 1993). Pinealectomy affected the amplitude of the circadian rhythm of electroretinographic response in *A. carolinensis* and *I. iguana,* suggesting an involvement of the pineal in the modulation of this rhythm (Shaw et al. 1993; Miranda-Anaya et al. 1999)

The behavioral effects of pinealectomy are likely to be mediated by melatonin, because of the following observations: (i) pinealectomy greatly reduces the amount of circulating melatonin and abolishes its circadian rhythmicity (review in: Tosini 1997); (ii) daily injections of exogenous melatonin can entrain activity rhythms (Underwood and Harless 1985; Vivien-Roels et al. 1988; Bertolucci and Foà 1998); (iii) Melatonin administration lengthens the period of circadian rhythms in *S. olivaceous* and *S. occidentalis* (Underwood 1979, 1981), in *D. dorsalis* (Janik and Menaker 1990) and in *P.sicula* (Foa et al. 1992b). Melatonin administration also entrained intact (Underwood and Harless 1985) and pinealectomized *S. occidentalis* (Hyde and Underwood 1995); (iv) a phase response curve to melatonin has been described in lizards (Underwood 1986) and (v) finally, melatonin administration altered the circadian rhythm of body temperature selection in *I. iguana* (Tosini and Menaker 1996).

Parietalectomy did not affect locomotor rhythms in *A. carolinensis* (Underwood 1983) and in *P. sicula* (Innocenti et al. 1993), while in *I.iguana* it produces slight changes in the circadian rhythms of locomotor activity and body temperature (Tosini and Menaker 1998). In *P. sicula* parietalectomy temporarily abolishes (one week) the circadian rhythm in body temperature selection (Innocenti et al. 1993).

Pineal and seasonality

Investigations in the Iguanid lizard *A. carolinensis* and *T. rugosa* demonstrated that 24-h cycles of both light and temperature can entrain the pineal melatonin rhythm and that differences in length of daily photoperiod or thermoperiod affect the phase, amplitude and duration of this rhythm (Underwood 1985; Underwood and Calaban 1987; Firth et al. 1999). Hence, the current ambient lighting and temperature conditions (and their seasonal change) are readily translated into an internal cue in the form of the pineal melatonin rhythm. This cue can be used to regulate both the daily and annual physiology of lizards (Underwood 1985). In the Lacertid lizard *P.sicula* the pineal was shown to be involved in the seasonal reorganization of the circadian system that is typical of this lizard (Foà et al. 1994; Innocenti et al. 1994). In constant temperature and constant darkness pinealectomy in *P. sicula* actually induces an immediate transition from the typical circadian locomotor behavior of summer, characterized by a marked bimodal pattern, short τ and long α, to the typical circadian locomotor behavior of autumn, characterized by an unimodal pattern, a long τ and short α (Fig. 2C). Again, the behavioral effects of chronic implants of exogenous melatonin (in silastic capsules) were found to be the same as those of pinealectomy in summer: the abolition of the bimodal pattern after application of the implants was always associated with a lengthening in τ and shortening in α (Foà et al. 1992b). Robust circadian rhythms of blood-borne melatonin expressed by intact *P. sicula* in late summer become heavily disrupted or abolished in response to either pinealectomy or melatonin implants (Foà et al. 1992a). Taken together, these results strongly support the view that the transition from a summer locomotor pattern to an autumn-winter one in response to both pinealectomy and melatonin implants is due to the concomitant suppression of circadian melatonin rhythms in the blood. Accordingly, in contrast to the situation in summer, in autumn and winter circadian rhythms of blood-borne melatonin do not seem to be required for the expression of the locomotor pattern typical of these seasons, and therefore in autumn-winter the behavioral effects of pinealectomy are expected to be substantially reduced with respect to those observed in summer. While results of comparative studies concerning 24h blood-borne melatonin profiles at different times of the year are still preliminary in *P. sicula* (Bertolucci et al., in preparation), previous work in the tortoise *T. hermanni* showed the occurrence of annual changes in melatonin rhythms, with maximal amplitude of these rhythms in summer and their complete disappearance in winter (Vivien-Roels et al. 1979). The results of an investigation which compared systematically the effects of pinealectomy on circadian locomotor behavior of *P. sicula* at different times of the year actually confirmed the existence of marked annual changes in the role of the pineal in the circadian organization of this lizard (Innocenti et al. 1996). Changes in τ in response to pinealectomy were found to be significantly greater in summer than in winter, spring and autumn, and α changed significantly only in spring and summer (Fig. 2A, B). Furthermore, while pinealectomy was effective in altering the locomotor rhythms of all individual lizards tested in summer, the same surgery left locomotor rhythmicity of many lizards tested in autumn and winter completely undisturbed (Innocenti et al. 1996). Further investigations demonstrated that daily injections of exogenous melatonin are capable of entraining circadian locomotor rhythms of *P. sicula* exclusively during the summer (Fig. 3; Bertolucci and Foà 1998). Altogether, these findings demonstrate that the pineal—via its hormonal product melatonin—is centrally involved in determining the circadian organization of the Lacertid lizard *P. sicula* in summer, while it is only marginally (or not at all)

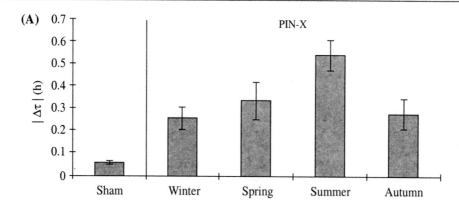

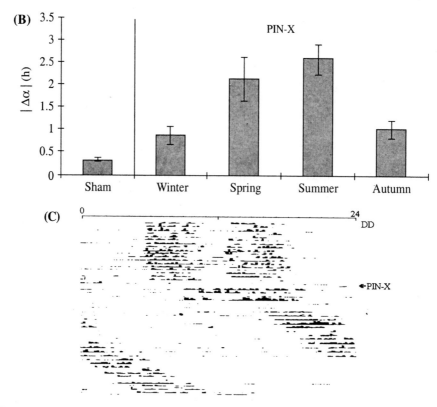

Fig. 2   Seasonal differences in the behavioral effect   f pinealectomy in the lacertid lizard *Podarcis sicula*. A, B. Means (± S.E.M.) of the absolute changes in the freerunning period of locomotor rhythms ($|\Delta\tau|$) and circadian activity time ($|\Delta\alpha|$) induced by pinealectomy (PIN-X) in different seasons and by sham pinealectomy (SHAM). Pinealectomy was effective in altering the freerunning period (τ) in all seasons, changes in τ were significantly greater in summer than in winter, spring and autumn. Circadian activity time (α) was found to change significantly in response to pinealectomy only in spring and summer. Since no seasonal differences in $|\Delta\tau|$ and $\Delta\alpha|$ were found among the four seasonal groups of SHAM, the data were pooled. C. Locomotor activity record of one lizard subjected to pinealectomy (PIN-X) in summer. Each horizontal line is a record of one day's activity, and consecutive days are mounted one below the other. Pinealectomy markedly lengthens τ, shortens α and abolishes the bimodal locomotor pattern.

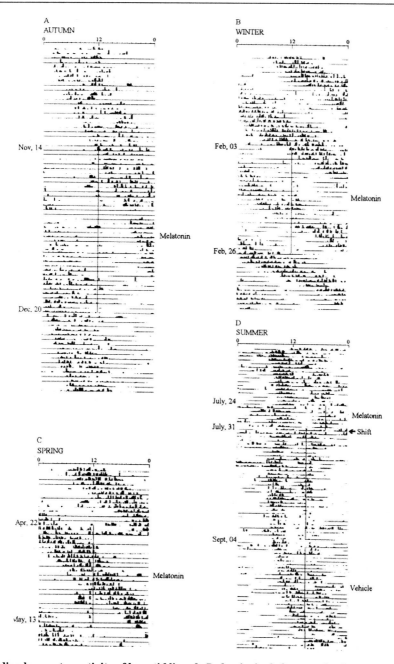

Fig. 3 Circadian locomotor activity of lacertid lizards *Podarcis sicula* freerunning in constant temperature (29°C) and darkness (DD). Lizards were collected and subjected to daily melatonin injections in autumn (A), winter (B), spring (C) and summer (D), respectively. A-C. Starting and ending dates of melatonin treatment are shown on the left of each record. The vertical line drawn through each record shows the time of day of melatonin injections during the whole injection period. D. On July 31st the time schedule of melatonin injections was advanced (Shift) from 7:00 p.m. to 3:00 p.m. On September 4th melatonin was replaced with vehicle solution.

involved in the other seasons. As mentioned before, the effects of pinealectomy on circadian locomotor behavior of lizards may vary consistently depending on the species (arrhythmicity, period changes, no effects). On the other hand, because of the seasonal differences in the behavioral effects of pinealectomy we have found in *P. sicula,* it seems reasonable to doubt that the differences among lizards are completely interspecific in nature. Instead they may, at least in part, depend on the particular season in which the behavioral effects of pinealectomy have been examined in each different species. The Lacertidae, as well as many Iguanidae, inhabit temperate zones, i.e. zones in which seasonal changes in circadian organization are likely to have evolved in response to the regular seasonal fluctuations in photoperiod and/or thermoperiod experienced by the lizards throughout the year (Underwood 1992; Foà et al. 1994). Hence, before deciding about interspecific differences, one should verify whether, for instance, *A. carolinensis* and *G. galloti* were tested in a season when the behavioral effects of pinealectomy are maximal and whether *D. dorsalis* was tested in a season when these effects are minimal.

Although the locomotor rhythms of the lizard tested in summer entrain successfully to the 24h period of melatonin injections, the locomotor rhythms of lizards tested in all other seasons do not entrain to the 24h period of the injections.

Role of the retina in the circadian system

The retinas of reptiles can participate in circadian function not only as photosensory input to the clock, but also as loci of circadian oscillators: in *I. iguana* the retina isolated in culture drives circadian rhythms of melatonin synthesis (Tosini and Menaker 1998). Bilateral ocular enucleation under constant bright light (LL) was found to induce a marked shortening in τ in *S. olivaceus, S. occidentalis*, and in some case arrhythmicity in *S. olivaceus* (Underwood and Menaker 1976; Underwood 1981). Enucleation has a modest effect on the circadian rhythms of body temperature and locomotor activity in *I. iguana*, however enucleation plus pinealectomy abolished both rhythms in the 30% of the animals tested (Bartell et al. 1999). Bilateral retinalectomy induces a marked shortening in τ in *P. sicula* kept in constant darkness (DD). These data suggest that the retinae may either (*S. olivaceus*) be a component of the primary pacemaker that drives locomotor rhythms, or (*S. occidentalis, P. sicula*) at least play an important modulating role, that is independent of light perception (*P. sicula*), on this primary pacemaker (Underwood 1981; Menaker 1982; Foà 1991). In *P. sicula* electrolytic lesions of both optic nerves at the level of the optic chiasma in DD produce the same behavioral effects as bilateral retinalectomy (Foà 1991; Minutini et al. 1994). This demonstrates that the influence of the retinae on the circadian system of *P. sicula* is neurally mediated. Accordingly, in the iguanid lizard *S. occidentalis* the influence of the retinae on the circadian system appears to be neurally mediated, since bilateral optic nerve section induces marked changes in shape of the phase-response curve to light (Underwood 1985).

The retinas play also a role in entrainment of circadian rhythms to LD cycles, since the light threshold for entrainment is lower in sighted than in blinded *S. olivaceus* (Underwood 1973). Intact retinas, however, are not necessary for entraining behavioral rhythms of lizards to LD cycles.

Several investigations made it clear that extraretinal photoreceptors participate in mediating entrainment of circadian rhythms of reptiles to LD cycles. In nine species of lizards, representing four different taxonomic families (Iguanidae, Gekkonidae, Xantusidae, Lacertidae) the locomotor rhythms can be entrained to LD cycles after enucleation of lateral eyes (Hoffmann 1970; Underwood and Menaker 1970, 1976; Underwood 1973, 1985).

Studies carried out in the iguanid lizard *S. olivaceus* and *P. sicula* showed that ablation of all known photoreceptive structures (lateral eyes, pineal, and parietal eye) in the same individual animal does not prevent entrainment of their circadian rhythms of locomotor activity to light (Foà et al. 1993; Underwood and Menaker 1976). Furthermore, shielding the brains of blinded-pinealectomized *S. olivaceus* entrained to LD cycles causes them to freerun. All this demonstrates the existence of brain photoreceptors mediating entrainment of locomotor rhythms to LD cycles. Reptile extraretinal photoreceptors must be quite sensitive because blind lizards can be entrained to an LD cycle as dim as 1 lux (Underwood 1973). Encephalic photoreceptors mediating entrainment have been documented in *Alligator missisipiensis* (Kavaliers and Ralph 1981).

Recent attempts to localize photoreceptors in the deep brain by using antibodies were successful. In the iguanid lizards *A. carolinensis* and *I. iguana* anti-opsins antibodies labeled neurons in the basal region of the lateral ventricles (Foster et al. 1993; Grace et al. 1996). A brain rhodopsin was recently cloned in *P. sicula*, but its location in the brain is unknown (Pasqualetti et al. 1997).

Two different mechanisms have been proposed for entrainment (Aschoff 1960; Pittendrigh 1981): in one, only the transitions from light to dark and from dark to light are considered effective for entrainment to 24 hr LD cycles (non-parametric entrainment); in the other, light and darkness are assumed to exert a more o less continuous action on the velocity of circadian oscillators (parametric entrainment). The observation that the velocity of oscillators changes by changing the intensity of LL (Aschoff's rule) supports the model of parametric entrainment. In diurnal animals, for example, the light portion of the 24 hr LD cycle may increase the velocity of the oscillators and the dark portion decrease their velocity, with the net effect of entraining period of the zeitgeber. Hence, it may be interesting to examine what array of photoreceptors mediate the response of the circadian system to changes in LL intensities. Underwood and Menaker (1976) investigated this aspect of circadian organization in *S. olivaceus* (Iguanidae) and *P. sicula* (Lacertidae), by testing the locomotor behavior of these diurnal lizards exposed to different levels of LL intensities after bilateral enucleation. When intact, both *S. olivaceus* and *P. sicula* obey Aschoff's rule for diurnal animals (Hoffmann 1960; Underwood and Menaker 1976). After blinding, *S. olivaceus* continues to obey to Aschoff's rule, while *P. sicula* cannot discriminate among different levels of LL and between LL and DD (Underwood and Menaker 1976). This led to conclude that in *S. olivaceus* extraretinal photoreceptors can mediate the response of the circadian system to changes in level of LL, while in *P. sicula* this function is accomplished only by the retinae of the lateral eyes. In contrast with this, new investigations in *P. sicula* showed clearly the existence in this lizard of extraretinal photoreceptors which are clearly capable of mediating the effects of changing in level of LL on circadian locomotor behavior (Fig. 4; Foà et al. 1993). Such a disagreement may depend on genetic differences between animals, since the *P. sicula* used in the experiments by Underwood and Menaker were collected in north-east Italy and Croatia, whereas the *P. sicula* used by Foà et al. were collected in central Italy (about 500–700 km apart). Even if this interpretation is correct, it is still unclear why some lizards can use extraretinal photoreception to discriminate among different intensities of LL, while others have only the retinae available to accomplish this function.

Role of hypothalamic areas in circadian organization

The suprachiasmatic nuclei of hypothalamus (SCN) are known to be the master circadian pacemaker of mammals and their lesion abolishes all circadian rhythms. During the last decade the SCN have been recognized to play a role also in the circadian system of lizards (Janik et al. 1990; Minutini et al. 1994). First of all, the lizard SCN are topographically similar to the mammalian SCN. As in mammals, the lizard SCN lies just dorsal to the optic chiasma and adjacent to the third ventricle, in

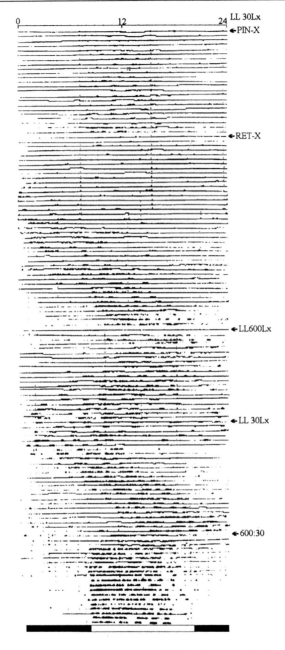

Fig. 4 Locomotor activity record of a lacertid lizard *Podarcis sicula* subjected to pinealectomy (PIN-X) and then to retinalectomy (RET-X) under LL 30 Lx. This lizard can discriminate between different levels of LL (Aschoff's rule) after PIN-X-RET-X: τ shortens under LL 600 Lx and lengthens under LL 30 Lx. The final part of the record shows entrainment of the activity rhythm of the PIN-X-RET-X lizard to a 24 h light cycle. These data show the existence of extrapineal-extraretinal photoreceptors in *P. sicula*.

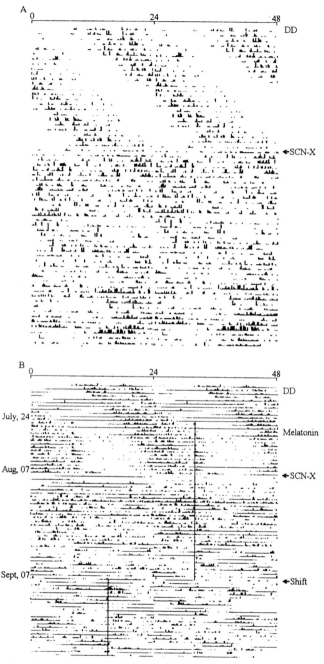

Fig. 5   **Effects of complete electrolytic lesions to the SCN (SCN-X) on circadian locomotor rhythms in the lacertid lizard *Podarcis sicula*. Locomotor activity records were double-plotted to aid in interpretation. A. Record of a lizard freerunning in constant darkness (DD), which became arrhythmic in response to SCN-X. B. While intact, a lizard tested during the summer entrained to the 24 h period of melatonin injections. After SCN-X this lizard became behaviorally arrhythmic. Melatonin injections continued after surgery, and their schedule was shifted on September 7th from 11:00 am to 03:00 p.m. The SCN-X lizard remained arrhythmic during the whole injection period.**

the region of transition from the preoptic area to the hypothalamus (Casini et al. 1993). Furthermore, as in rodents, the SCN of the Iguanid lizard *D. dorsalis* were shown to bind antibodies raised against arginine-vasopressine and neuropeptide Y (Janik et al. 1994). Collectively, these data strongly support the contention that the SCN of lizards are homologous to the SCN of mammals. Electrolytic lesions to 90% or more of the SCN (SCN-X) were found to abolish circadian rhythms of locomotor activity both in *D. dorsalis* and *P. sicula* (Fig. 5A; Janik et al. 1990; Minutini et al. 1994). Since except SCN lesions, no experimental treatment or lesion has so far succeeded in abolishing circadian locomotor rhythmicity in both *P. sicula* and *D. dorsalis*, it seems likely that in both lizard species the SCN contain the primary circadian pacemaker driving locomotor rhythms. Nevertheless, whether the SCN of reptile species contain circadian oscillators is still unknown.

Other experiments confirmed the role of the SCN as primary pacemaker in the *P. sicula* circadian system. Daily injections of exogenous melatonin entrain locomotor rhythms of intact, pinealectomized and unilaterally SCN-lesioned *P. sicula*, but are incapable of restoring rhythmicity in subjects previously rendered arrhythmic by SCN-X (Fig. 5B; Bertolucci and Foà 1998).

The fact that the circadian rhythm of behavioral temperature selection of *P. sicula* is not definitely abolished after both parietalectomy and pinealectomy suggests that the SCN or neighboring hypothalamic areas may be involved in driving this rhythm (Innocenti et al. 1993). Also in *I. iguana* experimental evidence suggests a role of the SCN in the circadian system, since removal of all the known circadian components (the retinas, the pineal and the parietal eye) does not abolish circadian rhythms of locomotor activity (Tosini and Menaker 1998). Interestingly, in *D. dorsalis* the daily bout of voluntary hypothermia disappears after lesion to the periventricular preoptic area of the hypothalamus (Berk and Heath 1975).

Conclusions

The available data on the circadian system of reptiles indicate that: (i) the retinas, pineal, parietal eye and the SCN may contain a circadian pacemaker; (ii) the role each of these structures play in circadian organization varies interspecifically and, moreover, it may fluctuate seasonally; (iii) circadian photoreceptors are located in the lateral eyes, the pineal, the parietal eye and in the brain (deep brain photoreceptors) and (iv) light and temperature are the primary entraining stimuli.

The study of circadian organization of reptiles also reveals that this taxonomic group provides a useful comparative model to study vertebrates' circadian organization. The circadian system of reptiles is multioscillatory and multiphotoreceptive, in which the contribution of each singular component and the interactions among the different component can be "easily" dissected.

References

Aschoff, J. (1960) Exogenous and endogenous components in circadian rhythms. Cold Spr. Harb. Symp. Quant. Biol. **25**: 11–28.

Bartell, P., Miranda-Anaya, M., Menaker, M. (1999) Effects of light, pinealectomy and enucleation on the circadian organization of the green iguana. International Congress on Chronobiology Abs.: 49.

Berk, M.L., Heath, J.E. (1975) Effects of preoptic, hypothalamic and telencephalic lesion on thermoregulation in lizards, *Dipsosaurus dorsalis*. J. Therm. Biol. **1**: 65–78.

Bertolucci, C., Foà, A. (1998) Seasonality and role of the SCN in entrainment of lizard circadian locomotor rhythms to daily melatonin injections. Am. J. Physiol. **274(43)**: R1004–R1014.

Casini, G., Petrini, P., Foà, A., Bagnoli, P. (1993) Pattern of organization of primary visual pathways in the European lizard *Podarcis sicula* Rafinesque. J. Hirnforsch. **34**: 361–374.

Cassone, V.M. (1990) Effetcs of melatonin on vertebrate circadian systems. Trends in Neurosci. **13**: 457–464.

Cowgell, G., Underwood, H. (1979) Behavioral theroregulation in lizards: circadian rhythm. J. Exp. Zool. **210**: 189–194.

Firth, B.T., Kennaway, D.J., Rozenbilds, M.A.M. (1979) Plasma melatonin in the scincid lizard, *Thachydosaurus rugosus* : diel rhythms, seasonality, and the effects of constant light and constant darkness. Gen. Comp. Endocrinol. **37**: 493–500.

Firth, B.T., Kennaway, D.J. (1980) Plasma melatonin levels in the scincid lizard *Thachydosaurus rugosus*: effect of constant and fluctuating temperature. Brain Res. **404**: 313–318.

Firth, B.T., Kennaway, D.J. (1987) Melatonin content of pineal, parietal eye and blood plasma of the lizard, *Thachydosaurus rugosus*: effect of constant and fluctuating temperature. Brain Res. **404**: 313–318.

Firth, B.T., Thompson, M.B., Kennaway, D.J. (1989) Thermal sensitivity of reptilian melatonin rhythms "cold" tuatara vs. "warm" skink. Am. J. Physiol. **256**: R1160–R1163.

Firth, B.T., Belan, I., Kennaway, D.J., Moyer, R.W. (1999) Thermocyclic entrainment of lizard blood plasma melatonin rhythms in constant and cyclic photic environments. Am. J. Physiol. **277**: R1620–R1626.

Foà, A. (1991) The role of the pineal and the retinae in the expression of circadian locomotor rhythmicity in the ruin lizard, *Podarcis sicula*. J. Comp. Physiol. A **169**: 201–207.

Foà, A., Janik, D., Minutini, L. (1992a) Circadian rhythms of plasma melatonin in the ruin lizard *Podarcis sicula*: effetcs of pinealectomy. J. Pineal Res. **12**: 109–113.

Foà, A., Minutini, L., Innocenti A (1992b) Melatonin: a coupling device between oscillators in the circadian system of the ruin lizard *Podarcis sicula*. Comp. Biochem. Physiol. A **103**: 719–723.

Foà, A., Flamini, M., Innocenti, A., Minutini, L., Monteforti, G. (1993) The role of extraretinal photoreception in the circadian system of the ruin lizard *Podarcis sicula*. Comp. Biochem. Physiol. A **105**: 223–230.

Foà, A., Monteforti, G., Minutini, L., Innocenti, A., Quaglieri, C., Flamini, M. (1994) Seasonal changes of locomotor activity patterns in ruin lizards *Podarcis sicula*. I. Endogenous control by the circadian system Behav. Ecol. Sociobiol. **34**: 227–274.

Foà, A., Bertolucci, C., Marsanich, A., Innocenti, A. (1997) Pineal transplantation to the brain of pinealectomized lizards: effects on circadian rhythms of locomotor activity. Behav. Neurosc. **111**: 1123–1132.

Foster, R.G., Garcia-Fernandez, J.M., Provencio, I., DeGrip, W.J. (1993) Opsin localization and chromophore retinoids identified within the basal brain of the lizard *Anolis carolinensis*. J. Comp. Physiol. A **172**: 33–45.

Grace, M.S., Alones, V., Menaker, M., Foster, R.G. (1996) Light perception in the vertebrate brain: an ultrastructural analysis of opsin-and vasoactive intestinal polypeptide-immunoreactive neurons in iguanid lizards. J. Comp. Neurol. **367**: 575–594.

Hyde, L.L., Underwood, H. (1995) Daily melatonin infusions entrain the locomotor activity of pinealectomized lizards. Physiol. Behav. **58**: 953–951.

Hoffmann, K. (1960) Versuche zur Analyse der Tagesperiodik, I. Der Einfluss der Lichtintensitaet. Z. Vergl. Physiol. **43**: 544–566.

Hoffmann, K. (1970) Zur Syncronisation biologischer Rhythmen. Verh. Dtsch. Zool. Ges. pp. 166–273.

Innocenti, A., Minutini, L., Foà, A. (1993) The pineal and circadian rhythms of temperature selection and locomotion in lizards. Physiol. Behav. **53**: 911–915.

Innocenti, A., Minutini, L., Foà, A. (1994) Seasonal changes of locomotor activity patterns in ruin lizards *Podarcis sicula*. II. Involvement of the pineal. Behav. Ecol. Sociobiol. **35**: 27–32.

Innocenti, A., Bertolucci, C., Minutini, L., Foà, A. (1996) Seasonal variations of pineal involvement in the circadian organization of ruin lizards *Podarcis sicula*. J. Exp. Biol. **199**: 1189–1194.

Janik, D.S., Menaker, M. (1990) Circadian locomotor rhythms in the desert iguana I. The role of the eyes and the pineal. J. Comp. Physiol. A **166**: 803–810

Janik, D.S., Pickard, G.E., Menaker, M. (1990) Circadian locomotor rhythms in the desert iguana. II: Effects of electrolytic lesions to the hypothalamus. J. Comp. Physiol. A **166**: 811–816

Janik, D.S., Cassone, V.M., Pickard, G.E., Menaker, M. (1994) Retinohypothalamic projections and immunocytochemical analysis of the suprachiasmatic region of the desert iguana *Dipsosaurus dorsalis*. Cell Tiss. Res. **275**: 399–406.

Jarling, C., Scarperi, M., Bleichert, A. (1989) Circadian rhythm in the temperature preference of the turtle, *Chrysemys* (= *Pseudemys*) *scripta elegans*, in a thermal gradient. J. Therm. Biol. **14**: 173–178.

Kasal, C., Menaker, M., Perez-Polo, R. (1979) Circadian clock in culture: N-acetyltransferase activity of chick pineal glands oscillates *in vitro*. Science **203**: 656–658.

Kavaliers, M., Ralph, C.L. (1981) Encephalic photoreceptor involvement in the entrainment and control of circadian activity of young American alligators. Physiol. Behav. **26**: 413–418.

Menaker, M. (1982) The search for principles of physiological organization in vertebrate circadian system. In: Aschoff, J., Daan, S., Groos, G.A. (eds.) Vertebrate circadian system. Springer-Verlag, Berlin, pp. 1–12.

Menaker, M. (1985) Eyes–the second (and third) pineal gland? In: Evered, D., Clark, S. (eds.) Photoperiodism, melatonin and pineal. Pitman, London, pp. 39–52.

Menaker, M., Wisner, S. (1983) Temperature-compensated circadian clock in the pineal of *Anolis*. Proc. Natl. Acad. Sci. USA **80**: 6119–6121.

Menaker, M., Tosini, G. (1996). The Evolution of Vertebrate Circadian System. In: Honma, K., Honma, S. (eds.) Circadian Organization and Oscillatory Coupling. Hokkaido University Press, Sapporo, pp. 39–52.

Minutini, L., Innocenti, A., Bertolucci, C., Foà, A. (1994) Electrolytic lesions to the optic chiasma affect circadian locomotor rhythms in lizard. Neuroreport **5**: 525–527.

Miranda-Anaya, M., Bartell, P., Yamasaki, S., Menaker, M. (1999) Circadian rhythm of electroretinographich response (ERG) and effect of pinealectomy in *Iguana iguana*. Soc. Neurosci. Abs. **25**: 1132.

Molina-Borja, M. (1996) Pineal-gland and circadian locomotor-activity rhythm in the lacertid *Gallotia galloti eisentrauti*, pinealectomy induces arrhythmicity. Biol. Rhythm Res. **27**: 1–11.

Moyer, R.B., Firth, B.T., Kennaway, D.J. (1995) Effect of constant temperatures, darkness and light on the secretion of melatonin by pineal explants and retinas in the gecko *Christinus marmoratus*. Brain Res. **675**: 345–348.

Pasqualetti, M., Innocenti, A., Foà, A., Nardi, I. (1997) Cloning of a brain opsin from lizard *Podarcis sicula*. Biol. Rhythms Res. **28**: 127.

Pickard, G.E., Tang, W.X. (1993) Individual pineal cells exhibit a circadian rhythm in melatonin secretion. Brain Res. **627**: 141–146.

Pittendrigh, C.S. (1981) Circadian system: entrainment. In: Aschoff, J. (ed.) Handbook of Behavioural Neurobiology. Biological Rhythms. Plenum Press, New York, **4**: 95–124.

Refinetti, R., Susalka, S.J. (1997). Circadian rhythm of temperature selection in a nocturnal lizard. Physiol. Behav. **62**: 331–336.

Shaw, A.P., Collazo, C.R., Easterling, K., Young, C.D., Karwoski, C.J. (1993). Circadian rhythm in the visual system of the lizard *Anolis carolinensis*. J. Biol. Rhythms **8**: 107–124.

Tilden, A.R., Hutchison, V.H. (1993) Influence of photoperiod and temperature on serum melatonin in the diamondback water snake, *Nerodia rhombifera*. Gen. Comp. Endocr. **92**: 347–354.

Tosini, G. (1997) The pineal complex of Reptiles: physiological and behavioral roles: Ethol. Ecol. Evol. **9**: 313–333.

Tosini, G., Menaker, M. (1995) Circadian rhythm of body temperature in an ectotherm (*Iguana iguana*) J. Biol. Rhythms **10**: 248–255.

Tosini, G., Menaker, M. (1996) Pineal complex and melatonin affect the daily rhythm of temperature selection in the green iguana. J. Comp. Physiol. A **179**: 135–142.

Tosini, G., Menaker, M. (1998) Multioscillatory circadian organization in a vertebrate, *Iguana iguana*. J. Neurosc. **18**: 1105–1114.

Tosini, G., Moreira, L.F., Bartell, P., Menaker, M. (1998) Temperature compensation of circadian rhythms in a multioscillatory system. SRBR Meeting Abs. **6**: 175.

Underwood, H. (1973) Retinal and extraretinal photoreceptors mediate entrainment of the circadian locomotor rhythm in lizard. J. Comp. Physiol. **83**: 187–222.

Underwood, H. (1977) Circadian organization in lizards: The role of the pineal organ. Science **195**: 587–589.

Underwood, H. (1979) Melatonin affects circadian rhythmicity in lizard. J. Comp. Physiol. **130**: 317–323.

Underwood, H. (1981) Circadian organization in the lizard, *Sceloporus occidentalis*: the effects of blinding, pinealectomy and melatonin. J. Comp. Physiol. **141**: 537–547.

Underwood, H. (1983) Circadian organization in the lizard *Anolis carolinensis*: a multioscillatory system. J. Comp. Physiol. **152**: 265–274.

Underwood, H. (1985) Pineal melatonin rhythms in the lizard *Anolis carolinensis*: effects of light and temperature cycles. J. Comp. Physiol. A **157**: 57–65.

Underwood, H. (1986) Circadian rhythms in lizards: phase response curve for melatonin. J. Pineal Res. **3**: 187–196.

Underwood, H. (1990) The pineal and melatonin: regulators of circadian function in lower vertebrates. Experientia **46**: 120–128.

Underwood, H. (1992) Endogenous rhythms. In: Gans, C. (ed.) Biology of Reptilia. The University of Chicago Press, Chicago and London, Vol. 18, pp. 22–29.

Underwood, H., Menaker, M. (1970) Extraretinal light perception: entrainment of the biological clock controlling lizard locomotor activity. Science **170**: 190–193.

Underwood, H., Menaker, M. (1976) Extraretinal photoreception in lizard. Photochem. Photobiol. **24**: 227–243.

Underwood, H., Harless, M. (1985) Entrainment of the circadian activity rhythm of a lizard to melatonin injection. Physiol. Behav. **35**: 267–270.

Underwood, H., Calaban, M. (1987) Pineal melatonin rhythms in lizards *Anolis carolinensis*: I. Response to light and temperature cycles. J. Biol. Rhythms. **2**: 179–193.

Vivien-Roels, B., Arendt, J., Bradtke, J. (1979) Circaadian and circannual fluctuations of pineal indoleamines (serotonin and melatonin) in *Testudo hermanni* Gmelin (Reptilia, Chelonia) I. Under natural conditions of photoperiod and temperature. Gen. Comp. Endocr. **37**: 197–210.

Vivien-Roels, B., Pevet, P., Claustrat, B. (1988) Pineal and circulating melatonin rhythms in the box turtle, *Terrapene carolina triunguis*: effects of photoperiod, light pulse, and environmental temperature. Gen. Comp. Endocr. **69**: 163–173.

Yamazaki, S., Goto, M., Menaker, M. (1999) No evidence for extraocular photorecptors in the circadian system of the Syrian hamster. J. Biol. Rhythms **14**: 197–201.

Xiong, W.H., Solessio, E.C., Yau, K.W. (1998) An unusual cGMP pathway underlying depolarizing light response of vertebrate parietal-eye photoreceptors. Nature Neurosci. **1**: 359–365.

13. The Circadian Pacemaking System of Birds

R. Brandstätter*

Department of Biological Rhythms and Behaviour, Max-Planck
Research Centre for Ornithology,
Von-der-Tann-Straße 7, D-82346 Erling-Andechs/Germany

In contrast to the highly centralized circadian clock of mammals, where the major pacemaker is located in the hypothalamic suprachiasmatic nuclei and information about the photic environment is exclusively received from the eyes, the avian circadian pacemaking system is characterized by high complexity and diversity. Birds have the capacity to obtain information about the photic environment by retinal, pineal, and deep encephalic photoreceptors. There are at least three autonomous circadian oscillators involved in the regulation of circadian rhythms at the organismic level, the retina, the pineal gland, and the hypothalamic oscillator.

With one exception, the Japanese quail, all avian pineal glands yet studied contain autonomous circadian oscillators that regulate melatonin synthesis in a rhythmic fashion but the degree of self-sustainment is variable between species. The avian retina has also been shown to produce melatonin rhythmically, and this capacity is independent of extra-retinal oscillators. The presence of a hypothalamic oscillator has been demonstrated by disruptions of circadian activity rhythms after hypothalamic lesions in a variety of species but whether birds possess a nucleus homologous to the mammalian SCN is still under debate. Recent results demonstrating rhythmic clock-gene expression in the avian hypothalamus suggest a complex organization of the avian hypothalamic oscillator.

Melatonin originating from either the pineal gland (in passeriform species) or the pineal gland and the retina (in columbiform birds) plays an important role in the regulation of circadian activity rhythms in most avian species. However, melatonin-independent control of circadian activity rhythms may also occur (e.g. in some galliform species). Experimental manipulations of the melatonin signal have been shown to influence the persistence of circadian activity rhythms, the range of zeitgeber periods to which rhythmicity can be synchronized, and resynchronization times following phase shifts of exogenous zeitgebers. Changes and differences in specific features of the melatonin signal can also be found under a variety of natural environmental conditions. This variability may be functionally important for the temporal organization of birds and related to their high diversity of life-history strategies.

Introduction

In most cases, the temporal organization of animals is demonstrated by the sequential occurrence of activity and rest. Very rarely, animals are continuously active or distribute their activity uniformly over night and day. Most animals restrict their activity phases within a 24-hour day to either day ("diurnal" activity rhythm) or night ("nocturnal"), or to one or both of the twilight periods ("crepuscular"). Day/night rhythms of physiology and behaviour are coupled to a variety of abiotic and biotic environmental factors, which cycle with a periodicity of about 24 hours. The most fascinating feature

*E-mail: brandstätter@erl.ornithol.mpg.de

of these rhythms is that they persist even when organisms are shielded from 24-hour environmental variations. Thus, these rhythms are not only a direct reflection of the rotation of the earth on its axis (and the subsequent variations of certain environmental parameters, such as the daily change of light and dark) but have an endogenous nature within the organism. From prokaryotes to man, endogenous circadian rhythms have been described that control a great number of biochemical, physiological, and behavioural phenomena, including photosynthesis in plants, development of insects, and sleep in mammals (Aschoff 1981; Pittendrigh 1993).

At the most basic level, these rhythms are generated within cells that contain a particular molecular clockwork. Such "clock cells" use molecular loops that close within the cell borders and, thus, do not require cell-cell interactions to produce an intracellular circadian rhythm. However, to be effective for the organism, these molecular oscillations have to be transduced within the clock cell to change its activity and, in multicellular organisms, outside the cell to induce daily changes in behaviour or general physiology (e.g. Dunlap 1999). In multicellular organisms, clock cells form "functional units", so-called circadian oscillators or circadian systems. The best known examples of such circadian systems include the eyes of some marine mollusks (Eskin 1979; Block and McMahon 1984), the retina of amphibians (Besharse and Iuvone 1983), the mammalian suprachiasmatic nucleus (SCN) (Klein et al. 1991; Moore and Silver 1998), and the pineal gland of non-mammalian vertebrates, particularly those of birds (e.g. Takahashi et al. 1980; Robertson and Takahashi 1988a, b; Murakami et al. 1994; Brandstätter et al. 2000).

General organization of the avian circadian pacemaking system

Birds have the capacity to obtain information about the photic environment from retinal, pineal, and deep encephalic photoreceptors (Cassone and Menaker 1984; Menaker et al. 1997; Foster and Soni 1998; Silver et al. 1988; Kojima and Fukada 1999). There are at least three autonomous and anatomically distinct oscillators available in birds to organize circadian pacemaking at the organismic level: the retina (Binkley et al. 1980; Underwood 1994), the pineal gland (Gaston and Menaker 1968; Zimmerman and Menaker 1979; Brandstätter et al. 2000), and a hypothalamic oscillator (Takahashi and Menaker 1979a, b, 1982). Several lines of evidence suggest that these components interact with each other to produce a stable organismic circadian rhythmicity (Cassone and Menaker 1984; Gwinner 1989a; Cassone 1990; Gwinner and Hau 2000) (Fig. 1). Besides birds, such a "multi-oscillatory" organization of the circadian pacemaking system can also be found in lower vertebrates (e.g. Underwood 1989; Menaker et al. 1997). In contrast, the circadian pacemaking system of mammals is highly centralized. Photic input to the major pacemaker, the hypothalamic suprachiasmatic nucleus, is received exclusively from the retinae of the eyes and no other structure than the SCN has yet been identified as containing a circadian oscillator that may act as a pacemaker (Hastings and Maywood 2000).

The pineal gland

The function and the significance of the avian pineal gland have been investigated in a variety of studies. With a few exceptions, birds have well-developed pineal glands which rhythmically release melatonin into the blood stream (Gwinner and Hau 2000). Rhythmic melatonin in the blood can be found during synchronization, i.e., when animals are exposed to light/dark cycles (LD), as well as under constant conditions (continuous darkness [DD] or continuous dim light [dimLL]) (Kumar et al. 1999), demonstrating that melatonin production is regulated by endogenous circadian oscillators.

Information on the intrapineal nature of this rhythmic melatonin production comes from a variety of *in vitro* studies with isolated cultured pineal glands or primary cultured pineal cells (Deguchi 1979;

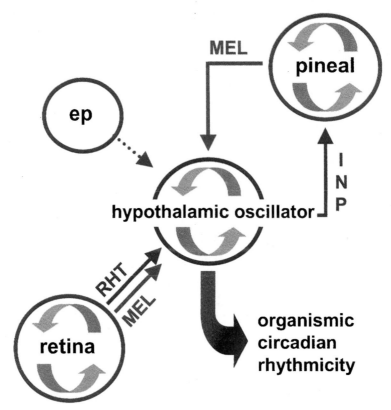

Fig. 1 Diagram of the components of the avian circadian pacemaking system. Oscillatory components are indicated by orange arrows, photoreceptive structures are encircled in green. Hormonal signal pathways are indicated by red arrows, neural pathways are indicated by blue arrows; ep = encephalic photoreceptors; INP = indirect neural pathways; MEL = melatonin; RHT = retino-hyphothalamic tract.

Takahashi et al. 1980; Takahashi and Menaker 1984). With one exception, the Japanese quail, all avian pineal glands yet studied contain autonomous circadian oscillators that regulate melatonin synthesis in a rhythmic fashion (e.g. Takahashi 1981; Robertson and Takahashi 1988a; Murakami et al. 1994; Brandstätter et al. 2000). Chicken pineal cell cultures express circadian oscillations of melatonin release for several cycles in constant darkness with a period close to 24 hours (Roberston and Takahashi 1988a). Advances or delays of the light/dark cycle produce a corresponding shift in the melatonin rhythm and the phase shifts persist after transfer to constant darkness, showing that the underlying circadian oscillator was entrained (Robertson and Takahashi 1988b). Autonomous circadian rhythms of melatonin production are maintained under constant conditions not only in whole-organ cultured chicken pineal glands (Takahashi et al. 1980) but also in dispersed pineal cells (e.g. Robertson and Takahashi 1988a, b), and even in isolated single pineal cells (Nakahara et al. 1997). Thus, like the isolated pineal gland as a whole, dissociated and individually cultured pineal cells contain (1) a photoreceptive input pathway, (2) a circadian oscillator or pacemaker that generates the rhythm, and (3) an output pathway resulting in the synthesis and release of melatonin.

Circadian rhythms of melatonin release were also found in cultured pineal glands of house sparrows

(Takahashi 1981; Brandstätter et al. 2000) and European starlings (Takahashi 1981), as well as in primary cultured pineal cells of pigeons and house sparrows (Murakami et al. 1994). In Japanese quail, cultured pineal cells (Murakami et al. 1994) and whole-organ cultured pineal glands (Brandstätter unpubl. data) have been shown to exhibit a diurnal oscillation of *in vitro* melatonin release under light/dark conditions, but the oscillation did not persist under constant conditions. These results suggest that the melatonin synthesis pathway in the Japanese quail pineal gland, as in other avian species, is coupled to a photoreceptive system. Coupling to an oscillator, as it has been found in all other species investigated until now, appears to be absent or to be very weak in this species.

There is considerable variation in the degree of persistence of the circadian melatonin rhythm of isolated pineal oscillators. Whereas the rhythm disappear or is strongly reduced in amplitude in cultured chicken pineal glands after several days in culture (Takahashi et al. 1980; Takahashi 1981), pineal glands from other species, such as the house sparrow, show high-amplitude rhythms of *in vitro* melatonin release for a week or more (Fig. 2).

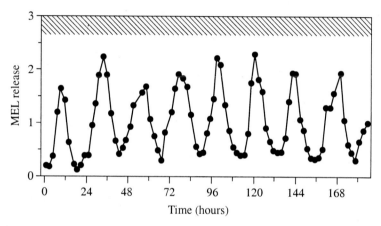

Fig. 2   Circadian rhythm of *in vitro* melatonin release from an isolated whole-organ cultured pineal gland of the house sparrow. Melatonin release is characterized by a high-amplitude rhythm during 8 cycles in continuous darkness (data shown as the relative amount of melatonin released into the culture medium; Brandstätter unpubl.).

The retina

The avian retina has so far been investigated in a few species only. Like the avian pineal gland, the retina of birds has been shown to produce melatonin rhythmically and there are several lines of evidence that this capacity is autonomous, i.e., independent of extra-retinal oscillators (Hamm and Menaker 1980; Thomas and Iuvone 1991; Bernard et al. 1997, 1999; Chong et al. 1998). Similar to the situation in mammals, the avian retina projects to the hypothalamus via the retinohypothalamic tract. In general, retinal fibres terminate in various hypothalamic locations but the anatomical distribution of retinal projections varies between species (e.g. Hartwig 1974; Cassone and Moore 1987; Norgren and Silver 1989; King and Follett 1997). Retinal fibres that could directly be involved in the transmission of photic information to the circadian pacemaking system can particularly be found in the lateral hypothalamic area and, 'to a lesser degree, close to the suprachiasmatic nucleus (e.g. Cassone and Moore 1987; Norgren and Silver 1989).

The avian "SCN"

In mammals, the "master clock" controlling circadian rhythmicity is located in the hypothalamic suprachiasmatic nuclei (SCN) which are characterized by distinct anatomical, physiological, and neurochemical features (Klein et al. 1991) as well as the presence of specific transcription factors and clock genes (Dunlap 1999). Until now, no comparable structure has been unambiguously described in the brain of any non-mammalian vertebrate. In birds, early anatomical studies described a suprachiasmatic nucleus (SCN) located in the anterior hypothalamus adjacent to the third ventricle (Crosby and Woodburne 1940; Oksche and Farner 1974) and experimental lesioning studies suggested that this suprachiasmatic nucleus might indeed represent the structure functionally equivalent to the mammalian SCN (Takahashi and Menaker 1979a, b, 1982). Since then, several hypothalamic cell groups have been assumed to represent the central nervous circadian pacemaker in birds, particularly the above-mentioned suprachiasmatic nucleus as well as a small cell group in the lateral hypothalamus, called either lateral hypothalamic retinorecipient nucleus (Norgren and Silver 1989) or "visual" SCN (Cassone and Moore 1987; King and Follett 1997).

Given that retinohypothalamic transmission of photic information to the SCN is a crucial step for the entrainment of the hypothalamic pacemaker in mammals, tract-tracing techniques were used in a certain number of studies in birds in an attempt to localize the avian hypothalamic oscillator. However, the presence of retinohypothalamic projections in birds does not allow conclusions about the localization of the oscillator (Hartwig 1974; Cassone and Moore 1987; Norgren and Silver 1989; King and Follett 1997). Although some neurotransmitters and neuropeptides which are rhythmically produced in the mammalian SCN were also found in the hypothalamus of birds, their anatomical localizations were as inconclusive as retinal projections (Panzica 1985; Cassone and Moore 1987; Norgren and Silver 1989, 1990; King and Follett 1997) and whether birds possess a nucleus homologous to the mammalian SCN remained controversial (Fig. 3).

Several reasons may have contributed to the lack of detailed knowledge about the avian equivalent of the mammalian hypothalamic circadian oscillator, including incomplete and inconsistent anatomical descriptions of the avian hypothalamus, confusion in delineation as well as terminology of cell groups, and the lack of a molecular approach defining key genes and/or transcription factors of circadian oscillators. Recently, two avian homologues of the drosophila period gene have been cloned in the Japanese quail and demonstrated to be rhythmically expressed in the retina and in the pineal gland (Yoshimura et al. 2000). Additionally, expression of a *Per* gene was found in the hypothalamus of the house sparrow showing marked diurnal rhythmicity in the suprachiasmatic nucleus as well as in the lateral hypothalamus (Brandstätter et al. 2001a). Thus, rhythmic clock gene expression in the avian hypothalamus is not restricted to a single cell group, as is the case in mammals, suggesting a rather complex organization of the avian hypothalamic circadian oscillator.

Functional significance of the individual circadian systems for the generation of circadian rhythms at the organismic level

In the house sparrow, the pineal oscillator is vital to the persistence of circadian rhythmicity *in vivo* (Fig. 4A). In birds kept under constant conditions, pinealectomy abolished the circadian rhythms of locomotor activity (Gaston and Menaker 1968), body temperature (Binkley et al. 1971), and feeding (Heigl and Gwinner 1994). There are several lines of evidence that these effects are primarily due to the lack of a rhythmic melatonin signal after removal of the pineal gland: disruptions of the neural connections of the pineal gland by either pineal stalk deflection or chemical sympathectomy did not

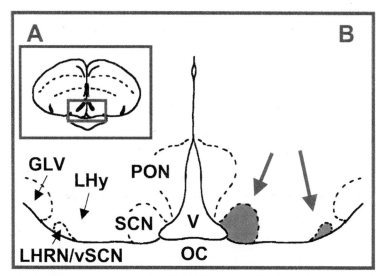

Fig. 3   Schematic drawing of a cross section through the anterior hypothalamus of the house sparrow brain. The insert (A) shows the anatomical position of the cell groups shown in B. The two cell groups that have been assumed to represent the functional equivalent of the mammalian hypothalamic oscillator are drawn in blue (right hemisphere) and indicated by red arrows. GLV = ventrolateral geniculate nucleus, LHRN/vSCN = lateral hypothalamic retinorecipient nucleus (also called visual SCN), LHy = lateral hypothalamic area, OC = optic chiasma, PON = preoptic nucleus, SCN = suprachiasmatic nucleus, V = third ventricle.

abolish free-running activity rhythms in constant darkness, unlike complete pinealectomy (Zimmerman and Menaker 1975).

Rhythmicity was restored when the pineal gland of a sparrow was implanted into the anterior chamber of the eye of an arrythmic pinealectomized host and the emerging rhythm had the phase of the rhythm of the donor bird, indicating that circadian clock properties were transplanted with the pineal gland (Zimmerman and Menaker 1979) (Fig. 5).

Additionally, rhythmicity was restored in arrhythmic pinealectomized sparrows when exogenous melatonin was rhythmically applied (Chabot and Menaker 1992; Heigl and Gwinner 1994). The dramatic effects of pinealectomy on circadian activity rhythms found in house sparrows have been confirmed in other passeriform birds, including white-crowned sparrows and white-throated sparrows (Gaston 1971; McMillan 1972). However, in another passeriform bird, the European starling, pinealectomy only disturbed locomotor and feeding activity rhythms but had no consistent effect on their persistence under constant conditions (Gwinner 1978).

Although most of the information on circadian oscillator properties has been obtained in the chicken, little is known about the effects of pineal removal in this species: pinealectomy has been reported to have no effect on circadian activity rhythms in only one study (MacBride 1973) but, for example, pinealectomy has been shown to shift or abolish daily rhythms of certain immune parameters which could be restored by applications of exogenous melatonin (Rosolowska-Huszcz et al. 1991). In the pigeon as well as in the Japanese quail, pinealectomy had no effects on free-running circadian activity rhythms (Simpson and Follett 1981; Ebihara et al. 1987; Chabot and Menaker 1992).

These species-specific differences in the effects of pinealectomy on circadian rhythmicity at the organismic level are, to a certain degree, related to the origin of circulating melatonin. In house

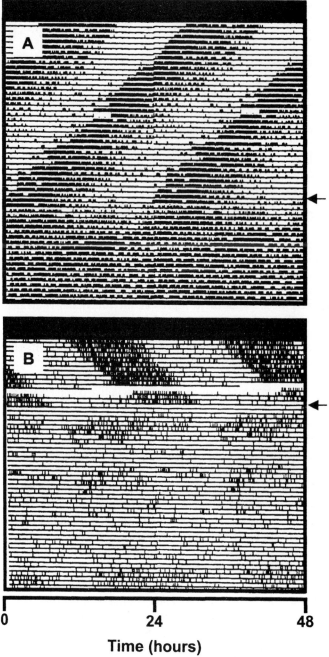

Fig. 4   Effects of pinealectomy (A) and electrolytic lesions of the suprachiasmatic nuclei (B) on circadian activity rhythms (double-plotted actograms) of house sparrows kept in DD. Surgeries have been performed at the time points indicated by the arrows on the right. Neither the pineal gland (B) nor the hypothalamic oscillator (A) maintain the rhythmicity of the bird when the other is removed A: modified from Gwinner et al. (1997), B: modified from Takahashi and Menaker (1982a).

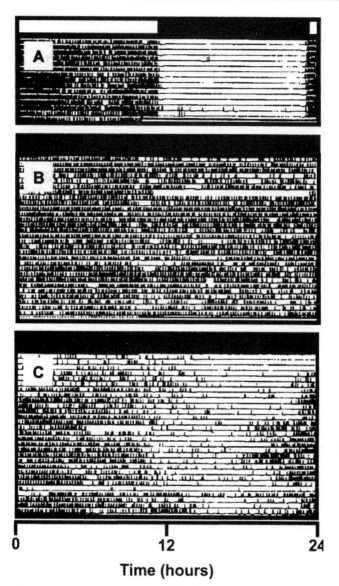

Fig. 5 Locomotor activity of house sparrows in LD (A), after pinealectomy in DD (B), and after transplantation of a pineal gland into the anterior eye chamber in DD (C). The light/dark schedule in A is indicated on top; continuous darkness is indicated by black horizontal bars on top of B and C (for details see text; modified from Zimmerman and Menaker 1979).

sparrows and other passeriform birds, the pineal gland is the only source of circulating melatonin (Fig. 6). Although the degree of disruption of circadian activity rhythms appears to be different between species, ranging from changes of the duration of the activity phase and the period of free-running rhythms to complete abolishment of circadian activity, elimination of the endogenous melatonin rhythm always has some effect on behavioural rhythms in passeriform birds. In non-passeriform birds, such as the chicken, the Japanese quail, and the pigeon, effects of pinealectomy on behavioural activity rhythms are either not detectable (Japanese quail, pigeon) or insufficiently investigated (chicken). However, as in passeriforms, pinealectomy completely abolished the plasma melatonin rhythm in the chicken indicating that circulating melatonin is derived exclusively from the pineal gland. Extrapineal melatonin, produced rhythmically in both the retina and the Harderian gland, did not contribute to circulating melatonin (Cogburn et al. 1987). Interestingly, pinealectomy did not eliminate the rhythm of circulating melatonin in either the Japanese quail or the pigeon (Fig. 6). In the pigeon, pinealectomy or enucleation reduced the amplitude of the rhythm of circulating melatonin but only the simultaneous removal of the pineal gland and the eyes eliminated the plasma melatonin rhythm (Ebihara et al. 1987). In the Japanese quail, pinealectomy or enucleation also reduced the amplitude of the melatonin rhythm in the blood but even after combined removal of both structures a low-amplitude rhythm of melatonin remained in the blood (Underwood and Siopes 1984). Similarly, neither pinealectomy nor blinding alone abolished behavioural rhythmicity in pigeons whereas birds that had been both pinealectomized and blinded showed no more circadian activity rhythms (Ebihara et al. 1987). Cyclic melatonin infusion into pinealectomized blinded pigeons restored activity rhythms (Oshima et al. 1989). These observations indicate that not only the pineal gland but also the eye is involved in circadian organization of the pigeon.

In the Japanese quail, it also requires the simultaneous elimination of the pineal gland and the eyes to eliminate most of the melatonin rhythm in the blood (Underwood and Siopes 1984) (Fig. 6). Interestingly, removal of the pineal gland had no effect on activity rhythms in either LD or DD but enucleation abolished activity rhythms in a majority of animals investigated in LD and in all animals in DD. Combined pinealectomy and enucleation resulted in arrhythmicity of all birds in both LD and DD. The dramatic effect of enucleation is obviously not due to the release of melatonin since enucleation was mimicked by optic nerve section, a treatment that disrupts neural signal transmission to the brain but leaves the plasma melatonin rhythm intact (Underwood et al. 1990; Underwood 1994).

In summary, melatonin originating from either the pineal gland or the pineal gland and the retina plays an important role in circadian pacemaking at the organismic level in most avian species. However, as exemplified by the above-mentioned results from the Japanese quail, melatonin-independent circadian rhythms may also occur. Additionally, there are several lines of evidence that a further circadian oscillator located in the hypothalamus is also involved in the regulation of circadian activity rhythms in birds. Pinealectomized enucleated pigeons are able to entrain to LD cycles and show residual rhythmicity for a while after transfer from LD cycles to dimLL (Ebihara et al. 1987). Following release into constant conditions, circadian activity rhythms of pinealectomized house sparrows usually do not disappear immediately but damp out over a series of transitional cycles (Gaston and Menaker 1968) suggesting that there remains at least one damped oscillator after removal of the pineal gland. Lesions of the suprachiasmatic hypothalamus resulted in severe disruptions of circadian activity rhythms in house sparrows (Takahashi and Menaker 1979a, b, 1982) (Fig. 4B). Thus, house sparrows bearing lesions of the suprachiasmatic hypothalamus were arrhythmic in spite of the presence of an intact pineal gland, and pinealectomized house sparrows could not maintain circadian activity rhythms although there was still an intact hypothalamic oscillator. Thus, neither the pineal gland nor the hypothalamic oscillator can maintain the rhythmicity of the bird when the other is removed. It is likely that the presence of a hypothalamic oscillator is a common feature of the avian circadian pacemaking system since similar effects of hypothalamic lesions were found in all further species investigated until now, including the Java sparrow, the pigeon, and the Japanese quail (Ebihara and Kawamura 1981; Simpson and Follett 1981; Ebihara et al. 1987).

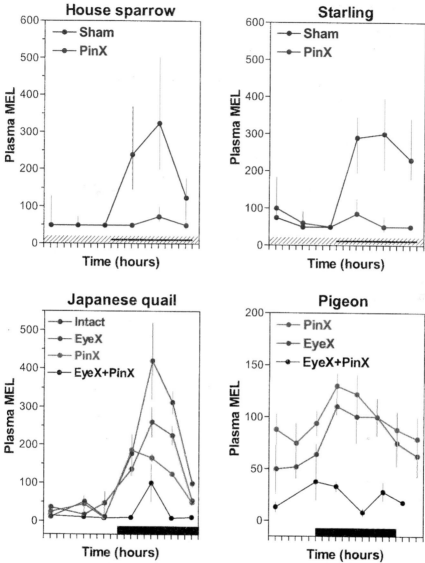

Fig. 6 Plasma melatonin rhythms in various bird species after removal of individual components of the circadian pacemaking system. Pinealectomy (PinX, drawn in green) abolished circadian rhythms of circulating melatonin in house sparrows (upper left graph) and European starlings (upper right graph) kept in DD as compared to sham-pinealectomized birds (drawn in blue). Continuous darkness is indicated by shaded bars, subjective nights are indicated by horizontal black lines. Data are shown as medians with quartiles (redrawn from Janik et al. 1992). In the Japanese quail (lower left graph), pinealectomy (PinX, green) or enucleation (EyeX, red) reduce plasma melatonin levels in animals entrained to LD 12:12 hours as compared to intact birds. A low-amplitude rhythm of circulating melatonin is still present after combined pinealectomy and enucleation (EyeX+PinX, black). Night is indicated by the horizontal black bar (redrawn from Underwood and Siopes 1984). In the pigeon (lower right graph), pinealectomy (PinX, green) or enucleation (EyeX, red) reduce the amplitude of the plasma melatonin rhythm which is abolished by combined pinealectomy and enucleation (EyeX+PinX, black). Night is indicated by the black horizontal bar (redrawn from Ebihara et al. 1987; control values of circulating melatonin not available). Plasma melatonin values are given in pg/ml in all graphs.

Plasticity and seasonality of the circadian pacemaking system

Besides the generating and maintaining daily rhythms, birds, like most organisms, have the ability to "read" the annual progression of certain environmental parameters and to utilize this information to regulate their annual biological rhythms. The most important zeitgeber for annual rhythms of physiology and behaviour, including gonadal activity, moult, and migratory behaviour, is the annual cycle of photoperiod (Gwinner 1989b) and it has been shown that the circadian system plays an important role in photoperiodic time measurement (Hamner 1963; Farner 1975; Farner and Gwinner 1980). However, the regulatory mechanisms for annual rhythms are less well understood than in mammals. In birds, as in mammals, seasonal changes in night length are reflected in corresponding changes in the duration of nocturnal melatonin production, as exemplified by studies in the Japanese quail (Underwood and Siopes 1985; Kumar and Follett 1993). In house sparrows, the duration of elevated melatonin has also been shown to parallel seasonal changes in night length to a certain degree, being shortest in summer and longest in winter (Brandstätter et al. 2001b). Additionally, there was a significant difference in peak amplitude levels of melatonin. Amplitudes were highest and similar in spring and summer, intermediate in autumn, and lowest in winter. In general, such changes in characteristic features of the nocturnal melatonin profile could be due to either differences in the amount of melatonin produced as a function of the underlying oscillator or as a consequence of a variety of exogenous environmental factors, such as intensity and duration of the light signal or temperature. Since circulating melatonin originates exclusively from the pineal gland in this species (Janik et al. 1992; Gwinner et al. 1997) and *in vitro* melatonin release of cultured pineal glands originating from summer and winter animals paralleled amplitude as well as duration differences found in the intact animals (Brandstätter et al. 2001b), these data suggest that the difference in these two parameters is indeed a function of the seasonal status of the pineal oscillator (Fig. 7).

Seasonal amplitude changes of circulating melatonin were also found in birds living at high latitudes. The amplitude of the 24-hour melatonin rhythm was drastically reduced around midsummer in penguins in Antarctica and Svalbard ptarmigans in the Arctic region (Miché et al. 1991; Cockrem 1991; Reierth et al. 1999). Furthermore, studies on captive garden warblers (Gwinner et al. 1993) have revealed that the amplitude of the plasma melatonin-rhythm becomes significantly reduced during the migratory seasons as compared to other times of the year. These changes in the amplitude of circulating melatonin (occurring spontaneously and concomitant with the circannual rhythm of migratory restlessness in birds kept in a constant 12-hour photoperiod) suggest an endogenous circannual rhythm of the amplitude of the circadian melatonin signal and, thus, point towards a tight coupling of certain mechanisms that regulate circadian and circannual rhythmicity.

The transduction of information about time into complex behaviour

The pineal melatonin rhythm is not only one of the components generating rhythms of behaviour in birds but, in addition, appears to be substantially involved in regulating complex circadian performance. This is exemplified by recent results indicating that photoperiodic information acquired by house sparrows and reflected in the pattern of melatonin release *in vivo* is retained for some time in isolated pineal glands cultured *in vitro* (Brandstätter et al. 2000). As in mammals, the duration of elevated melatonin in the plasma of sparrows has been found to be longer in birds kept under short-day (= long-night) conditions than in birds kept under long-day (= short-night) conditions (Fig. 8). In addition, the amplitude of the rhythm was low in long-night and high in short-night conditions. These different patterns persisted for at least two cycles in birds released from the two photoperiods into DD (Fig. 8). When pineal glands were explanted from house sparrows synchronized to distinct photoperiods,

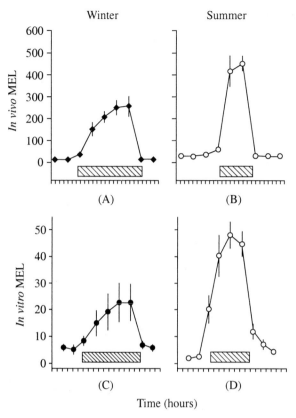

Fig. 7 **Seasonal variations of melatonin production in the house sparrow. 24-hour profiles of circulating melatonin (*in vivo* MEL in pg/ml plasma) in winter (A) and summer (B) as well as *in vitro* melatonin release from explanted cultured pineal glands (*in vitro* MEL in ng/ml culture medium) obtained from animals in winter (C) and summer (D) are characterized by elevated melatonin levels corresponding to the durations of darkness (shaded bars) and distinct amplitude levels (modified from Brandstätter et al. 2001b).**

in vitro melatonin release in DD was characterized by significant differences in the durations of elevated melatonin and significantly different night-time amplitude values (Brandstätter et al. 2000). These data indicate that one of the important components of the avian circadian pacemaking system, the pineal gland, can store and retain biologically meaningful information about time. Similar capacities have recently also been demonstrated in the mammalian SCN (Mrugala et al. 2000). The temporarily stored information might be used by birds to determine whether day length increases or decreases. It could also allow birds to buffer the effects of weather-dependent short-term variations in photoperiod which might impair precise photoperiodic time measurement. Moreover, the capacity of the pineal gland to retain day length-related information may be one of the mechanisms underlying photoperiodic aftereffects on activity patterns, i.e. the observation that day-active birds initially exposed to long photoperiods retain longer activity times when transferred to constant light or darkness than animals previously exposed to short photoperiods (e.g. Gwinner 1980). However, photoperiodic aftereffects on behaviour last longer than those on the patterns of melatonin release in isolated pineal glands

(Brandstätter et al. 2000). According to the hypothesis that the conservation of the duration of the daily activity time in constant conditions is the result of photoperiod establishing a set of phase relationships between two or more constituent oscillators of the circadian system, which are then able to retain this temporal pattern for a certain time after transfer to constant conditions (Pittendrigh and Daan 1976), the long-term retention of previous conditions *in vivo* could indeed result from the interaction of at least two components of the circadian system, such as the pineal gland and the hypothalamic circadian oscillator.

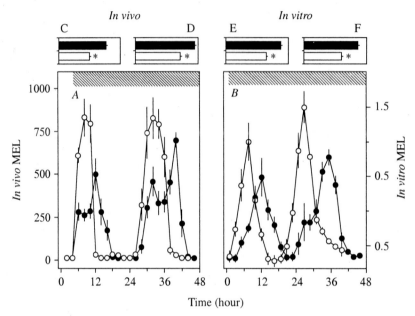

Fig. 8 *In vivo* and *in vitro* melatonin production in the house sparrow during constant conditions after previous synchronization to different photoperiods. A. *In vivo* melatonin profiles (*in vivo* MEL in pg/ml plasma) during 48 hours in constant conditions following synchronization with two distinct photoperiods. B. Profiles of relative melatonin release from isolated pineal glands (*in vitro* MEL) of the two photoperiodic groups during the first two cycles of culture in continuous darkness. Durations of elevated melatonin as well as amplitude values reflect the *in vivo* situation. Open circles: previous exposure to LD 16/8; black dots: previous exposure to LD 8/16. Hatched bars indicate constant darkness. Horizontal bars on top represent average durations of elevated melatonin during the first (C) and second (D) cycle *in vivo* as well as during the first (E) and second (F) cycle *in vitro*. * = P < 0.001 (modified from Brandstätter et al. 2000).

In summary, these recent results exemplify a high plasticity of the avian circadian pacemaking system as a whole as well as the capacity of an isolated component of the system to "internalize" information about the temporal organization of the environment and to transduce this information into a complex circadian performance at the organismic level.

Possible implications for ecology and behaviour

There is evidence that the pineal melatonin rhythm acts on at least one other oscillator within the circadian pacemaking system, which in turn feeds back onto the pineal gland (Cassone and Menaker 1984; Gwinner 1989a). Based on theoretical assumptions (Wever 1962, 1965), a change in the amplitude of the melatonin rhythm may have consequences for the persistence of circadian rhythmicity under constant conditions, the range of entrainment (i.e. the range of zeitgeber periods to which rhythmicity can be synchronized), and resynchronization times following phase shifts of exogenous zeitgebers. Some of these theoretical postulates have already been experimentally tested in birds in which the melatonin rhythm was abolished, either by surgically removing the pineal gland or by "masking" the endogenous rhythm of circulating melatonin with constant high levels of exogenous melatonin applied through implants. Under constant conditions, both pinealectomy and application of exogenous melatonin through silastic tubings resulted in loss of behavioural circadian rhythms in house sparrows (Turek et al. 1976; Gwinner et al. 1997). In synchronized conditions, the range of entrainment increased following pinealectomy, irrespective of whether the birds were exposed to a light or a food zeitgeber (Heigl and Gwinner 1995, 1999), and resynchronization times following phase shifts of food as well as light zeitgebers were reduced (Hau and Gwinner 1995; Abraham et al. 2000).

Experimental manipulations of the melatonin rhythm have also been performed in a more seasonal context in a variety of studies but, in contrast to the situation in mammals, neither pinealectomy nor treatment with exogenous melatonin had consistent influences on seasonal reproductive cycles in birds (Gwinner et al. 1981; Simpson et al. 1983; Follett et al. 1985). Nevertheless, several effects of such treatments have been described in birds, possibly indicating that certain aspects of the daily melatonin rhythm may play a role in the control of seasonal phenomena in this class of animals as well (Balasubramanian and Saxena 1972; Gwinner et al. 1981; McDonald 1982; Bentley et al. 1999; Guchhait and Haldar 1999). For example, doves have been shown to display greater nesting activity under long days than under short days. This difference was abolished by pinealectomy; the activity of long-day pinealectomized birds decreased to a level comparable to that of short-day birds (McDonald 1982). Interestingly, pinealectomy reduced plasma melatonin only to about 50% of that of sham-operated pigeons (Vakkuri et al. 1985; Ebihara et al. 1987; Foa and Menaker 1988). These observations suggest that an experimentally induced reduction of the amplitude of the endogenous melatonin rhythm may elicit a behavioural short-day response in a columbiform bird. Besides these results in pigeons, dawn and dusk injections of exogenous melatonin in Japanese quail transferred from short days to a stimulatory photoperiod did not prevent photoinduction but even resulted in a significantly stronger stimulation of the reproductive parameters investigated as compared to control animals (Juss et al. 1993). Since this treatment resulted not only in an extension of the duration of the nocturnal melatonin signal but also in an at least 10-fold increase in the amplitude of circulating melatonin, these data lead to the assumption that Japanese quail do not rely upon reading the duration of the melatonin signal as a hormonal transducer of day length but suggest a possible stimulatory effect of an increased melatonin amplitude. Recently Bentley et al. (1999) were able to demonstrate that exogenous melatonin attenuated the long-day-induced volumetric increase in song-control nuclei in the brain of the European starling regardless of the physiological state of the animals, suggesting a direct "seasonal" effect of melatonin in this passeriform species.

Under natural environmental conditions, changes in the amplitude of melatonin might also have several effects in a circadian as well as circannual context. A high-amplitude melatonin signal in spring and summer as it has been found in house sparrows (Brandstätter et al., 2001b) could be important for the stabilization of circadian rhythms and, thus, decrease susceptibility to photoperiodic noise that might negatively affect gonadal growth and reproductive performance. Seasonal changes in entrainment properties of circadian rhythms have indeed been described in canaries (Pohl 1994). The reduced melatonin amplitude found in house sparrows in winter could result in a reduction of the degree of self-sustainment of the pacemaking system as a whole. This, in turn, could facilitate adjustment to changing Zeitgeber conditions as the day length increases in late winter and early spring. The low-amplitude melatonin rhythms found around mid-summer in high-latitude birds (presumably caused by the high irradiance which varies only slightly in the course of a 24-hour day) might also represent a weakly self-sustained circadian pacemaking system with an increased range of entrainment. Such a mechanism might explain the fact that, in most cases, Arctic or sub-Arctic birds either exhibit 24-hour rhythms or are arrhythmic (Remmert 1965). Finally, the reduction of the amplitude of the melatonin rhythm might also be highly adaptive for migratory birds which actively change environmental zeitgeber conditions as a result of their own migratory flights. The advantage of this mechanism should be particularly pronounced in birds that migrate rapidly and along an east-westerly axis. Preliminary results of experiments with bramblings suggest that resynchronization times are in fact shorter during than before and after the migratory seasons (Pohl 2000).

Conclusions, speculations, and perspectives

Compared to mammals, birds are equipped with a rather complex and diverse circadian pacemaking system. It varies greatly between species and appears to be flexible within species and even within individuals. The latter is particularly obvious for the melatonin signal which, depending on season or latitude, is either characterized by a high- or a low-amplitude rhythm. The conclusion that these changes result in corresponding changes of the amplitude of the pacemaker as a whole is supported by a series of experimental results obtained in laboratory experiments (for recent reviews see Gwinner et al. 1994, 1997). Whether the variations in melatonin amplitude observed in free-living birds also reflect changes in overall pacemaker properties awaits rigorous experimental testing.

Ecologically, birds are extremely diverse and it is obvious that their immense radiation and their complex life history strategies are somehow reflected in the organization of their biological clocks. In migratory birds, the endogenous clock has to enable the animal to read the progression of the year very exactly and to temporally organize the animal so that it migrates on time. On the other hand, synchronization to changing environmental zeitgeber conditions has to be facilitated to prevent animals from being "jet-lagged" when migrating. Non-migratory birds may either experience dramatic changes of exogenous zeitgebers, such as day length and temperature, in the course of a year (in temperate zones or at high latitudes) or live in a more stable photic and thermal environment (in tropical regions) but, nevertheless, often show comparable daily and seasonal rhythms.

In summary, several common properties are detectable in the circadian pacemaking system of birds, such as the presence of a multiple photic input (from three distinct sources) and a certain functional compartmentalization (exemplified by the presence of at least three autonomous circadian oscillators). It is tempting to speculate that a system composed of several interacting components with distinct physiological properties, such as the retina, the pineal gland, and the hypothalamic oscillator, allows a higher degree of adaptation to changing environmental situations than a highly centralized system. However, besides flexibility, a system used for time measurement also needs stability and a high degree of self-sustainment to be able to regulate behaviour and to counteract environmental disturbances or zeitgeber "noise". Certain changes in pacemaker properties appear necessary to solve this problem and they are partly mediated by changes in specific features of the pineal melatonin rhythm. It is proposed that this variability is functionally important, for instance, either for increasing or decreasing the responsiveness to changes of the photic environment in temperate zone birds (Brandstätter et al. 2001b), for enabling high-latitude birds to remain synchronized with the low-amplitude zeitgeber conditions in midsummer, or for allowing birds quickly to adjust their behavioural rhythms to changing environmental conditions during migration (Gwinner and Brandstätter

2001). An example that changing environmental zeitgeber conditions are indeed "internalized" by the circadian pacemaking system is provided by photoperiodic aftereffects on behaviour and it has recently been shown that the circadian oscillator in the pineal gland is capable of translating relevant information about time into specific features of the melatonin signal and that this information can be retained after isolation from the animal and when shielded from exogenous zeitgebers.

We are far away from fully understanding all the demands of the environment on biological clocks as well as all aspects of their functional significance for behaviour and ecology. However, biological clocks are systems which have the capacity to detect and measure cyclic variations of the environment and to bring genetically determined endogenous rhythms into harmony with the outside world. It is fascinating to observe how similar and, at the same time, how diverse biological clocks are, from unicellular organisms to humans, and from house sparrows to pigeons.

Acknowledgements

I cordially thank Ebo Gwinner for his continuous support and innumerable fruitful discussions about birds and clocks since I came to Andechs. Thanks are also due to Gabi Wagner for critically reading and commenting on the manuscript. My dearest thanks belong to my wife Annette and my daughter Laura Katharina for all their love and patience.

References

Abraham, U., Gwinner, E., Van't Hof, T.J. (2000) Exogenous melatonin reduces the resynchronization time after phase shifts of a non-photic zeitgeber in the house sparrow (*Passer domesticus*). J. Biol. Rhythms **15**: 48–56.

Aschoff, J. (1981) A survey on biological rhythms. In: Aschoff, J. (ed.) Biological rhythms. Handbook of Behavioural Neurobiology, Vol. 4. Plenum Press, New York, pp. 3–10.

Balasubramanian, K.S., Saxena, R.N. (1972) Effect of pinealectomy and photoperiodism in the reproduction of Indian weaver birds, *Ploceus phillippinus*. J. Exp. Zool. **185**: 333–340.

Bentley, G.E., Van't Hof, T.J., Ball, G.F. (1999) Seasonal neuroplasticity in the songbird telencephalon: A role for melatonin. Proc. Natl. Acad. Sci. U.S.A. **96**: 4674–4679.

Bernard, M., Guerlotte, J., Greve, P., Grechez-Cassiau, A., Iuvone, M.P., Zatz, M., Chong, N.W., Klein, D.C., Voisin, P. (1999) Melatonin synthesis pathway: circadian regulation of the genes encoding the key enzymes in the chicken pineal gland and retina. Reprod. Nutr. Develop. **39**: 325–334.

Bernard, M., Iuvone, P.M., Cassone, V.M., Roseboom, P.H., Coon, S.L., Klein, D.C. (1997) Avian melatonin synthesis: photic and circadian regulation of serotonin N-acetyltransferase mRNA in the chicken pineal gland and retina. J. Neurochem. **68**: 213–224.

Besharse, J.C., Iuvone, P.M. (1983) Circadian clock in *Xenopus* eye controlling retinal serotonin N-acetyltransferase. Nature **305**: 133–135.

Binkley, S., Kluth, E., Menaker, M. (1971) Pineal function in sparrows: Circadian rhythms and body temperature. Science **174**: 311–314.

Binkley, S., Reilly, K.B., Hryshchyshyn, M. (1980) N-acetyltransferase in the chick retina. **1**. Circadian rhythms controlled by environmental lighting are similar to those in the pineal gland. J. Comp. Physiol. **139**: 103–108.

Block, G.D., McMahon, D.G. (1984) Cellular analysis of the *Bulla* ocular circadian pacemaker system. III. Localization of the circadian pacemaker. J. Comp. Physiol. **155**: 387–395.

Brandstätter, R., Abraham, U., Albrecht, U. (2001a) Initial demonstration of rhythmic per gene expression in the hypothalamus of a non-mammalian vertebrate, the house sparrow. Neuro Report **12**: 1167–1170.

Brandstätter, R., Kumar, V., Abraham, U., Gwinner, E. (2000) Photoperiodic information acquired *in vivo* is retained *in vitro* by a circadian oscillator, the avian pineal gland. Proc. Natl. Acad. Sci. USA **97**: 12324–12328.

Brandstätter, R., Kumar, V., Van't Hof, T.J., Gwinner, E. (2001b) Seasonal variation of *in vivo* and *in vitro* melatonin production in a passeriform bird, the house sparrow (*Passer domesticus*). J. Pineal Res. **31**: 120–126.

Cassone, V.M., Moore, R.Y. (1987) Retinohypothalamic projection and suprachiasmatic nucleus of the house sparrow, *Passer domesticus*. J. Comp. Neurol. **266**: 171–182.

Cassone, V.M. (1990) Melatonin: Time in a bottle. In: Mulligan, S.R. (ed) Oxford Reviews of Reproductive Biology, Vol. 12. Oxford University Press, Oxford, pp. 319–367.

Cassone, V.M., Menaker, M. (1984) Is the avian circadian system a neuroendocrine loop? J. Exp. Zool. **232**: 539–549.

Chabot, C.C., Menaker, M. (1992) Effects of physiological cycles of infused melatonin on circadian rhythmicity in pigeons J. Comp.Physiol. A **170**: 615–622.

Chong, N.W., Cassone, V.M., Bernard, M., Klein, D.C., Iuvone, P.M. (1998) Circadian expression of tryptophan hydroxylase mRNA in the chicken retina. Mol. Brain Res. **61**: 243–250.

Cockrem, J.F. (1991) Plasma melatonin in the Adelie penguin (*Pygoscelis adeliae*) under continuous daylight in Antarctica. J. Pineal Res. **10**: 2–8.

Cogburn, L.A., Wilson-Placentra, S., Letcher, L.R. (1987) Influence of pinealectomy on plasma and extrapineal melatonin rhythms in young chickens (*Gallus domesticus*) Gen. Comp. Endocrinol. **68**: 343–356.

Crosby, E.C., Woodburne, R.T. (1940) The comparative anatomy of the preoptic area and the *hypothalamus*. Res. Pub. Assoc. Res. Nerv. Ment. Dis. **20**: 52–169.

Deguchi, T. (1979) A circadian oscillator in cultured cells of chicken pineal gland. Nature **282**: 94–96.

Dunlap, J.C. (1999) Molecular bases for circadian clocks. Cell **96**: 271–290.

Ebihara, S., Kawamura, H. (1981) The role of the pineal organ and the suprachiasmatic nucleus in the control of circadian locomotor rhythms in the Java sparrow, *Padda oryizivora*. J. Comp. Physiol. **141**: 207–214.

Ebihara, S., Oshima, I., Yamada, H., Goto, M., Sato, K. (1987) Circadian organization in the pigeon. In Hoiroshige T., Honma, K. (eds.) Comparative aspects of circadian clocks. Hokkaido University Press, Sapporo, pp. 84–94.

Eskin, A. (1979) Circadian system of the *Aplysia* eye: Properties of the pacemaker and mechanisms of its entrainment. Fed. Proc. **38**: 2573–2579.

Farner, D.S. (1975) Photoperiodic controls in the secretion of gonadotrophins in birds. Am. Zool. **15** (suppl. **1**): 117–135.

Farner, D.S., Gwinner, E. (1980) Photoperiodicity, circannual and reproductive cycles. In: Epple A, Stetson, M.H. (eds.) Avian Endocrinology. Academic Press, New York, pp. 331–366.

Foa, A., Menaker, M. (1988) Blood melatonin rhythms in the pigeon. J. Comp. Physiol. A **164**: 25–30.

Follett, B.K., Foster, R.G., Nicholls, T.J. (1985) Photoperiodism in birds. Ciba Found. Symp. **117**: 93–105.

Foster, R.G., Soni, B.G. (1998) Extraretinal photoreceptors and their regulation of temporal physiology. Rev. Reprod. **3**: 145–150.

Gaston, S. (1971) The influence of the pineal organ on the circadian activity rhythm in birds. In: Menaker, M. (ed.) Biochronometry. National Academy of Sciences, Washington, D.C., pp 541.

Gaston, S., Menaker, M. (1968) Pineal function: The biological clock in the sparrow? Science **160**: 1125–1127.

Guchhait, P., Haldar, C. (1999) Regulation of pineal gland and gonadal functions of a tropical nocturnal bird, Indian spotted owlet, *Athene brama*, following different 5-methoxyindoles treatments. Biogen. Amin. **15**: 263–273.

Gwinner, E. (1978) Effects of pinealectomy on circadian locomotor activity rhythms in European starlings, *Sturnus vulgaris*. J. Comp. Physiol. **126**: 123.

Gwinner, E. (1980) Relationship between circadian activity patterns and gonadal function: evidence for internal coincidence? In Proceedings of the 17th Ornithological Congress, Berlin Germany, Verlag der deutschen Ornithologischen Gesellschaft, Berlin, pp. 409–416.

Gwinner, E. (1989a) Melatonin in the circadian system of birds: model of internal resonance. In: Hiroshige, T., Honma, K. (eds.) Circadian clocks and ecology. Hokkaido University Press, Sapporo, pp. 27–53.

Gwinner, E. (1989b) Photoperiod as a modifying and limiting factor in the expression of avian circannual rhythms. J. Biol. Rhythms **4**: 237–250.

Gwinner, E., Brandstätter, R. (2001) Complex bird clocks. Phil. Trans. B (in press).

Gwinner, E., Hau, M. (2000) The pineal gland, circadian rhythms, and photoperiodism. In Whittew, G.C. (ed.) Sturkie's Avian Physiology, fifth edition. Academic Press, New York, pp. 557–568.

Gwinner, E., Hau, M., Heigl, S. (1994) Phasic and tonic effects of melatonin on avian circadian systems. In: Hiroshige T., Honma K. (eds.) Circadian clocks and evolution. Hokkaido University Press, Sapporo, pp. 127–137.

Gwinner, E., Schwabl-Benzinger, I., Schwabl, H., Dittami, J. (1993) Twenty-four hour melatonin profiles in a nocturnally migrating bird during and between migratory seasons. Gen. Comp. Endocrinol. **90**: 119–124.

Gwinner, E., Hau, M., Heigl, S. (1997) Melatonin: Generation and Modulation of Avian Circadian Rhythms. Brain Res. Bull. **44**: 439–444.

Gwinner, E., Wozniak, J., Dittami, J. (1981) The role of the pineal organ in the control of annual rhythms in birds. In Oksche A, Pevet P. (eds.) The Pineal Organ: Photobiology-Biochronometry-Endocrinology. Elsevier/North-Holland Biomedical Press, Amsterdam, pp. 99–121.

Hamm, H.E., Menaker, M. (1980) Retinal rhythms in chicks: circadian variation in melantonin and serotonin N-acetyltransferase activity. Proc. Natl. Acad. Sci. USA **77**: 4998–5002.

Hamner, W.M. (1963) Diurnal rhythm and photoperiodism in testicular recrudescence of the house finch. Science **142**: 1294–1295.

Hartwig, H.G. (1974) Electron microscopic evidence for a retinohypothalamic projection to the suprachiasmatic nucleus of *Passer domesticus*. Cell Tiss. Res. **153**: 89–99.

Hastings, M., Maywood, E.S. (2000) Circadian clocks in the mammalian brain. Bioessays **22**: 23–31.

Hau, M., Gwinner, E. (1995) Continuous melatonin administration accellerates resynchronization following phase shifts of a light-dark cycle. Physiol. Behav. **58**: 89–95.

Heigl, S., Gwinner, E. (1994) Periodic melatonin in the drinking water synchronizes circadian rhythms in sparrows. Naturwissenschaften **81**: 83–85.

Heigl, S., Gwinner, E. (1995) Synchronization of circadian rhythms of house sparrows by oral melatonin: effects of changing period. J. Biol. Rhythms **10**: 225–233.

Heigl, S., Gwinner, E. (1999) Periodic food availability synchronizes locomotor and feeding activity in pinealectomized house sparrows. Zoology **102**: 1–9.

Janik, D., Dittami, J. and Gwinner, E. (1992) The effect of pinealectomy on circadian plasma melatonin levels in house sparrows and European starlings. J Biol. Rhythms **7**: 277–286.

Juss, T.S., Meddle, S.L., Servant, R.S., King, V.M. (1993) Melatonin and photoperiodic time measurement in Japanese quail (*Coturnix coturnix japonica*). Proc. R. Soc. Lond. B **254**: 21–28.

King, V.M., Follett, B.K. (1997) C-fos expression in the putative avian suprachiasmatic nucleus. J. Comp. Physiol. A **180**: 541–551.

Klein, D.C., Moore, R.Y. and Reppert, S.M. (1991) Suprachiasmatic nucleus: The mind's clock, Oxford University Press, Oxford.

Kojima, D., Fukada, Y. (1999) Non-visual photoreception by a variety of vertebrate opsins. Novartis Found. Symp. **224**: 265–279.

Kumar V, Follett BK (1993) The circadian nature of melatonin secretion in Japanese quail (*Coturnix coturnix japonica*). J. Pineal Res. **14**: 192–200.

Kumar, V., Gwinner, E., Van't Hof, T.J. (2000) Circadian rhythms of melatonin in the European starling exposed to different lighting conditions: Relationship with locomotor and feeding rhythms. J. Comp. Physiol. A **186**: 205–215.

MacBride, S.E. (1973) Pineal biochemical rhythms of the chicken (*Gallus domesticus*): Light cycle and locomotor activity correlates. PhD-Thesis, University of Pittsburgh, Pittsburgh.

McDonald, P.A. (1982) Influence of pinealectomy and photoperiod on courtship and nest-building in male doves. Physiol. Behav. **29**: 813–818.

McMillan, J.P. (1972) Pinealectomy abolishes the circadian rhythm of migratory restlessness. J. Comp. Physiol. **79**: 105.

Menaker, M., Moreira, L.F., Tosini, G. (1997) Evolution of circadian organization in vertebrates. Braz. J. Med. Biol. Res. **30**: 305–313.

Miche, F., Vivien-Roels, B., Pevet, P., Spehner, C., Robin, J.P., Le Maho, Y. (1991) Daily patterns of melatonin secretion in an antarctic bird, the emperor penguin, *Aptenodytes forsteri*: Seasonal variations, effect of constant illumination and of administration of isoproterenol or propranolol. Gen. Comp. Endocrinol. **84**: 249–263.

Moore, R.Y., Silver, R. (1998) Suprachiasmatic nucleus organization. Chronobiol. Int. **15**: 475–487.

Mrugala, M., Zlomanczuk, P., Jagota, A., Schwartz, W.J. (2000) Rhythmic multiunit neural activity in slices of hamster suprachiasmatic nucleus reflect prior photoperiod. Am. J. Physiol. **278**: 987–994.

Murakami, M., Nakamura, H., Nishi, R., Marumoto, N., Nasu, T. (1994) Comparison of circadian oscillation of melatonin release in pineal cells of house sparrow, pigeon and Japanese quail, using cell perfusing systems. Brain Res. **651**: 209–214.

Nakahara, K., Murakami, N., Nasu, T., Kuroda, H., Murakami, T. (1997) Individual pineal cells in chick possess photoreceptive, circadian clock and melatonin-synthezising capacities *in vitro*. Brain Res. **774**, 242–245.

Norgren, R.B., Silver, R. (1989) Retinohypothalamic projections and the suprachiasmatic nucleus in birds. Brain Behav. Evol. **34**: 73–83.

Norgren, R.B., Silver, R. (1990) Distribution of vasoactive intestinal peptide-like and neurophysin-like immunoreactive neurons and acetylcholinesterase staining in the ring dove hypothalamus with emphasis on the question of an avian suprachiasmatic nucleus. Cell Tiss. Res. **259**: 331–339.

Oksche, A., Farner, D.S. (1974) Neurohistological studies of the hypothalamo-hypophysial system of *Zonotrichia leucophrys gambelii* (Aves, Passeriformes) with special attention to its role in the control of reproduction. Adv. Anat. Embryol. Cell. Biol. **48**: 1–136.

Oshima, I., Yamada, H., Sato, K., Ebihara, S. (1989) The role of melatonin in the circadian system of the pigeon, *Columbia livia*. In: Hiroshige, T., Honma, K. (eds.) Circadian Clocks and Ecology. Hokkaido University Press, Sapporo, pp. 118–126.

Panzica, G.C. (1985) Vasotocin-immunoreactive elements and neuronal typology in the suprachiasmatic nucleus of the chicken and Japanese quail. Cell Tiss. Res. **242**: 371–376.

Pittendrigh, C.S. (1993) Temporal organization: reflections of a Darwinian clock-watcher. Ann. Rev. Physiol. **55**: 17–54.

Pittendrigh, C.S., Daan, S. (1976) A functional analysis of circadian pacemakers in nocturnal rodents: V. Pacemaker structure: a clock for all seasons. J. Comp. Physiol. **106**: 333–355.

Pohl, H. (1994) Entrainment properties of the circadian system changing with reproductive state and moult in the canary. Physiol. Behav. **55**: 803–810.

Pohl, H. (2000) Circadian control of migratory restlessness and the effects of exogenous melatonin in the brambling (*Fringilla montifringilla*). Chronobiol. Int. **17**: 471–488.

Reierth, E., Van't Hof, T.J., Stokkan, K.A. (1999) Seasonal and daily variations in plasma melatonin in the high-arctic Svalbard ptarmigan (*Lagopus mutus hyperboreus*). J. Biol. Rhythms **14**: 314–319.

Remmert, H. (1965) Über den Tagesrhythmus arktischer Tiere. Z. Morphol. Ökol. Tiere **55**: 142–160.

Robertson, L.M., Takahashi, J.S. (1988a) Circadian clock in cell culture: I. Oscillation of melatonin release from dissociated chick pineal cells in flow-through microcarrier culture. J. Neurosci. **8**:12–21.

Robertson, L.M., Takahashi, J.S. (1988b) Circadian clock in cell culture: II. *In vitro* photic entrainment of melatonin oscillation from dissociated chick pineal cells. J. Neurosci. **8**: 22–30.

Rosolowska-Huszcz, D., Thaela, M.J., Jagura, M., Stepien, D., Skwarlo-Sonta, K. (1991) Pineal influence on the diurnal rhythm of nonspecific immunity indices in chickens. J.Pineal Res. **10**: 190–195.

Silver, R., Witkovsky, P., Horvath, P., Alones, V., Barnstable, C.J., Lehman, M.N. (1988) Coexpression of opsin- and VIP-like-immunoreactivity in CSF-contacting neurons of the avian brain. Cell Tiss. Res. **253**: 189–198.

Simpson, S.M., Follett, B.K. (1981) Pineal and hypothalamic pacemakers: Their role in regulating circadian rhythmicity in Japanese quail. J. Comp. Physiol. A **144**: 381–389.

Simpson, S.M., Urbanski, H.F., Robinson, J.E. (1983) The pineal gland and the photoperiodic control of luteinizing hormone secretion in intact and castrated Japanese quail. J. Endocrinol. **99**: 281–287.

Takahashi, J.S. (1981) Neural and endocrine regulation of avian circadian systems. Dissertation Abstracts Int. **42–05**: 1724.

Takahashi, J.S., Hamm, H., Menaker, M. (1980) Circadian rhythms of melatonin release from individual superfused chicken pineal glands in vitro. Proc. Natl. Acad. Sci. USA **77**: 2319–2322.

Takahashi, J.S., Menaker, M. (1979a) Brain mechanisms in avian circadian systems. In: Suda, M., Hayaishi, O., Hachiro, N. (eds.) Biological rhythms and their central mechanism. Elsevier, Amsterdam, pp. 95–109.

Takahashi, J.S., Menaker, M. (1979b) Physiology of avian circadian pacemakers. Fed. Proc. **38**: 2583–2588.

Takahashi, J.S., Menaker, M. (1982) Role of the suprachiasmatic nuclei in the circadian system of the house sparrow, *Passer domesticus*. J Neurosci **2**: 815–828.

Takahashi, J.S., Menaker, M. (1984) Multiple redundant circadian oscillators within the isolated avian pineal gland. J. Comp. Physiol. **154**: 435–440.

Thomas, K.B., Iuvone, P.M. (1991) Circadian rhythm of tryptophan hydroxylase activity in chicken retina. Cell. Mol. Neurobiol. **11**: 511–527.

Turek, F.W., McMillan, J.P., Menaker, M. (1976) Melatonin: effects on the circadian locomotor rhythm of sparrows. Science **194**: 1441–1443.

Underwood, H. (1989) The pineal gland and melatonin: regulators of circadian function in lower vertebrates. Experientia **45**: 914–922.

Underwood, H. (1994) The circadian rhythm of thermoregulation in Japanese quail. I. Role of the eyes and pineal. J. Comp. Physiol. A **175**: 639–653.

Underwood, H., Barrett, R.K., Siopes, T. (1990) Melatonin does not link the eyes to the rest of the circadian system in quail: a neural pathway is involved. J. Biol. Rhythms **5**: 349–361.

Underwood, H., Siopes, T. (1984) Circadian organization in Japanese Quail. J. Exp. Zool. **232**: 557–566.

Underwood, H., Siopes, T. (1985) Melatonin rhythms in quail: regulation by photoperiod and circadian pacemakers. J. Pin. Res. **2**: 133–143.

Vakkuri, O., Rintamaki, H., Leppaluoto, J. (1985) Plasma and tissue concentrations of melatonin after midnight light exposure and pinealectomy in the pigeon. J. Endocrinol. **105**: 263–268.

Wever, R. (1962) Zum Mechanismus der biologischen 24-Studen-Periodik. Kybernetik **1**: 139–154.

Wever, R. (1965) Pendulum versus relaxation oscillation. In Aschoff, J. (ed.) Circadian Clocks. North-Holland Publ. Inc, Amsterdam, pp. 74–83.

Yoshimura, T., Suzuki, Y., Makino, E., Suzuki, T., Kuroiwa, A., Matsuda, Y., Namikawa, T., Ebihara, S. (2000) Molecular analysis of avian circadian clock genes. Mol. Brain Res. **78**: 207–215.

Zimmerman, N.H., Menaker, M. (1975) Neural connections of sparrow pineal: role in circadian control of activity. Science **190**: 477–479.

Zimmerman, N.H., Menaker, M. (1979) The pineal gland: a pacemaker within the circadian system of the house sparrow. Proc. Natl. Acad. Sci. USA **76**: 999–1003.

14. Neurochemical Aspects of the Entrainment of the Mammalian Suprachiasmatic Circadian Pacemaker

H.D. Piggins*, A.N. Coogan, D.J. Cutler and H.E. Reed

School of Biological Sciences, 3.614 Stopford Building, Oxford Road, University of Manchester, Manchester, UK M13 9PT

The localisation of the mammalian circadian pacemaker to the suprachiasmatic nucleus of the hypothalamus (SCN) some 27 years ago has lead to an explosion of interest in this area of the brain. Today, research is aimed at determining how *zeitgeber* information is relayed to the SCN circadian pacemaker, including the identification of the neural pathways via which this information is communicated, the neurochemicals contained in these pathways, the receptors mediating the actions of these neurochemicals, second messenger factors implicated in mediating the phase-resetting effects of neurochemicals, and finally the sequence of intracellular events that lead to changes in clock gene expression.

Through tract tracing methodologies, the pathways between the SCN and other structures in the brain have been identified, while immunohistochemical techniques have resolved some of the neurochemicals contained in these projections. The use of in situ hybridisation has enabled researchers to identify the neurochemicals, and their associated receptors, that are synthesised in SCN neurones. These techniques have been used to elucidate neurochemical subdivisions within the SCN, and to determine if these transcripts fluctuate over the circadian cycle. In this chapter, we provide a brief overview of some of the significant findings concerning the neurochemical basis for the entrainment of the mammalian circadian pacemaker.

Introduction

In mammals, the neural pathways conveying photic information to the SCN are well characterised. These include a direct, monosynaptic retinal projection to the SCN, called the retinohypothalamic tract (RHT), and a polysynaptic pathway via retinally innervated cells of the intergeniculate leaflet (IGL) and ventral lateral geniculate nucleus (LGN) of the thalamus, which project to the SCN as well as to the contralateral LGN complex (Figure 1).

The RHT originates from a subset of retinal ganglion cells of the gamma or Perry type III class, whose axons primarily innervate the ventral division of the rodent SCN (Figure 2A), as well as other hypothalamic structures. In the rat, the RHT forms mainly Gray type I synapses with SCN cells, although an estimated 25% are Gray type II synapses; Gray type I and type II synapses are presumed to be excitatory and inhibitory, respectively. Consistent with this neuroanatomical data, *in vivo* electrophysiological studies have found that retinal illumination evokes increases in the electrical activity of rat SCN neurones, although a small proportion of responsive cells are suppressed. Considerable evidence indicates that the RHT utilises glutamate as the main transmitter (Ebling 1996). Glutamate

---

*E-mail: hugh.piggins@man.ac.uk

immunoreactivity (-ir) is found in RHT terminals in the SCN, and glutamate-immunopositive vesicles have been detected, at the electron microscopic level, in the thickened synaptic apposition of synaptic bulbs contacting neurones and astrocytes of the rodent SCN. Further, stimulation of the optic nerve evokes the release of tritiated aspartate and glutamate into the SCN. *In vitro* neurophysiological studies have shown that such stimulation excites retinorecipient SCN neurones, while N-methyl-D-aspartate (NMDA) and non-NMDA ionotropic glutamate receptor antagonists can block these excitatory effects.

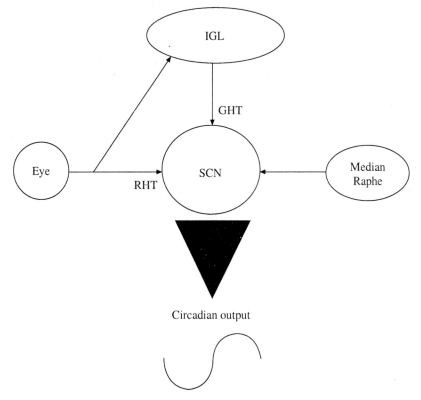

Fig. 1   Main neural inputs to the mammalian SCN. IGL = intergeniculate leaflet, GHT = geniculohypothalamic tract, RHT = retinohypothalamic tract. See text for further details.

Subsequent *in vitro* studies have determined that a circadian pattern in the spontaneous discharge rate of rodent SCN neurones persists when maintained in a brain slice *in vitro*. The peak in this firing rate rhythm is a well-established indicator of the phase of the SCN circadian pacemaker. Investigations by Ding et al. (1994) showed that glutamate and ionotropic glutamate receptor agonists phase-dependently phase-reset the time of peak of this rhythm in a manner resembling the effects of light pulses on rodent activity rhythms. These *in vitro* phase shifts were blocked by selective ionotropic glutamate receptor antagonists, demonstrating for the first time that glutamate could mimic photic phase-shifting.

The contribution of these ionotropic glutamate receptors to mediating the effects of light on the rodent circadian pacemaker *in situ* was shown by a series of *in vivo* studies which investigated the well-characterised photic induction of immediate early genes (IEGs) such as *c-fos*, and its associated protein Fos, as well as phase-shifts in wheel-running behaviour. In these pioneering studies, Abe and

Rusak (1994) found that peripherally injected MK-801, a non-competitive NMDA receptor antagonist, blocked the photic induction of Fos-ir in the ventral portion of the hamster SCN and attenuated the phase-resetting actions of light on hamster wheel-running rhythms. Subsequent studies, in which NMDA and non-NMDA receptor antagonists were microinjected into the SCN region, showed that the effects of these agents on gene expression and wheel-running rhythm phase occurred at the level of the SCN. Further, Mintz and colleagues (1999) have recently demonstrated that microinjection of the ionotropic glutamate receptor agonist, NMDA, into the SCN region phase-dependently phase-reset hamster behavioural rhythms in a manner resembling the phase-shifting actions of light. The phase-shifting actions of NMDA were blocked by pretreatment with an NMDA receptor antagonist. When considered together, these findings firmly implicate glutamate as the main neurotransmitter that conveys photic information from the RHT to the SCN, and that NMDA and non-NMDA classes of glutamate receptors mediate these effects.

There are, however, exceptions to these general findings. For example, the induction of Fos-ir in the dorsal region of the hamster SCN by light pulses is resistant to antagonists of both types of ionotropic glutamate receptors. Further, glutamate receptor blockade can be overcome by high intensity light pulses. The recent neuroanatomical and physiological demonstrations of metabotropic glutamate receptor subtypes in the rodent SCN raise the possibility that other glutamatergic processes play a role in photic entrainment. Moreover, these anomalies also indicate that other neurotransmitters, such as substance P and pituitary adenylate cyclase activating polypeptide, and their associated receptors, are involved in mediating RHT input to the SCN circadian pacemaker.

Other RHT neurotransmitters

Substance P
In addition to glutamate, a number of other neurotransmitters have been localised to the rodent RHT. Immunohistochemical studies have suggested that substance P (SP) is found in the RHT of rodents (Figure 2B), non-human primates, and humans, while SP-immunopositive retinal ganglion cells have been found in rabbits. Some studies have indicated that removal of the eyes (optic enucleation) abolishes most SP-ir terminals in the rat SCN. However, other investigators have failed to replicate these findings and instead have found no significant changes in SP-ir in the rat SCN following optic enucleation. These apparent discrepancies are presumably attributable to differences in immunohistochemical procedures or the duration of post-enucleation recovery time. Functional studies conducted *in vitro* on brain slice preparations containing the SCN, have shown that SP both activates rodent SCN neurones, and phase-shifts the cellular activity and metabolic rhythms of rat SCN cells in a temporal pattern resembling the effects of light on rodent behavioral rhythms (Shibata et al. 1992). These effects appear to be mediated via a SP receptor since they were blocked by the broad spectrum neurokinin receptor antagonist, spantide. Consistent with these data, it has also been found that intraventricularly injected spantide attenuated the effects of light on IEG expression in the hamster SCN. From the results of these studies, it is speculated that SP release from the RHT participates in conveying the effects of light to the rodent circadian pacemaker contained in the SCN.

However, there are a number of discrepant findings that collectively argue against such an interpretation. In the hamster, SP activates a small proportion of SCN cells *in vitro*, but has little phase-resetting effects when microinjected into the SCN region of hamsters free-running in constant conditions (Piggins et al. 1995b; Piggins and Rusak 1997). Further, immunohistochemical investigations have revealed that the hamster SCN contains very few SP-ir terminals, particularly in the ventral, retinally innervated region of the SCN. Moreover, examinations of the distribution for one of the

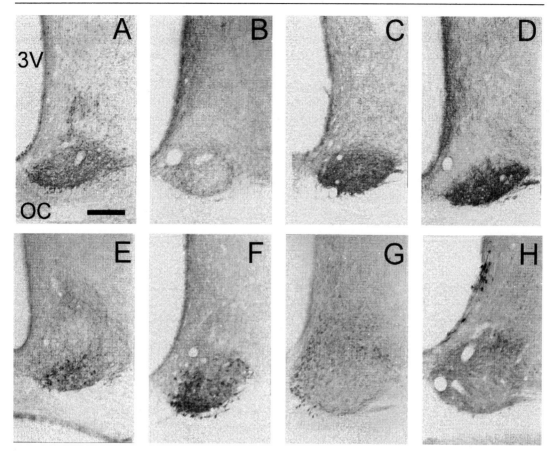

Fig. 2 **Neural inputs and intrinsic neurochemicals of the rat SCN. Photomicrographs show immunohistochemical markers in coronal sections through the rat SCN. (A) The retinohypothalamic tract input as demonstrated by the presence of the anterograde tracer cholera toxin subunit B, following injection of this tracer into the eye. (B) Substance P-ir terminals in the ventral region of the SCN. (C) Serotonin-ir terminals in the ventral SCN. (D) Neuropeptide Y-ir terminals in the ventral SCN. (E) Gastrin-releasing peptide-ir cells and fibres in the ventral and dorsal SCN, respectively. (F) Vasoactive intestinal polypeptide-ir cells and fibres in the ventral and dorsal SCN. (G) AVP cells in the medial SCN. (H) Lightly stained somatostatin-ir cells in dorsal SCN. Note the heavily labelled somatostatin-ir cells in the periventricular zone, dorsal to the SCN. 3V = third ventricle, OC = optic chiasm. Calibration bar = 100 μm.**

receptors through which SP is thought to influence neural activity, the neurokinin-1 (NK-1) receptor, have shown that in the hamster, NK-1-ir cells and cellular processes are found primarily along the dorsal and lateral borders of the SCN, with very few NK-1-ir dendrites traversing the retinally innervated core. Results from our laboratory show a similar restricted distribution for NK-1-ir and NK-1 mRNA in the rat SCN (unpublished observations). These data suggest that SP is unlikely to function as an RHT neurotransmitter in the SCN of the hamster, while its role in the rat SCN remains to be clarified. Further, the receptor mediating the cellular actions of SP on rodent SCN neurones is unlikely to be NK-1 since it is localised to regions of the SCN that do not receive a dense retinal input. The identification of the receptor mediating the actions of SP at the level of the SCN has been confounded by the use of compounds such as spantide that have poor selectivity for the NK-1 receptor. Further studies with agonists and antagonists that are selective for different SP receptors are required to clarify the role of SP in circadian rhythm processes.

Pituitary adenylate cyclase activating polypeptide

Another peptide that has been localised to the rat RHT is pituitary adenylate cyclase activating polypeptide (PACAP). Immunohistochemical studies have shown PACAP-ir retinal ganglion cells that project to the SCN, and co-localisation of PACAP and glutamate in fibres and terminals of the RHT projection to the SCN (Hannibal et al. 2000). *In situ* hybridisation studies demonstrate that mRNA for one of the receptors for PACAP, PAC_1, is heavily expressed in the rat SCN with an apparent circadian variation in its expression, such that higher levels are seen during the middle parts of the subjective day and late portions of the subjective night (Cagampang et al. 1998a). Although these anatomical data point to the potential involvement of PACAP in the regulation of SCN function, it has been reported that microinjection of PACAP into the SCN region of hamsters free-running in constant conditions, during the middle phases of the subjective day, does not phase-advance behavioural rhythms.

Despite these negative findings *in vivo*, PACAP has been shown to phase-reset the electrical activity rhythm of rodent SCN neurones maintained in brain slices *in vitro*, yet the results are controversial. Studies by Gillette and colleagues (Chen et al. 1999; Hannibal et al. 1997) have shown that PACAP (10^{-6} M) phase-advances the peak in neuronal activity of SCN cells *in vitro* during the subjective day, but not at night. This pattern of phase resetting is unlike that seen with light pulses and glutamate, and more closely resembles the phase-shifting effects of non-photic stimuli on rodent behavioural rhythms. This effect appears to involve activation of protein kinase A (PKA), since PKA inhibitors block these phase-resetting actions. Subsequent studies by this group have shown that PACAP accentuates the phase-delaying effects of glutamate *in vitro* and of light pulses *in vivo* when given during the early phases of the subjective night, while attenuating the phase-advancing effects of these stimuli during late subjective night. Harrington and colleagues (1999) have presented a very different interpretation. In their *in vitro* studies, PACAP tested at a similar range of concentrations (10^{-7} M to 10^{-6} M) phase-advanced hamster electrical activity rhythms during the subjective day, and was without effect during the subjective night. In contrast, a lower concentration of PACAP (10^{-9} M) had no phase-resetting effects during the middle portions of the subjective day, but caused photic-like phase shifts during the subjective night. *In vivo* studies by these investigators found that microinjection of PACAP at this lower concentration into the SCN region also phase-reset hamster rhythms during the subjective night. Therefore, PACAP shows an apparent marked concentration effect. The authors suggest that the concentration of PACAP released into the SCN may vary with time, and possibly acts via different mechanisms in a circadian-dependent fashion. However, as it is difficult to measure directly the amount of PACAP released from retinal terminals, it is not possible to say which doses correspond to physiological conditions at specific circadian time points.

It is apparent from the work of Piggins et al. (1996), as well as the photomicrographs in the studies of Chen et al. (1999), that PACAP-ir might be present in cells of the rat SCN, indicating that an additional source of PACAP in this nucleus may be from SCN interneurones. This raises the possibility that the actions of PACAP are more complex than originally thought. The task of determining the actions of PACAP on the SCN pacemaker, and the identity of the receptor(s) mediating these effects, is complicated further by possible *in vivo* versus *in vitro* differences, the lack of selective PACAP receptor tools, and the presence of splice variants of the PAC_1 receptor in the rat SCN. Additional research is required to elucidate fully the role of PACAP in the entrainment of the rodent SCN circadian pacemaker.

Non-photic inputs to the SCN

Serotonin

It was demonstrated over 25 years ago that the rodent SCN receives a dense serotonergic innervation. From the results of early neuroanatomical tract tracing studies, it was thought that the dorsal raphe was the source of this dense projection. However, recent investigations in hamster by Morin and colleagues using PHAL have conclusively demonstrated that it is neurones of the median raphe (MRN) that project to the SCN (Figures 1 and 2C), while cells in the dorsal raphe (DRN) innervate the IGL (Morin 1999). A similar study in the rat by Moga and Moore (1997) showed that the MRN is the exclusive source of the serotonin innervation of the SCN in this species.

Several research strategies have been devised to determine the role of serotonin in the regulation of circadian rhythms. These studies have revealed complex and occasionally apparently contradictory results. *In vivo* electrophysiological studies (Mason 1986; Ying and Rusak 1997) demonstrated that ionophoretically applied serotonin suppressed both the spontaneous activity and light-evoked increases in the activity of rodent SCN neurones. These results imply the presence of functional serotonin receptors in the rat SCN, and that activation of these serotonin receptors inhibits photic stimulation.

In vitro studies have found that electrical stimulation of the optic nerve activates populations of retinorecipient rat SCN neurones and that the serotonin receptor agonist, 8-OH-DPAT, suppressed these activational effects. This agent was also found to phase-advance the electrical activity rhythm of rat SCN neurones in a pattern that mimics the effects of non-photic stimuli on rodent activity rhythms. *In vivo*, peripherally administered 8-OH-DPAT causes phase advances in hamster wheel-running rhythms when given during the middle phases of the subjective day. Based on the pharmacology of 8-OH-DPAT, and the knowledge of serotonin receptors at the time, serotonin was thought to modulate SCN neuronal activity via the 5-HT_{1A} class of receptor. However, more recent studies indicate that 8-OH-DPAT is an agonist at the 5-HT_7 receptor, which is also present in the SCN. Taken together, these data imply that 8-OH-DPAT, acting via $5\text{-HT}_{1A/7}$ classes of receptors in the SCN, phase-resets the hamster circadian pacemaker in a non-photic manner.

However, it is unclear whether the phase-shifting actions of 8-OH-DPAT *in vivo* occur at the level of the SCN. A study by Mintz and colleagues (1997) reported that microinjection of 8-OH-DPAT into the SCN or IGL regions failed to phase-advance hamster wheel-running rhythms during the subjective day, whereas microinjection into the MRN phase-advanced this behavioural rhythm. Their data demonstrate that the MRN is the likely site of action of peripherally administered 8-OH-DPAT. Subsequent studies have failed to replicate these findings, and instead, have found that 8-OH-DPAT microinjection into either the SCN or IGL during the subjective day phase-advanced hamster activity rhythms. The sites of action of the phase-resetting actions of 8-OH-DPAT remain to be clarified.

An additional role of serotonin appears to be the modulation of photic input to the hamster SCN. 8-OH-DPAT has been shown to reduce both the phase-resetting actions and the effects on gene expression of light pulses in the hamster SCN during the subjective night. These studies indicate that serotonergic input to the rodent SCN may modulate photic input via the post-synaptic $5\text{-HT}_{1A/7}$ receptors.

Recent research by Pickard and colleagues (1999) has shown that 5-HT_{1B} receptors are present on RHT terminals in the rodent SCN, since optic enucleation reduced the level of expression of 5-HT_{1B} binding sites in the hamster SCN. Functional studies showed that microinjection of the 5-HT_{1B} receptor agonist, TFMPP, attenuated the phase-shifting effects of light on wheel-running rhythms and on IEG expression in the hamster SCN, while a 5-HT_{1B} receptor antagonist increased these actions of light. At no phase did TFMPP alone significantly alter gene expression in the SCN or the phase of hamster wheel-running rhythms.

Studies in which the serotonergic input to the hamster SCN was abolished through administration of the serotonergic neurotoxin 5,7-dihydroxytryptamine (5,7-DHT), into the ventricles or directly into the SCN or MRN, have shown that denervation of the serotonergic input to the SCN has few pronounced consequences on hamster behavioural rhythms (Morin 1999). The most robust effects of such a treatment include: an alteration in the phase-angle of entrainment such that the onset of wheel-running occurs some 30 min prior to lights off, lengthening of tau (τ) in constant conditions, and increased alpha without a corresponding increase in wheel-running. In addition, the wheel-running rhythm of 5,7-DHT-lesioned hamsters shows an increased probability of splitting in constant bright light conditions. Under constant dark, 5,7-DHT-lesioned animals show an increased magnitude of the phase delays evoked by light pulses, while the phase advances are similar to those seen in unlesioned hamsters. These studies suggest that the serotonergic input to the SCN is not necessary for the expression of wheel-running rhythms or for photic entrainment. However, serotonin seems to be important in preventing the disruptive effects of constant light on hamster behavioural rhythms.

These pharmacological studies show that serotonin has at least two sites of action within the rodent SCN, including pre-synaptic effects via 5-HT_{1B} receptors on RHT terminals and post-synaptic effects via $5\text{-HT}_{1A/7}$ receptors. At both pre- and post-synaptic sites, serotonin appears to dampen the photic input to the SCN circadian pacemaker. Further, although 8-OH-DPAT phase-advances the hamster SCN pacemaker *in vivo* during the middle portions of the subjective day, an intact serotonergic innervation of the SCN in this species is not necessary to mediate phase-shifts to some non-photic stimuli such as novel wheel-running, but is a requirement for others, including the benzodiazepine triazolam (which evokes phase-shifts during the middle phases of the subjective day). These latter data indicate that not all non-photic stimuli reset the SCN pacemaker by similar mechanisms and that they may utilise different neural substrates to evoke phase shifts.

A principal difficulty in evaluating the role of the MRN in circadian rhythm processes is the blinkered view that since MRN neurones synthesise serotonin, and because these serotonergic neurones innervate the rodent SCN, then all the effects of the MRN on the SCN circadian pacemaker must be mediated via serotonergic actions. However, serotonergic neurones represent only about 40% of the neurones in the rat MRN, indicating that other transmitters are likely to be contained in the MRN projection to the SCN. Further studies are necessary to determine the extent to which non-serotonergic raphe pathways to the SCN regulate rhythm phase.

Neuropeptide Y

The most prominent, non-retinal input to the SCN arises from retinally innervated cells of the IGL that subsequently project to the SCN to form the GHT (Figure 1). Early investigations revealed that neuropeptide Y (NPY) is an important neurochemical of this pathway (Figure 2D), although subsequent studies have indicated that other putative neurotransmitters, such as GABA, enkephalin, and possibly neurotensin, are also synthesised in the IGL (Harrington 1997). The receptor-based mechanisms mediating the actions of NPY on SCN neurones are complex, for example, *in vitro* microculture studies have shown that both pre-and post-synaptic receptors are involved. Application of NPY to SCN neurones *in vitro* potently suppresses their spontaneous electrical activity and phase-advances the neuronal activity rhythm during the subjective day. Microinjection of NPY into the SCN region phase-advances hamster wheel-running rhythms during the middle portions of the subjective day, but has negligible effects on phase during the subjective night. These phase-resetting effects of NPY appear to be mediated via the Y_2 receptor, both *in vivo* and *in vitro*. Further studies suggest that PKC plays a crucial role in mediating these *in vitro* and *in vivo* phase-resetting effects of NPY on the hamster SCN. One dichotomy that has emerged is that GABA, acting via $GABA_A$ receptors, appears to mediate the actions of NPY on the hamster SCN *in vivo*, but not *in vitro*. This disparity eludes to a potential difference in the manner by which NPY resets the clock in *in vivo* versus *in vitro* experimental settings, raising the possibility that the mechanisms that regulate firing rate rhythm phase *in vitro* may be different from those regulating the phase of wheel-running rhythms.

Electrical stimulation of the IGL also phase-shifts hamster activity rhythms in a manner similar to NPY microinjection. Lesions of the IGL remove most NPY-ir terminals in the hamster SCN, and abolish the phase-resetting effects of a number of non-photic stimuli including triazolam, novel wheel-running, and arousal. Further, central microinjection of NPY antisera, but not rabbit serum, blocks the phase-shifting effects of vigorous exercising in a novel running-wheel. These studies suggest that NPY release from GHT terminals in the SCN conveys non-photic information to the SCN circadian pacemaker.

GABA

The inhibitory transmitter GABA is found throughout the rodent SCN. GABA-ir is contained in the GHT input, and both GAD mRNA and GABA-ir are found in the majority of SCN neurones. The role of GABA in circadian rhythm processes has yet to be fully understood. Early electrophysiological studies showed that GABA and $GABA_A$ receptor agonists primarily suppress SCN cellular activity, while the $GABA_A$ receptor antagonist, bicuculline, either had no effect or evoked small increases in the spontaneous electrical activity of SCN neurones. The widespread distribution of GABA in the SCN has led some researchers to postulate that GABA is the main neurochemical of the rodent SCN pacemaker. An *in vitro* study by Wagner and colleagues (1997) has suggested that the effects of GABA on rodent SCN neurones varied across the circadian cycle such that $GABA_A$ receptor activation during the subjective day excited or increased cellular activity, but suppressed SCN neuronal activity during the subjective night. This temporal variation in the actions of GABA was speculated to be due to circadian changes in intracellular $[Cl^-]$ levels, with the model implying higher levels during the subjective day than the subjective night. An extensive *in vitro* study failed to replicate this finding, and instead, suggested that GABA inhibits SCN neuronal activity at all phases of the circadian cycle (Gribkoff et al. 1999). These results are consistent with those of Mason (1986) who observed that GABA inhibited rat SCN neuronal activity *in vivo* at all phases of the diurnal cycle. However, GABA does appear to be excitatory to developing hypothalamic neurones and consistent with this view, Liu and Reppert (2000) recently demonstrated that pulses of GABA applied in culture medium could entrain the electrical activity of foetal SCN neuronal cultures maintained on multi-electrode plates. These results suggest that GABA can play a role in the entrainment of the developing SCN circadian pacemaker.

GABA has been hypothesised to modulate photic inputs in the SCN. Recent studies by Gillespie and colleagues (1999) indicate that blockade of $GABA_A$ receptors with bicuculline potentiates the phase-shifting effects of light pulses on hamster behavioural rhythms. *In vitro* electrophysiological studies have also shown that $GABA_B$ receptors may modulate the activity of the excitatory effects of optic nerve stimulation on rat SCN neurones.

Peripheral administration of several types of benzodiazepines has been shown to phase-shift hamster behavioural rhythms when given during the middle portions of the subjective day. Benzodiazepines are thought to regulate the frequency of opening of the Cl^- ion channel of the $GABA_A$ receptor. Benzodiazepines suppress electrical activity in the rodent SCN *in vitro* and these effects appear to involve $GABA_A$ receptors. However, *in vivo* studies in which benzodiazepines were microinjected into a number of brain sites revealed that benzodiazepines do not appear to act at the level of the SCN to evoke phase shifts, but rather at a site upstream. Moreover, lesions of the GHT abolish the phase-resetting actions of benzodiazepines. Further, a recent study by Marchant and Morin (1999) has shown that structures of the subcortical visual system appear to be necessary for the phase-resetting effects of benzodiazepines to occur. Since many efferents of this visual system project to the IGL, it has been speculated that the IGL may function as a point of convergence for a number of non-photic inputs and, in turn, conveys this information to the SCN.

Nitric oxide

One of the most intriguing findings of the last decade was the demonstration by two laboratories that the gaseous neurotransmitter, nitric oxide (NO), modulated that phase-shifting actions of glutamate on rat SCN firing rate rhythms *in vitro* and the phase-resetting effects of light pulses on hamster wheel-running rhythms. These investigations showed that pharmacological agents that blocked the activity of nitric oxide synthases (NOS), the family of enzymes responsible for the synthesis of NO, or that removed extracellular NO (e.g. haemoglobin), blocked the phase-shifting effects of glutamate *in vitro* and light pulses *in vivo*, while agents that promoted the synthesis of NO phase-reset the rodent SCN circadian pacemaker. Some of these effects appear to involve the activation of protein kinase G (PKG) and cyclic GMP, since inhibition of PKG blocks the phase-advancing effects of light-pulses (Mathur et al. 1996). These studies were initially difficult to interpret since standard histochemical markers of NOS activity could not be visualised successfully in the rodent SCN. Subsequent development of antisera to one of the isoforms of NOS, neuronal NOS, has enabled researchers to conclusively demonstrate the presence of this enzyme in neurones of the rodent SCN, including a population of peptide-containing cells of the ventral SCN.

More recent studies with transgenic knockout mice lacking neuronal NOS have indicated that NO may not be necessary for light-evoked shifts in rodent behavioural rhythms (Kriegsfeld et al. 1999). These studies raise the possibility that other isoforms of NOS, such as endothelial NOS, located in the walls of blood vessels, may participate in modulating photic input to the mouse SCN circadian pacemaker. Further *in vivo* and *in vitro* studies with this animal model may help to determine a functional role for NO in photic entrainment processes.

Acetylcholine

Acetylcholine was initially hypothesised as the RHT neurotransmitter mediating the effects of light on the rodent SCN circadian pacemaker. However, subsequent immunohistochemical studies failed to find evidence for cholinergic markers in the RHT, and instead have shown that terminals containing choline acetyl transferase-ir in the rat SCN are the projections of cholinergic neurones located in the basal forebrain and brainstem (Bina et al. 1993). Investigations using immunohistochemical, *in situ* hybridisation, and radioligand receptor autoradiography techniques have shown that different types of muscarinic and nicotinic receptors are found in the SCN region of rodents. Both *in vitro* and *in vivo* studies have demonstrated that the muscarinic agonist carbachol phase-resets the rodent SCN circadian pacemaker in a pattern resembling the effects of light pulses on rodent behavioural rhythms. In the brain slice *in vitro*, these effects are blocked by a muscarinic receptor antagonist, but not by tetrodotoxin, indicating that carbachol acting via muscarinic receptors directly resets the rat SCN *in vitro*. These resetting actions of carbachol on the SCN *in vitro* are also blocked by a PKG inhibitor, indicating that following activation of acetylcholine receptors, PKG is involved in the intracellular pathways through which phase-resetting information is relayed to the molecular mechanisms of the SCN circadian pacemaker. However, *in vivo*, the effects of carbachol on the SCN have been shown to be abolished by the NMDA receptor antagonist MK-801, and both muscarinic and nicotinic receptor antagonists. These studies suggest that the mechanisms mediating the *in vivo* effects of carbachol are more complex than those found *in vitro*, presumably due to the presence of modulatory inputs to the SCN in the intact animal.

The role of acetylcholine in circadian rhythm processes has yet to be fully defined, however, the light-mimicking actions of carbachol suggest that muscarinic receptors can modulate the actions of light, and therefore the RHT input to the SCN. Since choline acetyl transferase-containing terminals arise from cells in the basal forebrain and brainstem—regions thought to regulate cortical arousal and the sleep-wake cycle—it seems likely that information regarding arousal is relayed to the SCN circadian pacemaker via these afferents.

Intrinsic neuropeptides of the SCN

Cells of the SCN have been shown to produce a number of neuropeptides, which, in turn, have been postulated to function as transmitters and modulators within the circadian system. Neurones of the retinally innervated ventrolateral portion of the SCN have been shown to produce vasoactive intestinal peptide (VIP), gastrin-releasing peptide (GRP), and peptide histidine isoleucine (PHI). VIP and PHI are always found to be co-localised, while a population of GRP-expressing SCN neurones also exhibit co-localisation with VIP/PHI. Neurones in the dorsomedial SCN, which express the neuropeptides arginine vasopressin and somatostatin, exhibit reciprocal innervation with GRP- and VIP/PHI-expressing neurones in the SCN. These findings all point to the likelihood that intrinsic peptidergic connections within the SCN form a complex network which is involved in both the conveyance of synaptic information within the SCN, and the regulation of SCN output.

Gastrin-releasing peptide

Gastrin-releasing peptide and neuromedin B (NMB), mammalian homologues of the anuran peptide bombesin, were initially found in gut tissue, but have subsequently been shown to have a wide distribution in the CNS of mammals. The presence of GRP-ir perikarya in the SCN of mammals (Figure 2E) has stimulated research into the elucidation of the role this peptide may play in circadian processes. In particular, a number of studies have focused on the putative role of GRP in processing retinal inputs to the SCN, as GRP-expressing neurones have been shown to directly receive retinal innervation (Tanaka et al. 1997). Further, application of light pulses at circadian times that lead to resetting of the clock also induces the expression of Fos in GRP-ir neurones (Romijn et al. 1996). Electrophysiological studies have indicated that application of GRP, and its related peptides bombesin and NMB, caused robust increases in firing rates of SCN neurones recorded in a slice preparation (Piggins et al. 1994). However, these data do not support those of another study that failed to show any significant effect on SCN neuronal firing rate of GRP alone, but indicated that a cocktail of GRP, VIP, and PHI was required to elicit substantial elevations in cell firing rate (Albers et al. 1991). Taken together, these findings suggest that there are functional BB_1 (NMB-preferring) and BB_2 (GRP-preferring) receptors in the rodent SCN. The presence of these receptor subtypes in the SCN has been demonstrated anatomically using radioligand receptor autoradiography and *in situ* hybridisation techniques.

The phase-shifting effect of GRP has been demonstrated *in vivo* in experiments where GRP was microinjected into the hamster SCN. Initial studies indicated that while application of GRP alone failed to elicit substantial phase shifts, a cocktail of VIP/GRP/PHI was necessary to produce appreciable phase delays in the early subjective night (Albers et al. 1991). However, Piggins and colleagues (1995a) reported that application of GRP (at doses similar to those used by Albers) produced robust phase shifts, with a phase response curve similar to that of light (i.e. no shift during the subjective day, phase delays following application during early subjective night and phase advances in response to application during late subjective night). There is no clear reason for the discrepancy between these findings. The possible physiological significance of the work of Albers is that GRP and VIP/PHI need to be co-released to cause phase-resetting of the clock. In both rat and hamster there are cells in the ventrolateral SCN that co-express GRP and VIP/PHI, although the numbers of these cells compared to those expressing only GRP or VIP/PHI are small (Aïoun et al. 1998). The physiological significance of co-release of GRP with VIP/PHI remains obscure. The role of GRP in phase-resetting of the clock was further investigated *in vitro* using the firing rate rhythm paradigm pioneered by Gillette and collaborators (e.g. Hannibal et al. 1997). GRP was found to produce large phase-dependent phase shifts in the firing rate rhythm of both rat and hamster SCN slices (McArthur et al. 2000). Again, the phase response curve to GRP strongly resembled that of light *in vivo* and that of glutamate *in vitro*. These findings, considered together, suggest strongly that GRP may be involved in light-induced resetting of the circadian pacemaker.

Although the phase-resetting properties of GRP on the circadian clock are well characterised, the mechanism that underpins the phase-dependency of the GRP-induced effects remains to be established. A number of studies have shown that animals kept under a light-dark cycle exhibit a diurnal rhythm in the level of GRP mRNA and GRP protein in the rat SCN. However, under free-running conditions, no rhythm in GRP levels was observed. Therefore, the phase response curve noted for the resetting effects of GRP do not appear to reflect alterations in endogenous levels of GRP. Further, the possibility that fluctuations in the levels of BB_1 and BB_2 receptors may underlie the GRP-induced effects in the SCN is unlikely since the expression of BB_1 and BB_2 receptor mRNA do not vary significantly over the day/night cycle (McArthur et al. 2000). Given the apparent constant levels of GRP and its receptors in the SCN, one possible mechanism which may give rise to this phase-dependent effect of GRP are oscillations in intracellular signalling factors, such as adenylyl cyclase.

Vasoactive intestinal polypeptide

Vasoactive intestinal polypeptide is a 28 amino acid peptide, widely distributed throughout the peripheral and central nervous systems, where it functions as a neuromodulator. High levels of VIP-ir have been demonstrated in cells of the retinally innervated, ventrolateral portion of the SCN (Figure 2F), and these cells give rise to extensive inter- and intra-SCN connections, as well as projecting to regions dorsal to the SCN (Watts and Swanson 1987). Some of these VIP-containing cells have been shown to express IEGs in response to light pulses (Romijn et al. 1996), indicating that they may play a part in mediating photic information in the SCN.

As noted above, VIP/PHI and GRP are co-localised in a proportion of cells of the ventrolateral SCN of the rat and it has been proposed that co-release of these peptides in varying ratios may have differential effects at different circadian phases. The importance of co-localisation of these peptides is however unclear based on the following discrepant findings. Albers and colleagues (1991) showed that microinjection of a cocktail of VIP, PHI and GRP directly into the SCN region of hamsters induced maximal phase shifts in locomotor activity in a manner similar to light pulses, whereas administration of each peptide alone was less effective. In contrast, Piggins and colleagues (1995a) found that microinjection of VIP alone into the SCN, during the subjective night, was able to significantly phase-shift hamster behavioural rhythms. This latter study is in agreement with data from our laboratory that show that VIP phase delays and phase advances the firing rate rhythm of rat SCN neurones when applied during the early and late subjective night respectively, with no effects during the subjective day (Reed et al. 2001). Taken together, these data indicate that VIP and related peptides have important roles in mediating responses to photic cues in the SCN.

There are currently three known receptors that bind VIP and the related, putative RHT peptide

PACAP (Harmar et al. 1998). The PAC_1 receptor binds PACAP with 1000 times greater affinity than VIP and has been shown, using *in situ* hybridisation histochemistry, to be present in the ventrolateral SCN (Hannibal et al. 1997). The $VPAC_1$ and $VPAC_2$ receptors bind VIP and PACAP with similar affinities but so far only $VPAC_2$ has been found in the SCN (DM region; Sheward et al. 1995). Our *in vitro* data, employing selective agonists to the above receptors, suggest that the phase-resetting actions of VIP are mediated by the $VPAC_2$ receptor (unpublished observations). *In situ* hybridisation has demonstrated that the mRNA for the $VPAC_2$ receptor cycles under constant darkness (Cagampang et al. 1998b), and it is possible that temporal variation in the availability of this receptor may account for the phase-dependent actions of VIP on the SCN clock.

Arginine vasopressin

Arginine vasopressin (AVP) is synthesised and released from neurones of the dorsomedial region of the rodent SCN (Figure 2G). This population of cells is responsible for the circadian rhythm in AVP in the cerebrospinal fluid. Radioligand and *in situ* hybridisation studies have shown radiolabelled AVP binding sites and heavy expression of AVP_{1a} receptor mRNA in the rodent SCN. AVP has been shown to have predominantly activational effects on rat SCN neurones, and these actions appear to be regulated via the AVP_{1a} class of receptor. Although AVP can alter the electrical activity of rodent SCN cells, it does not appear to reset the SCN circadian clock. Instead, AVP release is an output of the SCN circadian pacemaker and the activity of the AVP promoter has been shown to be directly regulated by molecular components of the circadian pacemaker.

Somatostatin

Somatostatin is synthesised in neurones of the medial SCN (Figure 2H), but its actions on the rodent SCN circadian clock have not been extensively researched. One study has shown that somatostatin resets the firing rate rhythm of rat SCN cells *in vitro* in a manner mimicking the phase-shifting actions of glutamate on this SCN output rhythm. *In vivo* in rats, depletion of somatostatin, via cysteamine, reveals a circadian rhythm in the VIP content in the SCN. Since there is evidence for neural connections between neurones in the dorsomedial and ventrolateral SCN, somatostatin neurones may function to dampen the rhythmic synthesis of peptides in the ventral SCN. The physiological role of somatostatin in circadian rhythm processes is necessarily speculative in the absence of functional studies.

Other neural inputs

Neurotensin

The presence of neurotensin-ir in the SCN, predominately in the retinorecipient ventrolateral portion (Watts and Swanson 1987), raises the possibility that neurotensin may have effects on the SCN circadian pacemaker. The content of neurotensin in the SCN does not exhibit a circadian rhythm under constant conditions, and also appears to be insensitive to the light-dark cycle. The potential role for neurotensin as a modulator of SCN function is further supported by the presence of the G protein-coupled NTR1 and NTR2 neurotensin receptors in the rat and mouse SCN, similarly distributed in the ventrolateral region. A third neurotensin receptor, NTR3, recently cloned but with no known physiological actions, is reportedly induced and inserted into the membrane following activation of the other neurotensin receptors (Vincent et al. 1999).

The role neurotensin may play in the regulation of the SCN is unclear but results from our laboratory demonstrate that this peptide stimulates the firing rat of rat SCN neurones, and these responses appear to be mediated via both NTR1 and NTR2 receptors (Coogan et al. 2001). It remains to be determined if these effects of neurotensin are direct or if neurotensin acts via the modulation of other neurotransmitter systems. There is evidence to support the latter possibility as neurotensin bindings sites in the rat SCN have been localised to VIP neurones and serotonin axons; neurotensin has also been demonstrated to regulate the excitatory effects of glutamate at other hypothalamic sites. The contribution of neurotensin to the functioning of the SCN requires further investigation but, based on the localisation of neurotensin and neurotensin receptors to the ventrolateral SCN, it is likely that neurotensin is involved in the phase-resetting effects of light on the SCN clock.

Orexin/Hypocretin

The orexins, orexin-A and orexin-B, and the structurally similar hypocretins, are a recently discovered family of neuropeptides. *In situ* hybridisation and immunohistochemical techniques have detected orexin-producing neurones in the brain of many species, including human, which are localised exclusively to the lateral hypothalamic area. Orexin-ir fibres are distributed widely throughout the brain and spinal cord (Cutler et al. 1999), and together with the broad but differential localisation of the two currently known receptors for the orexins, OX-1 and OX-2 (Trivedi et al. 1998), suggest that the orexins participate in a host of neuroendocrine, autonomic and behavioural systems.

The potential importance of the orexins in the modulation of the circadian timing system has been highlighted by the findings that intraventricular injection of the orexins increase vigilance/attention states and locomotor activity in rats, behaviours that are under clear circadian control (Williams et al. 2000). Moreover, the stimulation of feeding by the orexins shows a circadian dependency, as food intake in rats is increased only when the orexins are given during the subjective day, i.e. at times when feeding behaviour is normally diminished. The putative role of the orexins in arousal and the sleep-wake cycle has been emphasised by the discoveries that knockout of the mouse orexin gene induces a narcolepsy-like state, and that narcolepsy in dogs is caused by a mutation affecting the OX-2 receptor. Furthermore, there is evidence to indicate that human narcolepsy is associated with decreased orexin-A content in the cerebrospinal fluid (Nishino et al. 2000)

It is presently unclear whether the actions of the orexins on various circadian-based activities occur via a direct action at sites associated with arousal/sleep-wakefulness (e.g. the locus coeruleus, thalamic paraventricular nucleus) or feeding (e.g. lateral hypothalamic area, arcuate nucleus)—many of which express mRNA for the orexin receptors—or via actions on the central circadian clock. There are anatomical data to support the contention that the orexins may influence the circadian system: the rat and hamster SCN and IGL exhibit sparse to moderate quantities of orexin fibres, the rat SCN contains moderate levels of the orexin peptides, and Fos is stimulated in the rat SCN following intraventricular administration of orexin-A (Williams et al. 2000). Although the diurnal rhythm of prepro-orexin mRNA and orexin-A content, which in the anterior hypothalamus peaks at ZT0 and are minimal at ZT12-14, indicates that orexin levels in some brain regions are under circadian control, it remains to be established how the orexins contribute to the regulation of sleep and arousal and what role, if any, they play in the SCN.

Growth hormone-releasing factor

Growth hormone-releasing factor (GRF) exhibits structural similarity to VIP and is classically known for its role in growth hormone release from the pituitary. The pattern of growth hormone secretion is affected by the light/dark cycle and is clearly regulated by the SCN, as destruction of the SCN abolishes this rhythm. The diurnal variation in hypothalamic GRF mRNA levels, which are highest at ZT0 and lowest at ZT13-21, also reflects the influence of the SCN on GRF function. GRF has been implicated in two other processes that have a strong circadian component: the sleep-wake cycle and feeding behaviour (Krueger et al. 1999). GRF reportedly promotes sleep (especially non-REM sleep) and blockade of the endogenous GRF system disrupts spontaneous sleep. With respect to feeding, injection of GRF into the rat or hamster SCN-medial preoptic area (but not in other brain regions or peripherally) stimulates food intake but only when administered during the subjective day, when feeding activity is normally quiescent (Vaccarino et al. 1995). Further studies by these researchers have shown that antagonism of GRF activity in the rat, with an antibody against GRF, selectively attenuated feeding at the beginning of the subjective night (ZT12). These findings demonstrate convincingly that GRF is an important mediator of the rhythm of feeding behaviour and raises the possibility that GRF has actions on the SCN circadian system.

The notion that GRF may function as a neuromodulator in the SCN is supported by the presence of GRF-ir axons in the dorsal portion of the rat SCN (Sawchenko et al. 1985). The significance of this GRF innervation of the SCN is unknown, as GRF receptors have not been detected in this nucleus. Moreover, microinjection of GRF into the hamster SCN phase-shifts the free-running rhythm of locomotor activity in a non-photic-like fashion. From these observations, it was speculated that GRF modulates the SCN indirectly via its actions on feeding behaviour, a well-established regulator of SCN function. It is interesting to note, however, that the SCN is enriched with mRNA for the growth hormone secretagogue receptor, a receptor that does not bind GRF but similarly stimulates growth hormone secretion when activated (Guan et al. 1997). The endogenous ligand for this receptor is unknown at present but the possibility that it may influence SCN function cannot be discounted.

Corticotropin-releasing factor

The primary function of corticotropin-releasing factor (CRF) is to control the synthesis and release of adrenocorticotropin hormone from the pituitary. However, the extensive distribution of CRF neurones throughout the hypothalamus, and elsewhere in the central nervous system, is consistent with the diverse range of systems reportedly influenced by CRF, including the sleep-wake cycle. There are moderate levels of CRF-ir in the rat SCN, as measured by radioimmunoassay techniques, and CRF-containing neurones have been detected throughout the SCN with a greater concentration in the ventrolateral portion. The existence of mRNA for one of the CRF receptors, CRF2, in the rat SCN (Chalmers et al. 1995), supports the possibility that CRF may also function as a neurotransmitter in the SCN.

The role of CRF in the modulation of the circadian system is purely speculative; intraventricular injection of CRF has been shown to decrease the amplitude of the circadian rhythm of hamster locomotor activity, but only when administered during the early part of the light phase (Seifritz et al. 1998). Moreover, when rats maintained under constant dark conditions are given a light pulse during the early subjective night—a time when light is capable of shifting the clock—Fos is stimulated in neurones of the ventrolateral SCN, and some of these cells were found to be immunoreactive for CRF. These data suggest that CRF may serve to modulate the responsiveness of the SCN to light, although it is unknown whether these putative effects are direct or if they arise from interactions with VIP, a peptide that is found in the majority of CRF-containing neurones in the SCN.

Future studies

As the above review demonstrates, many neurotransmitters are known to alter the activity of rodent SCN neurones and to influence the phase of the mammalian circadian pacemaker. It is apparent that the acute actions of these neurochemicals at the cellular level are not necessarily mirrored by changes in circadian rhythm phase. The neurochemicals whose actions are best characterised are glutamate, NPY, and to a lesser extent, serotonin. However, even among these, there are obvious discrepancies. For example, the effects of ionotropic glutamate receptor blockade on photic phase-resetting can be overcome by bright light, and some hamster SCN neurones continue to express IEGs following a light pulse despite the presence of such antagonists. These data suggest that novel glutamate receptors may mediate, in part, the effects of light on this population of SCN neurones, however, the involvement of other neurotransmitter systems in photic phase-shifting cannot be disregarded. Studies that have provided insight into the acute and phase-resetting actions of these neurochemicals have not yet been matched by a complete understanding of the molecular mechanisms that convey this information to recently identified putative clock genes, such as *mper*1-3. The identification of the second messenger systems thought to be involved in these pathways should yield important information to address these issues.

Another problem that is emerging in SCN research is the apparent redundancy in the phase-resetting actions of neuroactive substances on the SCN clock. As summarised in this chapter, many neurochemicals have been shown to reset the rodent SCN *in vitro* and *in vivo*. In general, these actions are consistent between the two experimental settings, but it is difficult to resolve the apparent redundant overlap in their effects. Additionally, there is still much to be learned about the neural substrates that underlie the actions of a range of stimuli whose phase-resetting actions are well documented. For example, the recent finding that subcortical visual nuclei likely play a more significant role in the integration of phase-resetting information than was previously thought, highlights the fact that our understanding of SCN function and the mechanisms of entrainment is incomplete. Research into this region of the visual system, and further elucidation of the contribution of various neurotransmitter systems that project to, or are contained within the SCN, will undoubtedly be the subject of considerable research over the next decade.

References

Abe, H., Rusak, B. (1994) Physiological mechanisms regulating photic induction of Fos-like protein in hamster suprachiasmatic nucleus. Neurosci. Biobehav. Rev. **18**: 531–536.

Aïoun, J., Chambille, I., Peytevin, J., Martinet, L. (1998) Neurons containing gastrin-releasing peptide and vasoactive intestinal polypeptide are involved in the reception of the photic signal in the suprachiasmatic nucleus of the Syrian hamster: an immunocytochemical ultrastructural study. Cell Tiss. Res. **291**: 239–253.

Albers, H.E., Liou, S.Y., Stopa, E.G., Zoeller, R.T. (1991) Interaction of colocalized neuropeptides: functional significance in the circadian timing system. J. Neurosci. **11**: 846–851.

Bina, K.G., Rusak, B., Semba, K. (1993) Localization of cholinergic neurons in the forebrain and brainstem that project to the suprachiasmatic nucleus of the hypothalamus in rat. J. Comp. Neurol. **335**: 295–307.

Cagampang, F.R., Piggins, H.D., Sheward, W.J., Harmar, A.J., Coen, C.W. (1998a) Circadian changes in PACAP type 1 (PAC1) receptor mRNA in the rat suprachiasmatic and supraoptic nuclei. Brain Res. **813**: 218–222.

Cagampang, F.R., Sheward, W.J., Harmar, A.J., Piggins, H.D., Coen, C.W. (1998b) Circadian changes in the expression of vasoactive intestinal peptide 2 receptor mRNA in the rat suprachiasmatic nuclei. Mol. Brain Res. **54**: 108–112.

Chalmers, D.T., Lovenberg, T.W., De Souza, E.B. (1995) Localization of novel corticotropin-releasing factor receptor (CRF2) mRNA expression to specific subcortical nuclei in rat brain: comparison with CRF1 receptor mRNA expression. J. Neurosci. **15**: 6340–6350.

Chen, D., Buchanan, G.F., Ding, J.M., Hannibal, J., Gillette, M.U. (1999) Pituitary adenylyl cyclase-activating peptide: a pivotal modulator of glutamatergic regulation of the suprachiasmatic circadian clock. Proc. Natl. Acad. Sci. USA **96**: 13468–13473.

Coogan, A.N., Rawlings, N., Luckman, S.M., Piggins, H.D. (2001) Effects of neurotensin on discharge rates of rat suprachiasmatic nucleus neurons *in vitro*. Neuroscience **103**: 663–672.

Cutler, D.J., Morris, R., Sheridhar, V., Wattam TAK, Holmes, S., Patel, S., Arch, J.R.S., Wilson, S., Buckingham, R.E., Evans, M.L., Leslie, R.A., Williams, G. (1999) Differential distribution of orexin-A and orexin-B immunoreactivity in the rat brain and spinal cord. Peptides **20**: 1455–1470.

Ding, J.M., Chen, D., Weber, E.T., Faiman, L.E., Rea, M.A., Gillette, M.U. (1994) Resetting the biological clock: mediation of nocturnal circadian shifts by glutamate and NO. Science **266**: 1713–1717.

Ebling, F.J.P. (1996) The role of glutamate in the photic regulation of the suprachiasmatic nucleus. Prog. Neurobiol. **50**: 109–132.

Gillespie, C.F., Van Der Beek, E.M., Mintz, E.M., Mickley, N.C., Jansow, A.M., Huhman, K.L., Albers, H.E. (1999) GABAergic regulation of light-induced c-Fos immunoreactivity within the suprachiasmatic nucleus. J. Comp. Neurol. **411**: 683–692.

Gribkoff, V.K., Pieschl, R.L., Wisialowski, T.A., Park, W.K., Strecker, G.J., de Jeu MTG, Pennartz, CMA, Dudek, F.E., (1999) A reexamination of the role of GABA in the mammalian suprachiasmatic nucleus. J. Biol. Rhythms **14**: 126–130.

Guan, X.M., Yu, H., Palyha, O.C., McKee, K.K., Feighner, S.D., Sirinathsinghji, D.J., Smith, R.G., Van der Ploeg, L.H., Howard, A.D. (1997) Distribution of mRNA encoding the growth hormone secretagogue receptor in brain and peripheral tissues. Mol. Brain Res. **48**: 23–29.

Hannibal, J, Ding, J.M., Chen, D., Fahrenkrug, J., Larsen, P.J., Gillette, M.U., Mikkelsen, J.D. (1997) Pituitary adenylate cyclase-activating peptide (PACAP) in the retinohypothalamic tract: a potential daytime regulator of the biological clock. J. Neurosci. **17**: 2637–2644.

Hannibal, J., Moller, M., Ottersen, O.P., Fahrenkrug, J. (2000) PACAP and glutamate are co-stored in the retinohypothalamic tract. J. Comp. Neurol. **418**: 147–155.

Harmar, A.J., Arimura, A., Gozes, I., Journot, L., Laburthe, M., Pisegna, J.R., Rawlings, S.R., Robberecht, P., Said, S.I., Sreedharan, S.P., Wank, S.A., Waschek, J.A. (1998) Nomenclature of receptors for vasoactive intestinal polypeptide and pituitary adenylate cyclase-activating polypeptide. Pharmacol. Rev. **50**: 265–270.

Harrington, M.E. (1997) The ventral lateral geniculate nucleus and the intergeniculate leaflet: interrelated structures in the visual and circadian systems. Neurosci. Biobehav. Rev. **21**: 705–727.

Harrington, M.E., Hoque, S., Hall, A., Golombek, D., Biello, S. (1999) Pituitary adenylate cyclase-activating peptide phase shifts circadian rhythms in a manner similar to light. J. Neurosci. **19**: 6637–6642.

Kriegsfeld, L.J., Demas, G.E., Lee, S.E., Dawson, T.M., Dawson, V.L., Nelson, R.J. (1999) Circadian locomotor analysis of male mice lacking the gene for neuronal nitric oxide synthase (nNOS-/-). J. Biol. Rhythms **14**: 20–27.

Krueger, J.M., Obal, F., Fang, J. (1999) Humoral regulation of physiological sleep: cytokines and GHRH. J. Sleep Res. **8(1)**: 53–59.

Liu, C., Reppert, S.M. (2000) GABA synchronizes clock cells within the suprachiasmatic circadian clock. Neuron **25**: 123–128.

Marchant, E.G., Morin, L.P. (1999) The hamster circadian system includes nuclei of the subcortical visual shell. J. Neurosci. **19**: 10482–10493.

Mason, R. (1986) Circadian variation in sensitivity of suprachiasmatic and lateral geniculate neurones to 5-hydroxytryptoamine in the rat. J. Physiol. **377**: 1–13.

Mathur, A., Golombek, D.A., Ralph, M.R. (1996) cGMP-dependent protein kinase inhibitors block light-induced phase advances of circadian rhythms *in vivo*. Am. J. Physiol. **270**: R1031–R1036.

McArthur, A.J., Coogan, A.N., Aipru, S., Sugden, D., Biello, S., Piggins, H.D. (2000) Gastrin-releasing peptide phase-shifts suprachiasmatic nuclei neuronal rhythms *in vitro*. J. Neurosci. **20**: 5496–5502.

Mintz, E.M., Gillespie, C.F., Marvel, C.L., Huhman, K.L., Albers, H.E. (1997) Serotonergic regulation of circadian rhythms in Syrian hamsters. Neurosci. **79**: 563–569.

Mintz, E.M., Marvel, C.L., Gillespie, C.F., Price, K.M., Albers, H.E. (1999) Activation of NMDA receptors in the suprachiasmatic nucleus produces phase shifts of the circadian clock *in vivo*. J. Neurosci. **19**: 5124–5130.

Moga, M.M., Moore, R.Y. (1997) Organization of neural inputs to the suprachiasmatic nucleus in the rat. J. Comp. Neurol. **389**: 508–534.

Morin, L.P. (1999) Serotonin and the regulation of mammalian circadian rhythmicity. Ann. Med. **31**: 12–33.

Nishino, S., Ripley, B., Overeem, S., Lammers, G.J., Mignot, E. (2000) Hypocretin (orexin) deficiency in human narcolepsy. Lancet **355**: 39–40.

Pickard, G.E., Smith, B.N., Belenky, M., Rea, M.A., Dudek, F.E., Sollars, P.J. (1999) 5-HT1B receptor-mediated presynaptic inhibition of retinal input to the suprachiasmatic nucleus. J. Neurosci. **19**: 4034–4045.

Piggins, H.D., Antle, M.C., Rusak, B. (1995a) Neuropeptides phase shift the mammalian circadian pacemaker. J. Neurosci. **15**: 5612–5622.

Piggins, H.D., Cutler, D.J., Rusak, B. (1994) Effects of ionophoretically applied bombesin-like peptides on hamster suprachiasmatic nucleus neurons in vitro. Eur. J. Pharmacol. **271**: 413–419.

Piggins, H.D., Cutler, D.J., Rusak, B. (1995b) Ionophoretically applied substance P activates hamster suprachiasmatic nucleus neurons. Brain Res. Bull. **37**: 475–479.

Piggins, H.D., Rusak, B. (1997) Effects of microinjections of substance P into the suprachiasmatic nucleus region on hamster wheel-running rhythms. Brain Res. Bull. **42**: 451–455.

Piggins, H.D., Stamp, J.A., Burns, J., Rusak, B., Semba, K. (1996) Distribution of pituitary adenylate cyclase activating polypeptide (PACAP) immunoreactivity in the hypothalamus and extended amygdala of the rat. J. Comp. Neurol. **376**: 278–294.

Reed, H.E., Meyer-Spasche, A., Cutler, D.J., Coen, C.W., Piggins, H.D. (2001) Vasoactive intestinal polypeptide (VIP) phase shifts the suprachiasmatic nucleus clock *in vitro*. Eur. J. Neurosci. **13**: 839–843.

Romijn, H.J., Sluiter, A.A., Pool, C.W., Wortel, J., Buijs, R.M. (1996) Differences in colocalization between Fos and PHI, GRP, VIP and VP in neurons of the rat suprachiasmatic nucleus after a light stimulus during the phase delay versus phase advance period of the night. J. Comp. Neurol. **372**: 1–8.

Sawchenko, P.E., Swanson, L.W., Rivier, J., Vale, W.W. (1985) The distribution of growth-hormone-releasing factor (GRF) immunoreactivity in the central nervous system of the rat: an immunohistochemical study using antisera directed against rat hypothalamic GRF. J. Comp. Neurol. **237**: 100–115.

Seifritz, E., Klemfuss, H., Montes, J.M., Britton, K.T., Ehlers, C.L. (1998) Effects of corticotropin-releasing factor on circadian locomotor rhythm in the golden hamster. Pharmacol. Biochem. Behav. **60**: 855–862.

Sheward, W.J., Lutz, E.M., Harmar, A.J. (1995) The distribution of vasoactive intestinal peptide2 receptor messenger RNA in the rat brain and pituitary gland as assessed by *in situ* hybridization. Neurosci. **67**: 409–418.

Shibata, S., Tsuneyoshi, A., Hamada, T., Tominaga, K., Watanabe, S. (1992) Effect of substance P on circadian rhythms of firing activity and the 2-deoxyglucose uptake in the rat suprachiasmatic nucleus in vitro. Brain Res. **597**: 257–263.

Tanaka, M., Hayashi, S., Tamada, Y., Ikeda, T., Hisa, Y., Takamatsu, T., Ibata, Y. (1997) Direct retinal projections to GRP neurons in the suprachiasmatic nucleus of the rat. Neuro. Report **8**: 2187–2191.

Trivedi, P., Yu, H., MacNeil, D.J., Van der Ploeg, L.H.T., Guan, X.M. (1998) Distribution of orexin receptor mRNA in the rat brain. FEBS Lett. **438**: 71–75.

Vaccarino, F.J., Sovran, P., Baird, J.P., Ralph, M.R. (1995) Growth hormone-releasing hormone mediates feeding-specific feedback to the suprachiasmatic circadian clock. Peptides **16**: 595–598.

Vincent, J.P., Mazella, J., Kitabgi, P. (1999) Neurotensin and neurotensin receptors. Trends Pharmacol. Sci. **20**: 302–309.

Wagner, S., Castel, M., Gainer, H., Yarom, Y. (1997) GABA in the mammalian suprachiasmatic nucleus and its role in diurnal rhythmicity. Nature **387**: 598–603.

Watts, A.G., Swanson, L.W. (1987) Efferent projections of the suprachiasmatic nucleus: II. Studies using retrograde transport of fluorescent dyes and simultaneous peptide immunohistochemistry in the rat. J. Comp. Neurol. **258**: 230–252.

Williams, G., Harrold, J.A., Cutler, D.J. (2000) The hypothalamus and the regulation of energy homeostasis: lifting the lid on a black box. Proc. Nutr. Soc. **59**: 385–396.

Ying, S.W., Rusak, B. (1997) 5-HT7 receptors mediate serotonergic effects on light-sensitive suprachiasmatic nucleus neurons. Brain Res. **755**: 246–254.

Biological Rhythms
Edited by V. Kumar
Copyright © 2002, Narosa Publishing House, New Delhi, India

15. Photoperiodism in Plants

P. J. Lumsden*

Department of Biological Sciences, University of Central Lancashire, Preston, UK

In photoperiodism induction occurs if a sufficiently long period of darkness or light is experienced. Rhythmicity can be demonstrated as responses to the timing of interruption of darkness by light (short-day plants, SDP), or additions of far-red light to constant white light (long-day plants, LDP). The resulting photoperiodic response rhythm (PRR) has a period of about 24 h, shows temperature compensation, and is entrained by light. Light also has a direct ('acute') response, inhibiting or promoting induction at particular phases, and so giving rise to the external coincidence model for control of induction.

In SDP the PRR has a constant phase relationship with the end of the light period; this could indicate that the underlying oscillator is stopped or suspended in continuous light, being released at the onset of darkness. However, if the length of the light period preceeding a partially inductive a dark period is varied, a rhythm in the degree of induction occurs, which is a function of the length of the photoperiod. It is therefore postulated that the oscillator continues to oscillate in continuous light, around a light limit cycle located on a particular isochron from where, at the onset of darkness it would return to a dark limit cycle. In LDP the PRR appears to be coupled to the light on signal. There is evidence that at least in some species both types of rhythm co-exist.

Introduction

Light is probably the most important environmental factor controlling plant growth and development. Three major categories of response to light can be identified. Firstly, plants can respond to the presence or absence of light (e.g. light acting through the photoreceptor phytochrome controls the germination of many seeds, and the subsequent greening of the seedling following emergence from the soil). Secondly, the quality of light due to shading and reflection by neighbours can elicit a response. Shade light and reflected light have a much reduced ratio of red : far-red light (R:FR) compared with sunlight. This again is detected by the receptor phytochrome and, depending on species the plant may demonstrate a shade-avoidance response, by increasing stem elongation to outgrow competitors and reach a better light climate. Finally, the daily duration of the daily photoperiod controls many aspects of development. This is the phenomenon of photoperiodism, a term resulting from the original work of Garner and Allard, who first definitively associated changes in plant development with changes in day length.

The discovery of photoperiodism

Photoperiodism is a widespread and important phenomenon, found in all classes of higher organism. With the invention of the incandescent lamp in the 19th century it was possible for researchers to

*E-mail: pjlumsden@uclan.ac.uk

artificially control the duration of daylengths. As a result, the Frenchman Julien Tournois and the German Hans Klebs suggested, independently, that the duration, rather than the quantity of light, is a major factor in plant development (Tournois 1912, 1914; Klebs 1913). Tournois found that plants of *Humulus* (hop) and *Cannabis* (hemp) kept under glass in winter flowered precociously; in fact plants given only six hours of daylight flowered most rapidly, even though they grew more slowly. Significantly he also concluded that it was the length of the night rather than the shortness of the day which was crucial. At about the same time Klebs observed that rosettes of *Sempervivum funkii* (house leek) flowered in the middle of winter if they were given a few days of constant light, while rosettes under natural (short) days always remained vegetative. He concluded that "in nature, flowering is probably determined by the fact that from the equinox (21 March) the length of the day increases and when it reaches a certain length, flowering is initiated. Light probably acts as a catalytic rather than a nutritive factor".

However, it was Garner and Allard (1920, 1923), who clearly demonstrated that flowering and many other responses in plants could be accelerated either by long days (LD) or short days (SD), depending on the plant. Their findings were based on observations with the late maturing strain 'Biloxi' of *Glycine max* (soybean) and the 'Maryland Mammoth' variety of *Nicotiana tabacum* (tobacco), being used in breeding programmes at the time. Plants growing in pots under glass in winter and early spring flowered while still quite small, compared with their continued vegetative growth outdoors in summer. After eliminating temperature and light intensity as possible seasonal factors, Garner and Allard tested the hypothesis that the causal factor was the relative length of the day and night. They did this by transferring plants of *Glycine max* cv Peking and *Nicotiana tabacum* cv. Maryland Mammoth to a darkened hut every day during summer, keeping the daily exposure of light to seven hours. Plants under the experimental short day regime flowered rapidly, while those grown in natural summer day lengths remained vegetative.

Garner and Allard proposed that the tobacco and soya plants would only flower if the duration of the daily light period was sufficiently short, and they introduced the terms *photoperiod* and *photoperiodism*. Following further experimentation they classified plants into photoperiodic groups: Short-day Plants (SDP) are those which flower or in which flowering is accelerated in days which are shorter than a "critical" day length. Long-day plants (LDP) are plants which flower or in which flowering is accelerated when the daily light period exceeds a critical day length. Plants which flower at the same time regardless of day length are called day-neutral plants (DNP). Plants that respond to day length can be further sub-divided into obligate (or qualitative) types, where a particular day length is essential for flowering, or facultative (or quantitative) types, where a particular day length accelerates but is not essential for flowering.

The word 'Photoperiodism' is derived from the Greek roots for 'light' and 'duration of time' and can be defined as the response to the changes in the daily duration of light and darkness that enable organisms to adapt to seasonal changes in their environment. A response occurs when the day or night length is longer or shorter than a threshold (critical) value. Therefore photoperiodic responses require a time-measuring mechanism, to which is closely coupled a photoperception system. Further, the timekeeping mechanism must operate very precisely and it must be insensitive to unpredictable variations in the environment. Finally, the seasonal range and rate of change of day length is a function of latitude, being lower in the tropics than at higher latitudes. In lower latitudes, timekeeping needs to be more precise than at higher latitudes in order to locate a seasonal event with the same degree of accuracy.

Selective advantage

It should not be surprising that organisms have developed the ability to detect and respond to changes in daylength. Such a response could clearly confer selective advantage because it provides a means of anticipating, and so preventing, the adverse effects of a particular seasonal environment. For example, the shortening day length during autumn precedes low winter temperatures, at high latitudes at least. The shortening days act as a signal to trigger bud dormancy and cold hardiness in plants (and to induce diapause in certain insects), responses which enable them to survive the unfavourable winter environment (Thomas and Vince-Prue 1996). Other survival strategies which develop in response to day length include the formation of storage organs such as bulbs or tubers. Summer conditions can also be unfavourable and bud dormancy is induced by the long-day conditions in some desert species (Schwabe and Naschmony-Bascombe 1963).

The other main type of photoperiodically controlled response is the switch to reproductive development (induction of flowering). Examples of plants flowering in response to lengthening days of spring or shortening days of autumn are common. Photoperiodic plants are even common in tropical latitudes where the seasonal day length changes are small and day length is used to synchronize reproductive or other activities with seasonal events such as dry or rainy periods. For flowering it also means that members of a population will flower at the same time, which increases the chances of outbreeding and thus genetic recombination. If a pollinating insect's behaviour is also photoperiodically controlled, this further improves the chance of successful pollination. Another example of the value of seasonal timing of flowering is that woodland plants can flower and seed before the dense leaf canopy is formed. Because of the ease of experimentation, induction of flowering is the response for which there is most experimental data.

Perception and induction occurs in the leaves

Regardless of whether the particular plant is a LDP or SDP, the main site for the perception of day length is the leaf, even though the observed responses usually take place elsewhere in the plant. This was demonstrated by Knott (1934) for flowering in spinach, which is a LDP. If the leaves, but not the apex, were exposed to long photoperiods this resulted in the development of floral primordia at the shoot apex, while if the apical bud alone was exposed to long days, plants remained vegetative. Similar results were obtained for vegetative responses. Tuberisation, a SD response, occurred in Jerusalem artichoke (*Helianthus tuberosus*) when the aerial stem tips were covered to give SD, provided that leaves longer than two inches were included (Hamner and Long 1939). Day length therefore is perceived in leaves and results in a localised change of properties, which constitutes a state of induction. A developmental change then occurs elsewhere as a result of a signal transmitted from the leaves to the responding organs.

Induction requires measurement of day or night length

Under the normal diurnal 24-h cycle, length of day and length of night are absolutely related. Therefore plants could detect the critical day length by measuring either the duration of light, or of darkness, or even their relative durations.

Although Tournois had originally suggested that in SDP it was the length of darkness which was crucial, it was Hamner and Bonner who first demonstrated this unambiguously. Using the SDP *Xanthium strumarium* (Cocklebur) they varied the duration of light and dark independently; flowering was only induced when the dark period was longer than a 'critical' value of 8.5 h, irrespective of the relative durations of light and darkness (Hamner and Bonner 1938; Hamner 1940). They also found that the promotive effect of a long night could be reduced or abolished if a light interruption or pulse (of a few minutes duration) was given in the middle of it, apparently confirming the importance of the dark period. Other studies on the timing mechanism revealed that, if the dark period was artificially lengthened, and interruptions (or 'night breaks') were given at regular intervals (a Bünsow protocol), a rhythm in the inhibition of flowering was revealed, with a period of about 24 h (Carr 1952). In several SDP, varying the duration of the dark period also results in a rhythm in the level of flowering, the period of the rhythm being about 24 h. (This is sometimes referred to as a Nanda-Hamner protocol). There is good evidence that the two phenomenas are manifestations of the same rhythmic process.

These developments vindicated the earlier idea of Erwin Bünning, that photoperiodism involved alternating phases of sensitivity to light, the photophile and scotophile, with the photophile beginning at the onset of the light (Bünning 1960). This model was modified to one in which there existed a rhythm of sensitivity to light, at particular phases of which light inhibited induction. Photoperiodic induction would occur if a particular inducible phase of a rhythm fell in darkness, the phase of the rhythm being a function of the duration of the photoperiod (and not simply running from dawn as in Bünning's model).

This idea constitutes an external coincidence model, in which there are two actions of light; firstly, setting the phase of the rhythm of sensitivity to light, and secondly in coincidence with the photoinducible phase, to prevent flowering, or other developmental phenomenon, such as tuberisation (Fig. 1). In

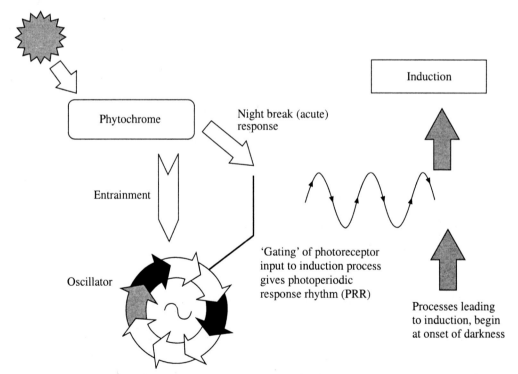

Fig. 1   **External coincidence model for the photoperiodic control of induction in SDPs. Induction occurs if light is not given at or before the critical phase, when the 'gate' is open. If light is given at this inducible phase, induction is prevented. Light inputs are shown from the receptor phytochrome, of which more than one molecular species may be involved.**

most SDP the phasing of the rhythm is very similar; after a sufficiently long photoperiod the point at which a night break results in maximum inhibition (NB_{max}) occurs 8–9 h into the dark period. Implicit in this observation is the fact that timing of the duration of darkness is achieved by the rhythm being coupled to and running from the onset of the dark period. The critical night length (CNL) generally corresponds in time to the time of NB_{max}, although under some circumstances the CNL is longer than the time to NB_{max}, but this is due to environmental effects on processes other than the operation of the circadian system.

The situation in LDP is rather different, where it is the duration of the light period which is usually measured. A brief interruption of darkness (which inhibits induction in SDP) is not usually sufficient to promote flowering in LDP; longer exposures (hours) are usually required for the plant to interpret the schedule as a long day. Rhythms in response to extended night breaks have been described in the LDPs *Sinapis*, *Hyoscyamus* and *Lolium* (Hsu and Hamner 1967; Kinet et al. 1973; Périlleux et al. 1994). A second difference from SDP is that the photoperiod should contain some far-red light, specifically, during the latter part of the day (i.e. the period whose duration is critical for the day to be interpreted as 'long'). In *Arabidopsis* a rhythm in response to 6 h periods of far-red light against a background of constant white in *Arabidopsis* (Deitzer 1984). This indicates that a rhythm of sensitivity operates, but that this is timed from the beginning of the light period, and continues into darkness. Induction occurs if the light period is long enough to illuminate a particular, sensitive phase of the rhythm.

The difference in measurement of dark and light is not strictly correlated with whether plants are SDP or LDP. For example, it is possible to identify certain SD plants, e.g. *Chrysanthemum*, where light quantity and quality does have some effect or where a long night break is required to prevent flowering. To reflect this situation plants can be classified as "light dominant" or "dark dominant", depending on whether they show the long-day or short-day type of perception response. An interesting recent example, which elegantly illustrates this distinction, is the phenomenon of bud dormancy in Norway spruce (*Picea abies*). Budset is a short-day response which requires one or a few cycles of short days and long nights (Dormling et al. 1968). However, considerable differences in sensitivity of response exist between populations of different latitudinal provenance. An arctic population of *Picea abies* will set terminal buds in response to a single long night of 16 hours, whereas a Romanian population requires 4 cycles of 8 h light/16 h dark (Qamaruddin et al. 1995). The CNL also differs between populations; from 7-10 hours for lowland populations in Romania to as little as 2–3.5 hours for populations from the arctic circle. Although in both populations bud set occurs in response to shortening days, the northern population appears to measure the (shortening) duration of the light period, while the southern population measures the (lengthening) duration of the dark period. Specifically, when plants from the two populations were subjected to day extensions of varying light quality, and were tested for their rhythmic responsiveness to night breaks, it was found that the southern population displayed dark dominant properties, i.e. the length of the dark period was being measured, while the northern population displayed light dominant features, i.e. it was measuring the duration of the light period. The inescapable conclusion is that both types of mechanism co-exist within the plant, and that selection pressures act to push populations towards one or the other response types. In this case the northern population experiences very short nights, under which conditions dark timekeeping may be a rather imprecise mechanism for measuring the duration of darkness and so may be selected against in populations originating at high latitudes.

Control of the photoperiodic response rhythm (PRR)

The rhythm in sensitivity to light is referred to as the photoperiodic response rhythm, or PRR

(Lumsden 1991), and satisfies the criteria for a circadian rhythm in having a period of about 24 h, and showing temperature compensation. A third criterion, that the phase of the rhythm can be altered by light, is the critical requirement for it to be involved in time measurement. In SDP the rhythm must be entrained or coupled to a light signal (or absence of light signal) at the end of the day, such that timing is initiated at the start of darkness.

As with other circadian rhythms, it is now generally accepted that the PRR is driven by some underlying circadian oscillator. Light which affects the phase of the rhythm does so by changing the phase of the oscillator. A range of responses to this action of light can then be identified.

(a) The PRR can be phase-shifted by light during darkness.

Plotting shift against phase at which light was given produces a phase response curve (PRC), the shape of which is common to all circadian rhythms: light early in the subjective night causes a phase delay, while light in the late subjective night or early subjective day causes a phase advance. The experiments of King and Cumming with *Chenopodium* generated a complete PRC, but required 6 h of light to generate phase shifts (King and Cumming 1972), which is rather inconvenient for investigating the photoreceptor involved. In *Pharbitis* seedlings, which are very sensitive to light, phase shifts have been induced by brief exposures to R, at certain phases of the rhythm only a few seconds of light were necessary (Lumsden and Furuya 1986).

(b) When light has been on for a longer period of time, a light-off signal rather than a phase shift is generated, and phase is constant from the end of the light period (or beginning of the dark period).

There are clear differences between species in sensitivity to this action of light. In *Chenopodium* seedlings 2 h of R had no effect on the PRR, 6 h caused phase shifts, and more than 12 h was needed for phase to be constant from the end of the light period (King and Cumming 1972). In *Pharbitis* seedlings phase shifting was achieved with as little as 6 secs R, and a light-off signal was generated by as little as 2 h of light (Lumsden and Furuya 1986). Where a constant relationship exists between the PRR and the end of the light period, the situation is more likely to be one of continuous entrainment, in which there is continuous input from photoreceptor to the oscillator rather than discrete phase shifting. Indeed this might intuitively be expected for day-light organisms, and is illustrated in a number of SDP where for photoperiods longer than a few hours NB_{max} occurs at a constant 8–9 h after the onset of darkness, for example in *Chenopodium* (Cumming et al. 1965), *Pharbitis* (Takimoto and Hamner 1965), and *Xanthium* (Papenfuss and Salisbury 1967). The implicit question 'After what duration does a light treatment cease to give a phase shift and generates a light-off signal instead' is, under most natural conditions, fairly academic. Since induction occurs as a result of daylength decreasing from the long days of summer, then in most cases the daylength will be sufficiently long for the circadian system to be subject to continuous entrainment rather than to repeating phase shifts. The phase of the PRR will therefore be constant from the onset of darkness, and when the CNL is exceeded, NB_{max} will fall in the dark period and induction will result.

(c) A rhythmic flowering response can also be seen in response to different durations of light preceeding a dark period close to the CNL.

If SDP are given a dark period just slightly longer than the CNL, induction will not be complete, and an intermediate level of flowering will result. If the length of the light period preceeding such a dark period is then varied, a rhythm in the degree of flowering occurs which is a function of the length of the photoperiod, and the period of this rhythmicity is very close to 24 h. This has been clearly seen in *Pharbitis nil* (Spector and Paraska 1973), and also in *Xanthium* (Papenfuss and Salisbury 1967) and *Chenopodium* (King 1975). Clearly, some rhythmic function is persisting during the light period.

All of these responses to light can be represented in a model in which the state variables of the oscillator occupy one of two limit cycles, a dark and a light limit cycle (Lumsden 1996). This is illustrated in Fig. 2. In constant light the oscillator moves from the dark to the light limit cycle, where it continues to oscillate, but around a fixed phase point (CT6). On return to darkness, the oscillator returns to the dark limit cycle, somewhere between CT5 and CT7, depending on the phase occupied at the time of lights off. The inducible phase, NB_{max}, then occurs about 9 h later. There will however be some variation in the time then taken for the oscillator to reach the inducible phase, which results in the observed rhythmic response to a near critical night length as a function of duration of the light period. This model is based on observations made by Peterson and Saunders (1980) for eclosion in the fly *Sarcophaga*. In darkness, the oscillator occupies the dark limit cycle, and brief exposures to light can bring about phase shifting, the extent and direction of the shift depending of course on the phase.

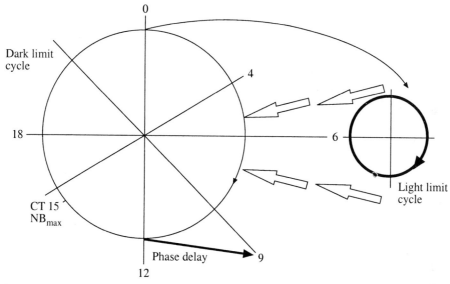

Fig. 2 **Limit cycle model for the oscillator controlling the PRR in a SDP. Lines are isochrons, representing circadian time (CT). In constant light, the oscillator moves to the light limit cycle, where it continues to oscillate between isochrons CT5 and 7. Open arrows indicate the trajectory of the oscillator following the onset of darkness, returning to the dark limit cycle. (See text for full description). The line at CT12 indicates the oscillator trajectory if a phase-shifting light interruption is given at CT12 during darkness; the oscillator moves to CT9, giving a phase delay of 3 h.**

In LDP rhythmicity is less easy to model. A well studied example is *Lolium*, where the effect of 8 h light given in a 40-h or 64-h dark period was investigated. In the 40 h dark period peaks of promotion of flowering were obtained when the interruptions were given at about 8 and 32 h but when the dark period was 60 h, only two peaks occurred, at about 4 h and 20–24 h. Light towards the end of the dark period at about 56 h was also strongly promotive (Périlleux et al. 1994). The results indicate that the response varies with the duration of the dark period, suggesting an interaction between the interrupting light period and either the preceding or following light period. A further difference between LDP and SDP is that LDP show a circadian rhythm in response to adding FR to a background of white fluorescent or red light; this rhythm runs during the light period and the phase is set by the dawn or dark to light transition. In *Hordeum* the phase of the rhythm in sensitivity to FR could itself be shifted by exposure to 6 h FR, consistent with the involvement of a circadian rhythm

(Deitzer et al. 1982), but also consistent with some kind of external coincidence model. The available evidence suggests that in LDPs the oscillator controlling the PRR occupies a single limit cycle, as compared with the light and dark limit cycle model for SDPs.

Actions of light; roles of photoreceptors

What of the photoperception system? We have established two separate actions of light, namely a direct effect on the PRR, and a phase controlling effect via the oscillator, in both SDPs and LDPs. With more than one input for light, there may be more than one pathway for a photoreceptor, and/or more than one photoreceptor involved. Several light-sensing pigments, phytochromes, blue/UV-A absorbing pigments, may be involved in the control of photoperiodic responses of higher plants. Phytochrome is now known to consist of a family chromoproteins, each of which can exist in a Pr form, which absorbs maximally in red light, or a Pfr form, which absorbs maximally in the far-red. Pr and Pfr can be reversibly interconverted by light, with biological activity usually associated with the Pfr form. The phytochrome family of proteins contains at least five different members—phytochromes A to E, encoded by separate genes. Individual phytochromes have been shown to have different biological roles as evidenced through the use of mutants and transgenic plants.

The effect of light as an interruption of darkness which inhibits flowering in SDPs or promotes flowering in LDPs has been referred to elsewhere as an 'acute' response, which would also include closure of leaflets of nyctinastic plants, enhancement of gene expression and opening of stomata. The response to a night break is clearly mediated by phytochrome, as shown by the early action spectra of the Beltsville group for the prevention of flowering in the SDPs *Glycine max* cv Biloxi and *Xanthium strumarium* (Parker et al. 1946) and for the promotion of flowering in the LDPs *Hordeum vulgare* (Borthwick et al. 1948) and *Hyoscyamus niger* (Parker et al. 1950). These are quite similar, with the most effective wavelengths of light being in the red region of the spectrum, between 600 and 660 nm. The response is R/FR reversible in a number of species (Thomas and Vince-Prue 1996), a further confirmation that phytochrome is the receptor.

Evidence that the analogous response in LDP, i.e. the promotion of flowering by light containing a high proportion of far-red, is mediated through phytochrome A comes from the work of Johnson et al (1994) who found that phytochrome A-deficient mutants of *Arabidopsis* are relatively insensitive to floral promotion by day-extensions with tungsten filament lamps, which produce light rich in far-red and which have a strong promoting effect on floral initiation in wild type plants. In *Triticum aestivum* (wheat), another LDP, wavelengths of light which were most promotive for flowering when given as day extensions also resulted in the highest amounts of phytochrome A as measured by ELISA (Carr-Smith et al. 1994). These data indicate a linkage between a specific photoreceptor and the photoperiodic mechanism in plants.

Interestingly, most LDPs do not respond to day extensions of blue light. However, *Arabidopsis,* and other members of the *Cruciferae* do respond (see Thomas and Vince-Prue, 1996). Although an involvement of phytcohrome cannot be completely excluded, the strong sensitivity of the members of the *Cruciferae*, but not other LDP, to induction by blue light, implies the action of a separate blue-absorbing photoreceptor. Support for this idea comes from studies of *Arabidopsis* where dichromatic irradiation with blue and monochromatic light at 589 nm was shown to be inductive, whereas irradiation with 589 nm light alone was ineffective (Mozley and Thomas 1995). It is not clear why the *Cruciferae* alone appear to use the blue light photoreceptor as part of their photoperiod mechanism. However it does suggest that plants are able to harness available photoreceptors into daylength sensing mechanisms, and build in redundancy in the mechanisms, presumably to provide some sort of fail safe system.

There is less direct evidence for phytochrome being the receptor for light which controls the phase of the PRR through impact on the oscillator. The main evidence is that red light (for SDP) or far-red light (for LDP) is most effective at changing the phase of the PRR (Thomas and Vince-Prue 1996).

Interestingly the picture is complicated by effects of phytochrome which do not directly involve the timing system. In *Sorghum*, a quantitative SDP, genotypes with the ma_3^R mutation have a much reduced sensitivity to photoperiod compared with the equivalent wild type, i.e. the mutants flower earlier under long days. The ma_3^R mutation has been found to be a deletion of the phytochrome B gene (Childs et al. 1997), indicating that lack of phytochrome B is responsible for the relative photoperiodic insensitivity of this mutant. In the photoperiodic potato species *Solanum tuberosum* ssp. *andigena*, which only tuberises in short days, transgenic plants that have reduced levels of phytochrome B mRNA and protein as a result of antisense inhibition are able to tuberise in long days, as well as short days. Here also the reduction in phytochrome B levels results in the loss of inhibition of tuberization that normally exists in wild-type plants in long days (Jackson et al. 1996). The antisense phytochrome B plants also tuberize at a much earlier developmental stage than normal, as with the early flowering phenotype of the *Sorghum* ma_3^R mutant. Early flowering is a common feature of phytochrome B mutants which is observed in LDP as well as SDP. *Arabidopsis* is a quantitative LDP which flowers earlier in long days than short days. The *phyB* mutant flowers earlier than wild-type in both short days and long days but nevertheless still shows a response to the different photoperiods. Phytochrome B thus appears to be correlated with the production of an inhibitory factor, normally present under long days but not short days, the removal of which allows photoperiodic responses to occur under long days.

Conclusions

It is clear that photoperiodism involves a circadian rhythm as the basis of time measurement. While most evidence is consistent with an external coincidence model, especially in SDP, suggestions have been made that photoperiodic mechanisms may operate according to an internal coincidence model. In this model there are two rhythms. One is initiated at the beginning of the light period, the other is initiated at the end of the light period or beginning of the dark period, and the phase relationship between these rhythms determines whether induction takes place or not (Thomas and Vince-Prue 1996). It is certainly worth noting that the dark dominant and light dominant types of response require the two types of rhythms (i.e. dusk-phased and dawn-phased respectively) which have been claimed to be required for internal coincidence models. It is possible to envisage situations where both mechanisms may operate together (as suggested by the studies in *Picea*), either as a precaution, should one mechanism be disabled, or to establish a dual key system, where both need to be satisfied to obtain a response.

Another interesting avenue for investigation is whether the PRR is controlled by the same oscillator which drives other rhythms such as leaf movement, stomatal aperture and *cab* gene expression. Evidence from mutants in *Arabidopsis* indicates a correlation between changes in the period length of the rhythms of *cab* gene expression and leaf movement and photoperiodic response (Millar 1999), consistent with a single oscillator controlling the photoperiodic system and the observed overt rhythm. In contrast, in SDP, the evidence indicates that certainly leaf movement and the PRR are controlled separately, as demonstrated in *Glycine* (Brest et al. 1971) and *Pharbitis* (Bollig 1977), strongly suggesting a separate oscillator for the photoperiodic response.

References

Bollig, I. (1977) Different circadian rhythms regulate photoperiodic flowering response and leaf movement in *Pharbitis nil* (L.) Choisy. Planta. **135**: 137–427.

Brest, D.E., Hoshizaki, T., Hamner, K.C. (1971) Rhythmic leaf movements in Biloxi soybeans and their relation to flowering. Plant Physiol. **47**: 676–681.

Bünning, E. (1960) Circadian rhythms and the time measurement in photoperiodism. Cold Spr. Harb. Symp. Quant. Biol. **25**: 249–256.

Borthwick, H.A., Hendricks, S.B., Parker, M.W. (1948) Action spectrum for the photoperiodic control of floral initiation of a long day plant, Wintex barley (*Hordeum vulgare*). Bot. Gaz. **110**: 103–118.

Carr, D.J. (1952) The photoperiodic behaviour of short-day plants. Physiol. Plant **5**: 70–84.

Carr-Smith, H.D., Thomas, B., Johnson, C.B., Plumpton, C., Butcher, G. (1994) The kinetics of type 1 phytochrome in green, light-grown wheat (*Triticum aestivum* L.). Planta **194**: 136–142.

Childs, K.L., Morgan, P.W., Miller, F.R., Pratt, L.H., Cordonnier-Pratt, M-M., Muller, J.E. (1997) The Sorghum bicolor photoperiod sensitivity gene, Ma_3, encodes a phytochrome B. Plant Physiol. **113**: 611–619.

Cumming, B.G., Hendricks, S.B., Borthwick, H.A. (1965) Rhythmic flowering responses and phytochrome changes in a selection of *Chenopodium rubrum*. Can. J. Bot. **43**: 825–853.

Deitzer, G.F., Hayes, R., Jabben, M. (1982) Phase shift in the circadian rhythm of floral promotion by far-red light in *Hordeum vulgare* L. Plant Physiol. **69**: 597–601.

Deitzer, G.F. (1984) Photoperiodic induction in long-day plants. In: Vince-Prue, D., Thomas, B., Cockshull, K.E. (eds.) Light and the Flowering Process. Academic Press, pp. 51–64.

Dormling, I., Gustafsson, Å., von Wettstein, D. (1968) The experimental control of the life cycle in *Picea abies* (L.) Karst. Silvae. Genet. **17**: 44–64.

Garner, W.W., Allard, H.A. (1920) Effect of the relative length of day and night and other factors of the environment on growth and reproduction in plants. J. Agric. Res. **18**: 553–606.

Garner, W.W., Allard, H.A. (1923) Further studies on photoperiodism, the response of plants to relative length of day and night. J. Agric. Res. **23**: 871–920.

Hamner, K.C. (1940) Interaction of light and darkness in photoperiodic induction. Bot. Gaz. **101**: 658–687.

Hamner, K.C., Bonner, J. (1938) Photoperiodism in relation to hormones as factors in floral initiation and development. Bot. Gaz. **100**: 388–431.

Hamner, K.C., Long, E.M. (1939) Localization of photoperiodic perception in *Helianthus tuberosus*. Bot. Gaz. **101**: 81–90.

Hsu, J.C.S., Hamner, K.C. (1967) Studies on the involvement of an endogenous rhythm in the photoperiodic response of *Hyoscyamus niger*. Plant. Physiol. **42**: 725–730.

Jackson, S.D., Thomas, B. (1997) Photoreceptors and signals in the photoperiodic control of development. Plant Cell Environ. **20**: 790–795.

Jackson, S.D., Heyer, A., Dietze, J., Prat, S. (1996) Phytochrome B mediates the photoperiodic control of tuber formation in potato. Plant J. **9**: 159–166.

Johnson, E., Bradley, M., Harberd, N.P., Whitelam, G.C. (1994) Photoresponses of light-grown *phyA* mutants of *Arabidopsis*. Plant Physiol. **105**: 141–149.

Kinet, J.M., Bernier, G., Bodson, M., Jacqmard, A. (1973) Circadian rhythms and the induction of flowering in *Sinapis alba*. Plant Physiol. **51**: 598–600.

King, R.W. (1975) Multiple circadian rhythms regulate photoperiodic flowering responses in *Chenopodium rubrum*. Can. J. Bot. **53**: 2631–2638.

King, R.W., Cumming, B.G. (1972) Rhythms as photoperiodic timers in the control of flowering in *Chenopodium rubrum* L. Planta **103**: 281–301.

Klebs, G. (1913) Über das Verhältnis der Aussenwelt zur Entwicklung der Pflanze. Sber. Akad. Wiss. Heidelberg **5**: 1–47.

Knott, J.E. (1934) Effect of a localized photoperiod on spinach. Proc. Am. Soc. Hort. Sci. **31**: 152–154.

Lumsden, P.J. (1991) Circadian rhythms and phytochrome. Ann. Rev. Plant Physiol. Plant Mol. Biol. **42**: 351–371.

Lumsden, P.J. (1996) A limit cycle model for the circadian clock in the photoperiodic control of flowering in short-day plants. The Flowering Newsletter **21**: 42–47.

Lumsden, P.J., Furuya, M. (1986) Evidence for two actions of light in the photoperiodic induction of flowering in *Pharbitis nil.* Plant Cell Physiol. **27**: 1541–1551.

Millar, A. (1999) Biological clocks in *Arabdopsis thaliana.* New Phytol. **141**: 175–197.

Mozley, D., Thomas, B. (1995) Developmental and photobiological factors affecting photoperiodic induction in *Arabidopsis thaliana* Henh. *Landsberg erecta.* J. Expt. Bot. **46**: 173–179.

Papenfuss, H.D. Salisbury, F.B. (1967) Aspects of clock resetting in flowering of *Xanthium.* Plant Physiol. **42**: 1562–1568.

Parker, M.W., Hendricks, S.B., Borthwick, H.A., Scully, N.J. (1946) Action spectrum for the photoperiodic control of floral initiation of short-day plants. Bot. Gaz. **108**: 1–26.

Parker, M.W., Hendricks, S.B., Borthwick, H.A. (1950) Action spectrum for the photoperiodic control of floral initiation of the long day Plant. *Hyoscyamus niger.* Bot. Gaz. **111**: 242–252.

Périlleux, C., Bernier, G., Kinet, J-M. (1994) Circadian rhythms and the induction of flowering in the long-day grass *Lolium temulentum* L. Plant Cell Environ. **17**: 755–761.

Peterson, E.L., Saunders, D.S. (1980) The circadian eclosion rhythm *in Sarcophaga argyrostoma*; a limit cycle representation of the pacemaker. J. Theor. Biol. **86**: 265–277.

Qamaruddin, M., Ekberg, I., Dormling, I., Norell, L., Clapham, D., Eriksson, G. (1995) Early effects of long nights on budset, dormancy, and abscisic acid content in two populations of *Picea abies.* Forest Genet. **2**: 207–216.

Schwabe, W.W., Naschmony-Bascombe, S. (1963) Growth and dormancy in *Lunularia cruciata* (L.) Dum. II. The response to daylength and temperature. J. Exp. Bot. **14**: 353–378.

Spector, C., Paraska, J.R. (1973) Rhythmicity of flowering in *Pharbitis nil.* Physiol. Plant. **29**: 402–405.

Takimoto, A., Hamner, K.C. (1965) Studies on red light interruption in relation to timing mechanisms involved in the photoperiodic response of *Pharbitis nil.* Plant Physiol. **40**: 852–854.

Thomas, B., Vince-Prue, D. (1996) Photoperiodism in Plants. Academic Press. London.

Tournois, J. (1912) Influence de la lumière sur la floraison du houblon japonais et du chanvre déterminées par des semis haitifs. C.R. Hebd. Séanc. Acad. Sci. Paris **155**: 297–300.

Tournois, J. (1914) Études sur la sexualité du houblon. Annis. Sci. Nat. (Bot.) **19**: 49–191.

Biological Rhythms
Edited by V. Kumar

16. Photoperiodism in Birds and Mammals

S.L. Meddle[1]*, G.E. Bentley[2] and V.M. King[3]

[1]Department of Biomedical Sciences, University of Edinburgh Medical School, George Square, Edinburgh EH8 9XD UK

[2]Department of Psychology, Behavioral Neuroendocrinology Group, Johns Hopkins University, 3400 North Charles Street, Baltimore, Maryland 21218, USA

[3]Department of Anatomy, University of Cambridge, Downing Street, Cambridge CB2 3DY, UK

Using the annual cycle of changing day length, photoperiodic animals restrict their reproductive efforts to a favorable time of year. Thus, the perception and measurement of day length are vital for maximal reproductive success. A clock mechanism to measure day length is necessary for photoperiodic responses. In one section, this chapter reviews the basic principles of the biological clock for measurement of and entrainment to a 24 h light: dark cycle. Some differences between birds and mammals and the ways in which they measure changing day length are highlighted; most notably, differences in photoreception and in the role for the pineal melatonin signal in the transduction of the light: dark signal. The mammalian clock has received a great deal of attention in recent years because of the identification of several clock genes and greater knowledge of how they interact. The clock in birds is less well understood. Birds measure day length in a circadian manner, but in contrast to mammals, pineal melatonin is not involved in photoperiodic time measurement. Seasonal breeding cycles of birds and the photoperiodic regulation of the hypothalamo-pituitary gonadal axis are discussed, as is a newly-identified role for melatonin in birds. Melatonin acts as an inhibitory hormone on seasonal neuroplasticity within the song control system of songbirds, acting in opposition to the stimulatory effects of gonadal steroids. Thus, there is a complex interaction between the circadian system, photoperiodic time measurement, activation and regulation of the neuroendocrine system regulating reproduction and hormonal actions upon the brain.

Photoperiodic regulation of seasonal reproduction

There are obvious advantages to seasonal breeding for animals that inhabit temperate zones. To ensure that offspring are produced at a time of year that maximises their best chance of survival these animals must anticipate the arrival of abundant resources and favourable weather conditions. As it takes weeks for aspects of the reproductive cycle (e.g. gonadal growth, gestation and incubation) to take place, and because the length of time for these physiological processes is fixed, this anticipation is a necessity. The evolutionary advantages gained from the ability to use predictable signals from the environment as a cue for the timed initiation of reproduction have resulted in a very powerful positive selective pressure. In middle and high latitudes the annual change in day length provides a robust and reliable indication to the organism of the time of year. Other important environmental cues such as temperature, food and mate availability also have profound effects on the onset of breeding. However

---

*E-mail: slmeddle@srv4.med.ed.ac.uk

here we will concentrate on how day length is perceived, measured and transduced into an endocrine signal. This chapter discusses the role of the circadian system in photoperiodic time measurement. It will review what is known at present about the neuroendocrine mechanisms controlling gonadotrophin releasing hormone (GnRH) release and finally a role for melatonin is discussed in relation to its effect on seasonal neuroplasticity in the song bird brain.

The role of the circadian clock in photoperiodic time measurement

Before any discussion of photoperiodic time measurement (PTM) in birds and mammals, it is useful to briefly review the structure of their circadian systems. The mammalian circadian system is made up of photoreceptive eyes, a circadian clock located in the suprachiasmatic nucleus (SCN) of the hypothalamus, and output pathways of clock controlled genes. The photic signal is perceived by the eyes and conducted via the retino-hypothalamic tract to the SCN where it acts to entrain the circadian clock to the external 24 h day/night cycle. The SCN is thought to drive all circadian oscillations of physiology and behaviour. One of the main cycles driven by the SCN, at least in terms of the circadian system itself, is the rhythm of melatonin secretion from the pineal gland. In the mammalian system melatonin is vital for the entrainment of foetal rhythmicity and for conveying photoperiodic information (Arendt 1997).

Birds too have photoreceptors, a circadian clock, and output pathways controlling physiological and behavioural rhythms, but there are notable differences as compared to mammals. Birds have three photoreceptors—the eyes, pineal and as yet unidentified 'deep brain photoreceptors', although the eyes and pineal are not requisite for the photoperiodic response. They have three circadian oscillators—the eyes, pineal and a hypothalamic pacemaker, the location of which is yet to be identified. Furthermore, both the eyes and pineal produce a circadian rhythm of melatonin secretion (see Gwinner et al. 1997 for review). In all bird species that have been examined, there is a robust daily rhythm in melatonin that is a function of night length [some examples are quail (*Coturnix japonica*), blackheaded buntings (*Emberiza melanocephala*) and garden warblers (*Sylvia borin*) (Kumar and Follett 1993; Kumar et al. 1996; Gwinner et al. 1993]. However in contrast to mammals, birds do not use this annual profile of nocturnal pineal melatonin secretion to convey photoperiodic time (Turek and Wolfson 1978; Juss et al. 1993).

Basic principles

In the early studies of photoperiodism by Bünning in 1960, PTM was imagined to be a photoperiodic reaction and was termed the "external coincidence" model. This model simply envisaged a photoinducible phase (Øi), occurring in the early subjective night, that when illuminated by light during the long days of late spring and summer resulted in a long day reproductive response. Earlier investigators had imagined this to be an hour-glass rhythm of photoinducibility, but in his model Bünning (1960) described a daily photoinducible phase generated by an endogenous circadian clock. This "external coincidence" model proposed two roles for light, one entraining the circadian system, and the other controlling photoperiodic induction by temporal coincidence with the photoinducible phase.

The first demonstration of a circadian basis for PTM was performed by Nanda and Hamner (1958) using resonance cycles on the Biloxi soybean. In these experiments the investigators used light/dark cycles of nT and $nT + 1/2T$, where $T = 24$ hours, n = no. of cycles and where the light pulse is equivalent to a short day e.g. 6 hours. Because the 6 hour light pulse is less than the critical day length required to produce long day responses (under normal circumstances), a long-day response in an hour-glass system will not occur in any of these photoperiodic paradigms. However, if a circadian oscillator controls the photoinducible phase then the cycles of nT result in a short-day photoperiodic response, while the cycles of nT + 1/2T cause a long-day photoperiodic response. Indeed, nT cycles produced short-day responses, while nT + 1/2T cycles produced long-day responses. Since this first experiment, photoperiodic induction has been observed in the Syrian hamster, *Mesocricetus auratus* (Elliott et al. 1972), house finch, *Carpodacus mexicanus* (Hamner 1963), blackheaded bunting, *Emberiza melanocephala* (Tewary and Kumar 1981), common Indian rosefinch, *Carpodacus erythrinus* (Kumar and Tewary 1982), brahminy myna, *Sturnus pagodarum* (Kumar and Kumar 1993), and European starling, *Sturnus vulgaris* (King et al. 1997) following nT + 1/2 T cycles. These results are best explained by the interpretation proposed by Hamner and Enright (1967) in which they measured both testicular size and activity rhythms in house sparrows (*Passer domesticus*). As predicted birds grew their testes when exposed to nT + 1/2T photocycles but not under nT. The real value of this experiment was that activity rhythms were used as a phase marker for the central circadian pacemaker and the testicular size was used as a marker for the reproductive response. This allowed the visualisation of the coincidence of every second light pulse in the nT + 1/2T cycles with the early subjective night and hence the circadian rhythm of photoinducibility resulting in long day photoperiodic responses.

A second hypothesis accounting for the circadian photoperiodic response is called the "internal coincidence" model and was proposed by Pittendrigh (1972). In this model the phase relationship between two circadian rhythms (one associated with dawn and the other associated with dusk) is both dependent upon and altered by changing daylengths. It is the nature of this changing phase relationship or the degree to which these two oscillations interact at different times of the year that drives the appropriate reproductive response. Perhaps the greatest weakness of this hypothesis is that it is very difficult to test, and to date no investigation has been successful in conclusively distinguishing it from the "external coincidence" model.

Other experiments, such as T cycles, have been designed to investigate the dual role of light in the external timing hypothesis, that is entraining the circadian system and in illuminating the photoinducible phase (Pittendrigh and Minis 1964). In this way it is possible using a light pulse shorter than the critical day length necessary to induce a long day response within a T cycle of the appropriate length, to entrain the clock with a light pulse coincident with the photoinducible phase. Indeed, Farner et al. (1977) successfully employed such a paradigm to induce gonadal growth in house sparrows. A relatively recent study of Kumar and Kumar (1995) shows that exposure to T-cycles alters critical day length for photoperiodic induction in the blackheaded bunting.

Night break experiments revealed the circadian nature of photoinducibility (Øi) in the white-crowned sparrow, *Zonotrichia leucophrys gambelii* (Follett et al. 1974) and in the blackheaded bunting (Kumar et al. 1996) (Figure 1). These 'night break' or modified Bünsow experiments are designed along similar lines as resonance and T experiments. Here the photoperiodic system is entrained with longer pulses of light, e.g. six or eight hours in duration, and shorter 'night break' pulses of approximately one or two hours placed at various times through the long dark period, testing for a persistent circadian rhythm of photoinducibility.

Birds

Since Hamner and Enright (1967) demonstrated that PTM is circadian in nature, there have been few advances in our understanding in the mechanisms involved in PTM in birds, let alone the biochemical pathways by which the photoperiodic information is perceived and decoded. It has been over 30 years since Menaker and Keats (1968) first demonstrated that locomotor rhythms were entrained to light-dark cycles by extraretinal photoreceptors in house sparrows. Testicular recrudescence and entrainment

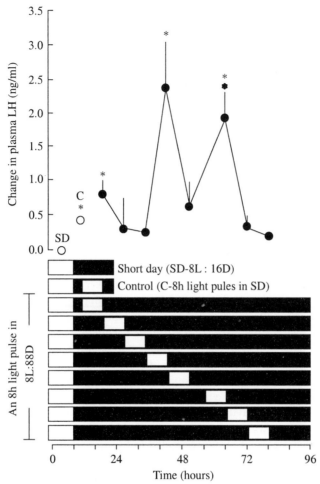

Fig. 1 The circadian nature of the photoperiodic response rhythm (Øi) in the blackheaded bunting (*Emberiza melanocephala*). In this experiment an 8 h light pulse was given at various times (indicated in the bottom bars: open portion—light period; closed portion—dark period) during 88 h of darkness (DD) and the photoperiodic effect assessed by the change in plasma concentrations of LH (filled circles). The capacity of the 8 h light pulse to cause induction clearly varies in a circadian fashion. Taken from Kumar et al. (1996).

of activity persisted in birds—that had been both enucleated and pinealectomised. These responses were abolished if light was prevented from penetrating the skull by the injection of India-ink under the scalp of these birds. Since then it has been established that the eyes are unnecessary for all photoperiodically driven processes such as nocturnal restlessness (or *zugunruhe*), migratory fat deposition and photorefractoriness. It is known that a 'deep brain photoreceptor' exists, since photoperiodic responses can still occur in birds with both the eyes and pineal removed (Wilson 1991), but the precise locations of these photoreceptors still await elucidation. Nonetheless, previous work in quail by Oliver and Bayle (1982) and Sicard et al. (1983) provided evidence for the hypothalamic infundibular region and the paraolfactory bulb to be critical areas for photoinduction. In the white-crowned sparrow, the ventromedial hypothalamus and tuberal region were shown to be sites for encephalic photoreception (Yokoyama et al. 1978). Immunocytochemical staining for opsin (an integral membrane photoreceptor protein) revealed opsin immunoreactivity in the lateral septum and tuberal hypothalamus of ring doves (*Streptopelia risoria*), ducks (*Anas platyrhynchos*), quail (Silver et al. 1988) and in songbirds (Saldanha et al. 1994). In quail it is known that these extra-retinal photoreceptors have a maximum spectral sensitivity around 492 nm, demonstrating the involvement of a rhodopsin-like photopigment (Foster and Follett 1985).

The existence of a paired suprachiasmatic nuclei (SCN) in the mammalian hypothalamus is indicative of an equivalent structure in birds. Initial attention was given to the *medial* SCN that lies bilaterally around the lateral recess of the third ventricle due to its homology to the mammalian SCN. Large lesions in this region disrupt circadian locomotor rhythms in quail (Simpson and Follett 1981). Nonetheless, the *medial* SCN is only weakly retinorecipient and does not express c-*fos* protein (Fos) in response to light (King and Follett 1997). Fos is a member of a family of immediate early genes that couple short-term signals received at the cell surface to long term cellular phenotype alteration (Morgan and Curran 1991). Another putative avian pacemaker was hypothesised to be located in the *visual* SCN, a more lateral structure that is highly retinorecipient (Wallman et al. 1994). King and Follett (1997) found this nucleus to express Fos following light stimulation in both quail and starlings. Despite this fact, they found that its expression was not dependent upon the circadian phase at which the light was given. To date, the identity of the avian pacemaker remains unresolved.

Although in most organisms studied so far PTM has a circadian basis, in a number of species it is not so clear cut. For example, both circadian and hour-glass type reproductive responses to photoperiodic schedules have been shown for the insect *Pieris brassica* (Dumortier and Brunnarius 1989) and the Japanese quail (see King et al. 1997). One possible explanation for these findings is that the hour-glass and circadian mechanisms are qualitatively the same with circadian pacemakers so highly damped that they can no longer support any rhythmicity i.e. an hour-glass pacemaker. The case of Japanese quail is an intriguing one, with a number of different studies showing hour-glass, or circadian type PTM with others showing heavily damped circadian rhythms. In quail rhythms of activity and photoinducibility are notoriously unstable (Juss et al. 1995). In fact, the existence of a circadian rhythm of Øi only becomes clear if birds are trained to live in darkness in photocycles such as LD 6:66 (one light pulse every 3 days). Even then, in these night break experiments Øi rapidly damps out (Follett et al. 1992). In an attempt to shed light onto these contradictory results King et al. (1997) examined the activity cycles and gonadal growth in quail and starlings when exposed to resonance paradigms. Both species are long day breeders but under LD 6:30 resonance cycles only starlings showed a typical long day response and grew their gonads. This suggests that starlings have an intact circadian system regulating PTM, while in quail it is an hour-glass timing system despite the fact that it is known that quail can produce circadian PTM responses (Follett et al. 1992). To reconcile these seemingly contradictory findings, the activity rhythm of these two species was also recorded. Quail and starlings both entrained normally to LD 6:18 as expected, but here the similarities stopped. Starlings entrained to LD 6:30 by maintaining 24 h periodicity (Figure 2) in exactly the same manner as the house sparrows in the classical Hamner and Enright (1967) study. In addition, when released into constant darkness (DD), the activity phase free-ran from the phase of the entrained activity. Quail, on the other hand, showed a consistent bout of activity during every light pulse, followed approximately 24 h later by a further bout of activity. When released into DD the activity cycle free-ran from the phase of the last pulse of light. This activity was consistent with a circadian system being reset to the same phase point by every 6 h light pulse in an hour-glass fashion. To investigate further, a phase response curve (PRC) to 6 h light pulses in DD was produced for both species. Starlings produced a standard type 1 PRC, while quail produced type 0. In order to establish if in quail each

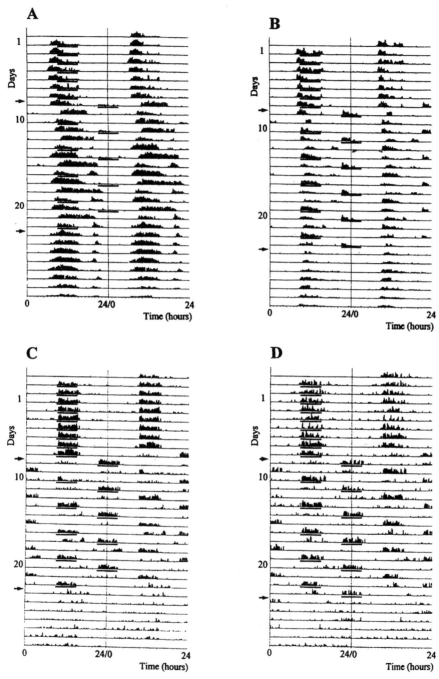

Fig. 2   Actograms of starlings (A, B) and quail (C, D) entrained to LD 6:18, followed by either 10 (A, C) or 11 (B, D) cycles of LD 6:30, then released into constant darkness and allowed to free-run. The horizontal black bars show the light phase of the LD cycle and lie immediately below the histogram of the day during which they fell; the arrows refer to the first and last light pulse of the LD 6:30 resonance paradigm. Note how the starlings maintain 24 h entrainment during exposure to LD 6:30, and free-run from the phase of the entrained activity. This contrasts to quail where no such 24 h entrainment occurs during exposure to LD 6:30. Instead, a small active phase is apparent approximately 24 h after each light pulse, and activity free-runs from the phase of the last pulse of light. Taken from King et al. (1997).

6 h light pulse caused resetting to a constant phase, the PRC was re-plotted with the circadian phase on the x-axis and the phase relationship between light and subsequent free-running activity on the y-axis. This produced a, horizontal line showing that, regardless of the circadian phase at which the light pulse fell, the phase relationship between the light pulse and subsequent activity was always the same. These results in conjunction with the various findings of the Follett group describe a circadian system that under certain (as yet unspecified) conditions will drive PTM but which under other circumstances operates as a highly damped circadian, or hour-glass oscillator.

More recently Zivkovik et al. (1999) argued that the circadian system, as reported by the core body temperature of the quail, is robust and that the LD 6:30 cycle is within the range of circadian entrainment of the quail circadian system. Nonetheless, the data from this experiment were largely consistent with the results of King et al. (1997).

Photoperiodism in mammals: recent developments

The role of the circadian clock in PTM is virtually understood in mammals. The clock is located in the SCN and drives the circadian cycle of melatonin production and release in the pineal gland. In contrast to birds and lower vertebrates, the mammalian pineal is not photoreceptive nor contains a functional circadian clock. Indeed, the role of the pineal in the mammalian circadian system is simplified to the production of melatonin, which in the context of reproduction conveys night length. This rhythm of melatonin production and secretion has several important features. First, melatonin is only expressed at night, and the duration of the peak expression reflects night duration. Secondly, the duration but not the phase of the melatonin signal is important for conveying photoperiodic information (see Bartness et al. 1993) suggesting that the melatonin signal does not act on the clock. Further, the duration of the melatonin-free interval is also critical to the development of the reproductive response (Maywood et al. 1991). These factors, that is high night-time melatonin concentrations, the duration (but not the phase) of the melatonin peak and the duration of the melatonin free interval are all required in the correct combination to elicit an appropriate reproductive response in mammals (see Arendt 1997 for review). The major questions left to answer in this field relate to the site of action of melatonin, how the information in the melatonin rhythm is decoded and how it drives the cascade of downstream events that eventually result in the appropriate reproductive responses.

The site of action of the photoperiodic melatonin signal is not known but there are three main candidates, the SCN, the dorsomedial nucleus of the hypothalamus (DMN) and the pars tuberalis (PT). The PT is the most likely site of action since melatonin binding studies across numerous photoperiodic species consistently show high binding in the PT with all other sites being variable between species. The DMN has also been identified as a potential site of action because there is high melatonin binding at this site in several species and DMN lesions block gonadotrophic responses to both melatonin and short day photoperiods in the Syrian hamster. The SCN is also favoured as a possible site of action as many mammalian species show melatonin binding here and lesions of the SCN in Siberian hamsters (*Phodopus sungorus*) can block the reproductive responses to melatonin infusions. However, this nucleus remains somewhat controversial as the site of action for melatonin in the photoperiodic reproductive response in mammals as SCN lesions studies using Syrian hamsters and sheep found no such effects (see Bartness et al. 1993 and Hastings et al. 1995 for review). Also, melatonin does not influence the expression of *c-fos* (a proto-oncogene of the family of immediate early genes, IEGs,) in the SCN of rats and hamster (Kumar et al. 1997).

More recently, the identification of numerous putative clock genes has allowed an important and potentially far-reaching advance in the understanding of both where and how the melatonin signal is transduced. Messager et al. (1999) investigated the expression of mPer1 and inducible cAMP repressor (ICER) mRNA in the SCN and PT in Syrian hamsters under long and short days. mPer1 mRNA expression was found to rise dramatically to a peak early in the light phase in both the SCN and PT. The authors concluded that photoperiod drives changes in mPer1 and ICER mRNA amplitude in the PT and it also altered the duration of the mPer1 peak in the SCN. However several further hypotheses can be derived. Firstly, it is likely that the inhibitory melatonin signal drives the rhythm of gene expression in the PT. This hypothesis is supported by the fact that in mouse strains that do not produce melatonin, no mPer1 is detected in the PT. Since mPer1 protein is a transcription factor and ICER is a transcription factor repressor, the photoperiodic control of their expression via the melatonin signal provides a feasible decoding and output mechanism for the photoperiodic signal. This raises the question of whether these genes (and by implication, the whole circadian oscillator) are also being driven by the melatonin signal, especially as Messager et al. (1999) indicated that they have unpublished data showing other putative clock genes are also expressed in the PT. If this is the case then this will be the first evidence for the existence and mechanism of the hypothesised 'slave oscillator' in the mammalian circadian system. Furthermore, it will demonstrate the existence of an hour-glass oscillator and show that it has the same basic mechanism as SCN self-sustaining circadian oscillators.

The neuroendocrine mechanisms regulating reproduction in birds

Photostimulation

Most birds, including those in the tropics, have the capability to respond to even the smallest increase in photoperiod (e.g. Hau et al. 1998; Bentley et al. 2000). However, the intensity of the light signal is vital as an alteration in day length perception occurs if light intensity is reduced (Bentley et al. 1998a). During photostimulation, once critical day length has been reached, there are immediate changes within the neuroendocrine system. Despite the fact that reproductive responses to photoperiod are well documented, very little is known about mechanism by which the photoperiodic cues are conveyed to the downstream neuroendocrine machinery that generates the reproductive response. The use of immediate early genes offers a novel approach by which to uncover the brain circuits involved in photoperiodism, particularly if the animal shows a rapid change to day length. In 1995 Meddle and Follett discovered that photoperiodic stimulation resulted in the appearance of Fos within the median eminence and basal tuberal hypothalamus of quail. Following on from this finding it was observed that Fos expression occurred by hour 18 from dawn of the first long day (Meddle and Follett 1997) that is, prior to the first rise in detectable luteinising hormone (LH) (Figure 3). This suggests therefore, that gene expression maybe involved in events preceeding photostimulation. This is especially intriguing, as we know Øi lies 12-16 hours after dawn (see Follett et al. 1998, for review) and that the endocrine signal (GnRH) rises significantly at the same time as LH, some 4 hours following Øi (Perera and Follett 1992). Fos activation in the median eminence was located within glial cells, providing more support to the view that GnRH release may be controlled at the cell terminals by glia. Another important feature of the photoperiodic response is that a single long day leads to a prolonged period of induction even if the quail are replaced back on short days. This feature has been termed "carry over" and also argues for a genomic component in the photoneuroendocrine machinery.

The role of gonadotrophin releasing hormone (GnRH) in the photoperiodic control of reproduction

Avian GnRH I was first purified in the early 1980's from chicken pituitaries (see Dunn and Millam 1998 for review). This was later followed by a discovery of a second form of GnRH, termed GnRH

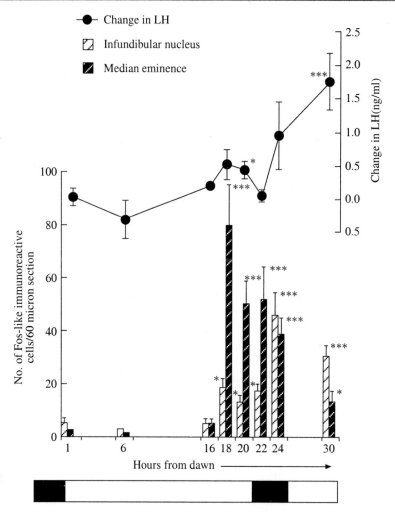

Fig. 3   Activation of two separate populations of brain cells within the basal tuberal hypothalamus in the brain of Japanese quail following transfer to a long day. Activation was demonstrated by counting the number of cells that expressed Fos immunoreactivity, indicative of immediate early genes being switched on. The first significant increase in Fos expression is 18 h from dawn, 2 hours prior to the first significant increase in LH. Statistical comparisons are shown against the values at 'dawn': *, $P < 0.05$; ***, $P < 0.001$. Taken from Meddle and Follett (1997).

II (Miyamoto et al. 1984). Both forms differ from the mammalian form of GnRH and their presence has been detected in all birds species studied. Neurones containing GnRH I are located within the preoptic region and those containing GnRH II are located more caudally in the midbrain (see Ball and Hahn 1997 for review). Both forms of GnRH stimulate gonadotrophin release, but only GnRH I is found within the median eminence suggesting that this peptide has a physiological gonadotrophin function. GnRH I fibres project from the cell bodies to terminate in the median eminence where GnRH is secreted into the hypothalamo-hypophysial portal system. The anterior pituitary is stimulated to release the gonadotrophins LH and FSH by GnRH and reproductive development ensues. Recent experiments by Baines et al. (1999) have demonstrated an increase in GnRH mRNA following photostimulation indicating that GnRH I neurones themselves become activated at this time.

Photorefractoriness

In seasonally breeding birds the end of the breeding season is heralded by rapid gonadal regression. This programmed gonadal regression occurs irrespective of photoperiod and has been termed photorefractoriness. Whilst in a photorefractory condition the response of the reproductive axis to long days changes and it is actively suppressed by the interaction of long days and thyroid hormones (Goldsmith and Nicholls 1984). In birds that are termed "absolutely photorefractory" the gonads will remain regressed as long as the birds experience long days. The recovery of photosensitivity only occurs following a period of short day exposure. This is in contrast to a "relatively refractory" state expressed by Japanese quail which is characterized by the lack of spontaneous gonadal regression under long photoperiods. The reproductive neuroendocrine system only becomes inactive when day length is shortened. We are only just starting to understand the mechanisms underlying photorefractoriness. We know that photorefractoriness occurs at the hypothalamic level as the pituitary remains sensitive to GnRH and the gonads remain responsive to LH and FSH year round (see Dawson 1999 for review). Previous studies have suggested that the decline in releasable GnRH I causes photorefractoriness, as levels of GnRH I are correlated with this condition. However this theory does not account for the finding that European starlings become absolutely photorefractory before levels of GnRH I in its precursor peptide proGnRH-GAP, decline in the median eminence (Parry et al. 1997). Certainly, photorefractoriness is associated with decrease in hypothalamic GnRH, an increase in prolactin (see Dawson 1999 for review), an increase in the number of synapses onto GnRH cell bodies (Parry and Goldsmith 1993) and an increase in hypothalamic prolactin-releasing hormone, vasoactive intestinal polypeptide (Deviche et al. 2000).

Evidence suggests a neural mechanism for photorefractoriness, but peripheral thyroid hormones and prolactin also play a part in the regulation of GnRH secretion. Prolactin itself does not appear to cause photorefractoriness (Dawson and Sharp 1998) but it is coincident with moult onset. Thyroid hormones too are required for the development and maintenance of photorefractoriness, and may be acting through expression of nerve growth factor which in turn causes synaptogenesis onto GnRH I cell bodies (Bentley et al. 1997).

Recently, Meddle et al. (1999) provided evidence that photorefractory birds retain the capacity to secrete GnRH by demonstrating that photorefractory white-crowned sparrows are able to elicit an LH rise following administration of the neuroexcitatory amino acid glutamate analogue *N*-methyl-D-aspartate (NMDA). This supports the hypothesis that reproductive quiescence of the reproductive axis in this species is not initially a consequence of an exhaustion in GnRH. In addition NMDA caused Fos expression in the basal tuberal hypothalamus, the same region that becomes activated following photoperiodic stimulation. Photorefractory individuals also exhibited no significant decrease in the amount of GnRH I within the hypothalamus. This reinforces the view that changes in GnRH I number are not central to changes in reproductive function and instead it is the neural regulation of GnRH I secretion that determines seasonality in this species.

The role of melatonin in seasonal neuroplasticity in songbirds: Its action as an inhibitory hormone

Coincident with changes in reproductive activity, seasonal neuroplasticity occurs within a discrete network of telencephalic nuclei that are involved in song learning and production. Increases in song control nuclei volume largely depend upon seasonal increases in circulating testosterone (T) and its metabolites (Nottebohm et al. 1987). There are, however, gonad- and testosterone-independent seasonal changes in song nuclei volume (Bernard et al. 1997). Until recently, it has been unclear as to what factors might be contributing to these T-independent neuronal changes. In a recent experiment,

Bentley et al. (1999) castrated European starlings to remove the neuromodulating activity of gonadal steroids and exposed them to different photoperiods to induce reproductive states characteristic of different seasonal conditions. Long days increased the volume of the song control nucleus HVC (or High Vocal Center) compared to its volume on short days. However, birds implanted with exogenous melatonin and exposed to long days did not exhibit volumetric increases in HVC. This effect was observed regardless of reproductive state or circulating steroid levels. This is the first direct evidence of a role for melatonin in functional plasticity within the central nervous system of vertebrates. It appears that melatonin is acting in an inhibitory fashion to "fine-tune" the more dramatic stimulatory effects of T on the song system, thereby precisely timing the volumetric changes to a specific time of the year.

Recently, it has been discovered (via quantification of melatonin receptor activity within the song control nuclei) that there is an interaction of reproductive state with the action of melatonin upon the song system at different stages during the annual cycle (Bentley and Ball 2000). Brains were sampled from photosensitive starlings exposed to short days, photostimulated starlings exposed to long days and photorefractory starlings also exposed to long days. Each condition contained groups of gonad-intact and castrated birds. Melatonin receptor (MelR) distribution was assessed *in vitro* by 125-Iodomelatonin (IMEL) receptor autoradiography and its distribution was similar to that described in other songbird species. However, there was a striking down-regulation of MelR in the song control nucleus Area X of intact and castrated photostimulated birds on long days, as compared to their photorefractory counterparts on the same long days and to the short-day groups (Figure 4). Down-regulation of MelR occurred independently of gonadal steroids, as the same general pattern was seen in castrated and intact starlings (but see Bentley and Ball 2000 for discussion). Furthermore, only changes in reproductive state (in intact birds and castrates), not changes in photoperiod or duration of melatonin signal, affected the density of MelR in Area X of photorefractory starlings. This conclusion can be drawn because the IMEL binding pattern was markedly different in Area X of the photostimulated and photorefractory groups, even though they were held on the same 18L:6D photoperiod. It appears that the only period during the starling reproductive cycle in which MelR are present in low densities within Area X is during the breeding season (photostimulated). As this dramatic change occurs even in castrated starlings it indicates that it is a consequence of a physiological change associated with reproductive state (i.e., whether the reproductive axis is "on" or "off"), rather than a result of changes in circulating gonadal steroids. This association with reproductive state is reminiscent of the suppression of cell-mediated immune function, which is also associated with photostimulation (Bentley et al. 1998b).

Such observations add weight to the idea that there are complex physiological events that occur when birds change reproductive states, and that these centrally-mediated changes are not limited to the hypothalamo-pituitary-gonadal axis (Nicholls et al. 1988). It appears that just as photorefractoriness in starlings is mediated by a central mechanism that is dependent upon the presence of thyroid hormones (Goldsmith and Nicholls 1984), then so is the up-regulation of melatonin receptors as starlings become photorefractory. The results of Bentley and Ball (2000) are consistent with the hypothesis that melatonin acts as an "inhibitory hormone" upon the song control system, possibly acting to reduce the energetic costs of maintaining large volumes of specific song control nuclei (Jacobs 1996). This is especially true when one considers that melatonin inhibits many second messenger systems (Vanecek 1998), including accumulation of the second messenger adenosine 3',5'-cyclic monophosphate (cAMP).

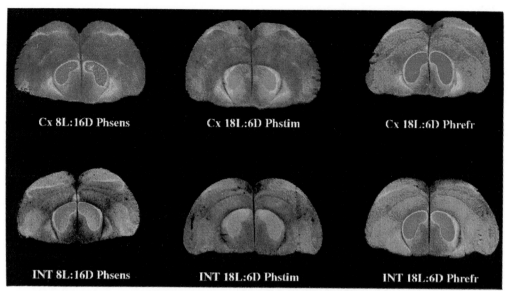

Fig. 4 Typical examples of pseudocolor-enhanced autoradiograms of [125]-Iodomelatonin (IMEL) receptor binding in Area X of castrated (Cx) or intact (INT) male European starlings in different reproductive conditions. Those that were exposed to short days and photosensitive = 8L:16D Phsens, those exposed to long days and photostimulated = 18L:6D Phstim and those that were exposed to long days but were photorefractory = 18L:6D Phrefr. Blue colour indicates low binding, followed by green, yellow, red, with white indicating the highest binding. Note the striking down-regulation of binding density in the Phstim birds, whether castrated or intact. Even though the Phstim and Phrefr birds are exposed to the same 18L:6D photoperiod, they exhibit marked differences in IMEL binding. Taken from Bentley and Ball (2000).

References

Arendt, J. (1997) The pineal gland, circadian rhythms and photoperiodism. In: Redfern, P.H. and Lemmer, B. (eds.) Physiology and Pharmacology of Biological Rhythms. Springer-Verlag, Berlin, Heidelberg, pp. 375–414.

Baines, E., Boswell, T., Dunn, I.C., Sharp, R.T., Talbot, R.T. (1999) The effect of photostimulation on the levels of gonadotrophin hormone releasing hormone (GnRH) mRNA in the hypothalamus of Japanese quail (*Coturnix coturnix japonica*). J. Reprod. Fert. Abs. Series **24**: 59.

Ball, G.F., Hahn, T.P. (1997) GnRH neuronal systems in birds and their relation to the control of seasonal reproduction. In: Parhar, I.S., Sakuma, Y. (eds.) GnRH Neurons: Gene to Behavior. Brain Shuppan, Tokyo, pp. 325–342.

Bartness, T.J., Powers, J.B., Hastings, M.H., Bittman, E.L., Goldman, B.D. (1993) The timed infusion paradigm for melatonin delivery: what has it taught us about the melatonin signal, its reception, and the photoperiodic control of seasonal responses? J. Pineal Res. **15**: 161–190.

Bentley, G.E., Ball, G.F. (2000) Photoperiod-dependent and -independent regulation of melatonin receptors in the forebrain of songbirds. J. Neuroendocrinol. (in press).

Bentley, G.E., Goldsmith, A.R., Juss, T.S., Dawson, A. (1997) The effects of nerve growth factor and anti-nerve growth factor antibody on the neuroendocrine reproductive system in the European starling *Sturnus vulgaris*. Gen. Comp. Endocrinol. **107**: 428–438.

Bentley, G.E., Goldsmith, A.R., Dawson, A., Briggs, C., Pemberton, M. (1998a) Decreased light intensity alters the perception of day length by male European starlings (*Sturnus vulgaris*). J. Biol. Rhythms **13**: 148–158.

Bentley, G.E., Demas, G.E., Nelson, R.J., Ball, G.F. (1998b) Melatonin, immunity and cost of reproductive state in male European starlings. Proc. R. Soc. B **265**: 1191–1195.

Bentley, G.E., Van't Hof, T.J., Ball, G.F. (1999) Seasonal neuroplasticity in the songbird telencephalon: a role for melatonin. Proc. Natl. Acad. Sci. USA **96**: 4674–4679.

Bentley, G.E., Spar, B.D., MacDougall-Shackleton, S.A., Hahn, T.P., Ball, G.F. (2000) Photoperiodic regulation of the reproductive axis in male zebra finches, *Taeniopygia guttata*. Gen. Comp. Endocrinol. (in press).

Bernard, D.J., Wilson, F.E., Ball, G.F. (1997) Testis-dependent and-independent effects of photoperiod on volumes of song control nuclei in American tree sparrows (*Spizella arborea*). Brain Res. **760**: 163–169.

Bünning, E. (1960) Circadian rhythms and the time measurement in photoperiodism. Cold Spr. Harb. Symp. Quant. Biol. **25**: 249–256.

Dawson, A. (1999) Photoperiodic control of gonadotrophin-releasing hormone secretion in seasonally breeding birds. In: Rao, P., Kluwer, P. (eds.) Neural Regulation in the Vertebrate Endocrine System. Academic/ Plenum Publishers, New York, pp. 141–159.

Dawson, A., Sharp, P.J. (1998) The role of prolactin in the development of reproductive photorefractoriness and postnuptial molt in the European starling (*Sturnus vulgaris*). Endocrinol. **139**: 485–490.

Deviche, P., Saldanha, C.J., Silver, R. (2000) Changes in brain gonadotropin-releasing hormone-and vasoactive intestinal polypeptide-like immunoreactivity accompanying reestablishment of photosensitivity in male dark-eyed juncos (*Junco hyemalis*). Gen. Comp. Endocrinol. **117**: 8–19.

Dumortier, B., Brunnarius, J. (1989) Diet-dependent switch from circadian to hour-glass-like operation of an insect photoperiodic clock. J. Biol. Rhythms **4**: 481–490.

Dunn, I.C., Millam, J.R. (1998) Gonadotropin releasing hormone: forms and function in birds. Poultry Avian Biol. Rev. **9**: 61–85.

Elliot, J.A., Stetson, M.H., Menaker, M. (1972) Regulation of testis function in golden hamsters: a circadian clock measures photoperiodic time. Science **178**: 771–773.

Farner, D.S., Donham, R.S., Lewis, R.A., Mattocks, P.W., Darden, T.R., Smith, J.P. (1977) The circadian component in the photoperiodic mechanism of the house sparrow, *Passer domesticus*. Physiol. Zoo. **50**: 247–268.

Follett, B.K., Mattocks, P.W., Farner, D.S. (1974) Circadian function in the photoperiodic induction of gonadotrophin secretion in the white-crowned sparrow. Proc. Natl. Acad. Sci. USA **71**: 1666–1669.

Follett, B.K., Kumar, V., Juss, T.S. (1992) Circadian nature of the photoperiodic clock in Japanese quail. J. Comp. Physiol. A **171**: 533–540.

Follett, B.K., King, V.M., Meddle, S.L. (1998) Rhythms and photoperiodism in birds. In: Lumsden, P.J., Millar, A.J. (eds.) Biological Rhythms and photoperiodism in plants. BIOS Scientific Publishers Ltd., Oxford, pp. 231–242.

Foster, R.G., Follett, B.K. (1985) The involvement of a rhodopsin-like photopigment in the photoperiodic response of the Japanese quail. J. Comp. Physiol. A **157**: 519–528.

Goldsmith, A.R., Nicholls, T.J. (1984) Thyroidectomy prevents the development of photorefractoriness and the associated rise in plasma prolactin in starlings. Gen. Comp. Endocrinol. **54**: 256–263.

Gwinner, E., Schwabl-Benzinger, I., Schwabl, H., Dittami, J. (1993) Twenty-four hour melatonin profiles in a nocturnally migrating bird during and between migratory seasons. Gen. Comp. Endocrinol. **90**: 119–124.

Gwinner, E., Hau, M., Heigl, S. (1997) Melatonin: generation and modulation of avian circadian rhythms. Brain Res. Bull. **44**: 439–444.

Hastings, M.H., Maywood, E.S., Ebling, F.J.P. (1995) The role of the circadian system in photoperiodic time measurement in mammals. In: Fraschini, F., Reiter, R.J. and Stankov, B. (eds.) The Pineal Gland and its Hormones. Plenum Press, New York, pp. 95–105.

Hamner, W.M. (1963) Diurnal rhythm and photoperiodism in testicular recrudescence of the house finch. Science **142**: 1294–1295.

Hamner, W.M., Enright, J.T. (1967) Relationships between photoperiodism and circadian rhythms of activity in the house finch. J. Exp. Biol. **46**: 211–227.

Hau, M., Wikelski, M., Wingfield, J.C. (1998) A neotropical forest bird can measure the slight changes in tropical photoperiod. Proc. R. Soc. Lond. B **265**: 89–95.

Jacobs, L.F. (1996) Sexual selection and the brain. TREE **11**: 82–86.

Juss, T.S. Meddle, S.L. Servant, R.S. King, V.M. (1993) Melatonin and photoperiodic time measurement in Japanese quail (*Coturnix coturnix japonica*). Proc. R. Soc. London B **254**: 21–28.

Juss, T.S., King, V.M., Kumar, V., Follett, B.K. (1995) Does an unusual entrainment of the circadian system under T36h photocycles reduce the critical day length for photoperiodic induction on Japanese quail? J. Biol. Rhythms **10**: 16–31.

King, V.M., Follett, B.K. (1997) c-*fos* expression in the putative avian suprachiasmatic nucleus. J. Comp. Physiol. A **180**: 541–551.

King, V.M., Bentley, G.E., Follett, B.K. (1997) A direct comparison of photoperiodic time measurement and the circadian system in European starlings and Japanese quail. J. Biol. Rhythms **12**: 421–442.

Kumar, V., Follett, B.K. (1993) The circadian nature of melatonin secretion in Japanese quail (*Coturnix coturnix japonica*). J. Pineal Res. **14**: 192–200.

Kumar, V., Tewary, P.D. (1982) Photoperiodic regulation of gonadal recrudescence in common Indian rosefinch: Dependence on circadian rhythm. J. Exp. Zool. **223**: 37–40.

Kumar, B.S., Kumar, V. (1993) Photoperiodic control of annual reproductive cycle in subtropical brahminy myna, *Sturnus pagodarum*. Gen. Comp. Endocr. **89**: 149–160.

Kumar, V., Kumar, B.S. (1995) Entrainment of circadian system under variable photocycles (T- photocycles) alters the critical daylength for photoperiodic induction in blackheaded buntings. J. Exp. Zool. **273**: 297–302.

Kumar, V., Goguen, D., Guido, M.E., Rusak, B. (1997) Melatonin does not influence the expression of c-*fos* in the suprachiasmatic nucleus of rats and hamster. Mol. Brain Res. **52**: 242–248.

Kumar, V., Jain, N., Follett, B.K. (1996) The photoperiodic clock in blackheaded buntings (*Emberiza melanocephala*) is mediated by a self-sustaining circadian system. J. Comp. Physiol. A **179**: 59–64.

Maywood, E.S., Lindsay, J.O., Karp, J., Powers, J.B., Williams, L.M., Titchener, L., Ebling, F.J.P., Herbert, J., Hastings, M.H. (1991) Occlusion of the melatonin-free interval blocks the short day gonadal response of the male Syrian hamster to programmed melatonin infusions of necessary duration and amplitude. J. Neuroendocrinol. **3**: 331–337.

Meddle, S.L., Follett, B.K. (1995) Photoperiodic activation of Fos-like immunoreactive protein in neurones within the tuberal hypothalamus of Japanese quail. J. Comp. Physiol. A **176**: 79–89.

Meddle, S.L., Follett, B.K. (1997) Photoperiodic driven changes in Fos expression within the basal tuberal hypothalamus and median eminence of Japanese quail. J. Neurosci. **17**: 8909–8918.

Meddle, S.L., Maney, D.L., Wingfield, J.C. (1999) Effects of N-methyl-D-aspartate on luteinizing hormone release and Fos-like immunreactivity in the male White-crowned sparrow (*Zonotrichia leucophrys gambellii*). Endocrinol. **140**: 5922–5928.

Menaker, M., Keats, H. (1968) Extraretinal light perception in the sparrow II. Photoperiodic stimulation of testis growth. Proc. Natl. Acad. Sci. USA **60**: 146–151.

Messager, S., Ross, A.W., Barrett, P., Morgan, P.J. (1999) Decoding photoperiodic time through Per1 and ICER gene amplitude. Proc. Natl. Acad. Sci. USA **96**: 9938–9943.

Miyamoto, K., Hasegawa, Y., Nomura, M., Igarashi, M., Kanagawa, K., Matsuo, H. (1984) Identification of the second gonadotropin-releasing hormone in the chicken hypothalamus: evidence that gonadotropin secretion is probably controlled by two distinct gonadotropin-releasing hormones in avian species. Proc. Natl. Acad. Sci. USA **81**: 3874–3878.

Morgan, J.I., Curran, T. (1991) Stimulus-transcription coupling in the nervous system: involvement of the inducible proto-oncogenes fos and jun. Ann. Rev. Neurosci. **14**: 421–451.

Nanda, K.K., Hamner, K.C. (1958) Studies on the nature of the endogenous rhythm affecting photoperiodic response of Biloxi soybean. Botanical Gazette (Chicago) **120**: 121–126.

Nicholls, T.J., Goldsmith, A.R., Dawson, A. (1988) Photorefractoriness in birds and comparison with mammals. Physiol. Rev. **68**: 133–176.

Nottebohm, F., Nottebohm, M.E., Crane, L.A., Wingfield, J.C. (1987) Seasonal changes in gonadal hormone levels of adult male canaries and their relation to song. Behav. Neural Biol. **47**: 197–211.

Oliver, J., Bayle, J.D. (1982) Brain photoreceptors for the photoinduced testicular responses in birds. Experientia **28**: 1021–1029.

Parry, D.M., Goldsmith, A.R. (1993) Ultrastructural evidence for changes in synaptic input to the hypothalamic luteinizing hormone-releasing hormone neurons in photosensitive and photorefractory starlings. J. Neuroendocrinol. **5**: 387–395.

Parry, D.M., Goldsmith, A.R., Millar, R.P., Glennie, L.M. (1997) Immunocytochemical localization of GnRH precursor in the hypothalamus of European Starlings during sexual maturation and photorefractoriness. J. Neuroendocrinol. **9**: 235–243.

Perera, A.D., Follett, B.K. (1992) Photoperiodic induction in vitro: the dynamics of gonadotropin-releasing hormone release from hypothalamic explants of the Japanese quail. Endocrinol. **131**: 2898–2908.

Pittendrigh, C.S. (1972) Circadian surfaces and the diversity of possible roles of circadian organization in photoperiodic induction. Proc. Natl. Acad. Sci. USA **69**: 2734–2737.

Pittendrigh, C.S., Minis, D.H. (1964) The entrainment of circadian oscillations by light and their role as photoperiodic clocks. Am. Nat. **98**: 261–294.

Saldanha, C.J., Deviche, P.J., Silver, R. (1994) Increased VIP and decreased GnRH expression in photorefractory dark-eyed juncos (*Junco hyemalis*). Gen. Comp. Endocrinol. **93**: 128–136.

Sicard, V., Oliver, J., Bayle, J.D. (1983) Gonadotrophic and photosensitive abilities of the lobus paraolfactorius: electrophysiological study in quail. Neuroendocrinol. **36**: 81–87.

Silver, R., Witkovsky, P., Horvath, P., Alones, V., Barnstable, C.J., Lehman, M.N. (1988) Coexpression of opsin- and VIP-like immunoreactivity in CSF-containing neurons of the avian brain. Cell Tiss. Res. **253**: 189–198.

Simpson, S.M., Follett, B.K. (1981) Pineal and hypothalamic pacemakers: their role in regulating circadian rhythmicity in Japanese quail. J. Comp. Physiol. A **145**: 391–398.

Tewary, P.D., Kumar, V. (1981) Circadian periodicity and the initiation of gonadal growth in blackheaded buntings (*Emberiza melanocephala*). J. Comp. Physiol. B. **144**: 210–203.

Turek, F.W., Wolfson, A. (1978) Lack of an effect of melatonin treatment via silastic capsules on photic-induced gonadal growth and the photorefractory condition in white-throated sparrows. Gen. Comp. Endocrinol. **34**: 471–474.

Vanecek, J. (1998) Cellular mechanisms of melatonin action. Physiol. Rev. **78**: 687–721.

Wallman, J., Saldanha, C.J., Silver, R. (1994) A putative suprachiasmatic nucleus of birds responds to visual motion. J. Comp. Physiol. A **174**: 297–304.

Wilson, F.E. (1991) Neither retinal nor pineal photoreceptors mediate photoperiodic control of seasonal reproduction in American tree sparrows (*Spizella arborea*). J. Exp. Zool. **259**: 117–127.

Yokoyama, K., Oksche, A., Darden, T.R., Farner, D.S. (1978) The sites of encephalic photoreception in photoperiodic induction of the growth of the testes in the white-crowned sparrow, *Zonotrichia leucophrys gambelii*. Cell Tiss. Res. **189**: 441–467.

Zivkovic, B.D., Underwood, G., Steele, C.T., Edmonds, K. (1999) Formal properties of the circadian and photoperiodic systems of Japanese quail: phase response curve and effects of T-cycles. J. Biol. Rhythms **14**: 378–390.

Biological Rhythms
Edited by V. Kumar
Copyright © 2002, Narosa Publishing House, New Delhi, India

17. Ultradian Rhythms

M.P. Gerkema*

Zoological Laboratory, University of Groningen, PO Box 14, Nl 9750 AA Haren, The Netherlands

A biological rhythm is called ultradian if its period is shorter than 24 hour. Ultradian rhythms have been observed in physiological functions, like cellular processes, respiraton, circulation, hormonal release and sleep stages, as well as in behavioral functions, often related to feeding patterns. Ultradian rhythms are characterized by diversity not only in period length (from hours to milliseconds) but also in mechanisms and functions. Besides homeostatic feedback loops at the behavioral level, several independent central nervous system (CNS) based ultradian pacemakers have been demonstrated. Attempts have been made to relate ultradian oscillations to each other. The Basic rest-activity Cycle (Brac) hypothesis supposes that the rhythm of rapid eye movement (REM) sleep episodes continues over the 24 hours of the day and is reflected in other physical and mental functions. Notwithstanding the resemblance of periodicity of REM cycles and some performances, the Brac concept cannot explain the diversity in frequency and mechanisms. The period length of many ultradian rhythms scale with body mass very similarly. Such allometry, without clarifying causal principles, indicates the potentials of synchronization and tuning of ultradian patterns. In general, functions of ultradian rhythms have been described in terms of energetic optimization and internal coordination. This may apply also to the example of ultradian feeding rhythms in voles. To obtain crucial savings in energy expenditure, however, voles furthermore have to synchronize their individual rhythms. Body contact is here essential, both in the entrainment mechanism and in the functional consequence of the synchronization process. In the absence of relevant geophysical cycles in the environment, synchronization with a biological external factor, i.e. with conspecifics, is characteristic of ultradian patterns in behavior and physiology.

Introduction

Rhythms in behavior and physiology that occur with a frequency higher than once per 24 hour are called ultradian (Halberg et al. 1965). Hence, by definition the period length may vary in the group of ultradian rhythms from several hours to milliseconds and even beyond. In overviews (Schulz and Lavie 1985; Wollnik 1989; Lloyd and Stupfel 1991; Gerkema 1992), the diversity among ultradian rhythms has been underlined, and, as pointed out by Aschoff and Gerkema (1985), this is true both for mechanisms as well as for the adaptive significance. Two major attempts have been made to connect various ultradian rhythms with each other. The Basic rest-activity cycle (Brac) hypothesis was originally formulated by Kleitmann (1961) and based on the assumption of a day and night ongoing alternation of sleep stages. Lindstedt and Calder (1981) started to compile allometric relations from literature, between body mass and period length of several physiological and behavioral rhythms.

 Perhaps the most well known ultradian rhythm is the cycle of sleep stages with Rapid Eye Movement (REM) (Aserinsky and Kleitman 1953). There are many other ultradian rhythms with

---

*E-mail: m.p.gerkema@biol.rug.nl

periods in the range of 0.1 to 5 h. Examples are the rhythms in concentration in cell metabolites (e.g. the glycolytic oscillator Boiteux et al. 1980). Several hormones are released in a pulsatile way, e.g. pituitary hormones (Van Cauter and Honinck 1985), insulin (Simon 1998) and leptin (Sinha et al. 1996). Higher frequencies are found in digestive physiology (e.g. gut beat), respiration, circulation and muscular twitches (references in Daan and Aschoff 1982). Behavioral rhythms are often related to feeding behavior (Daan and Aschoff 1981). This applies especially to herbivore animals in which fine tuned digestive processes impose cyclic activity patterns with a period shorter than 24 hour (Gerkema 1992). But also insectivorous species like shrews and moles show rather precise ultradian feeding patterns (Cowcroft 1954; Godfrey 1955). Also specific stages of reproduction are accompanied by short-term rhythms, e.g. courtship behavior and breeding (Daan and Aschof 1981). Ultradian rhythms reported for body temperature and oxygen consumption or carbondioxide production reflect in most cases also behavioral patterns that are accompanied by energetic costs (e.g. Stupfel et al. 1995). Besides, ultradian rhythms in general mobility and cognitive performances have been documented (Grau et al. 1995; Hatyashi et al. 1994; Neubauer and Freudenthaler 1995; Okudaira et al. 1984).

Mechanisms

Due to its broad definition, ultradian rhythms comprise at least four categories that should be distinguished. In its purest form, ultradian rhythms are continuous, ongoing during all phases of the 24 hours episode. Most animals show, however, polyphasic (and thus ultradian) activity patterns that are restricted to specific circadian times of the day. In addition, activity bouts can take place during very specific phases of the circadian cycle, e.g. bimodal or trimodal activity at the beginning, midpoint and ending of the circadian activity phase. Formally seen, these are ultradian patterns, because of the stable distance between such bouts (see, for instance, Poon et al. 1997). Whether such ultradian patterns are the result of circadian organization, and, in fact, are caused by circadian oscillators, can be distinguished rather easily by disruptions (e.g. lesions of the suprachiasmatic nucleus, SCN, in voles: Gerkema et al. 1990) or mutations (e.g. tau-mutant hamsters: Refinetti 1996) of the circadian system. In that case such patterns should disappear or at least be affected (Refinetti 1996). Otherwise, independent mechanisms have to be assumed (Gerkema et al. 1990). Finally, also circatidal rhythms are formally included in the definition of ultradian rhythmicity (see, for instance, Bornemann et al. 1998).

In several cases, ultradian pacemakers have been localized. This applies to spinal pacemakers, responsible for muscular activity related rhythms, and to the pontine oscillator of REM cycles. In the hypothalamus, GnRH rhythms have been established in the preoptic area and the arcuate nucleus, and the ultradian feeding oscillator in voles depends on the retrochiasmatic area (for references: see Gerkema 1992). Also homeostatic regulation of ultradian patterns has been demonstrated, for instance of nest-creeping behavior in the stickleback (Daan 1987). It should, however, be stressed that in each case the nature of the underlying ultradian mechanisms has to be clarified. Differentiation between homeostatic and clock like regulation asks for experimental manipulation of the periodicity. All examples given above have in common a negative feedback as an essential part of this mechanism, without however providing any clue to the physiological basis.

Basic rest-activity cycle

The Brac hypothesis was based originally on a periodicity of inter-feeding intervals in infants, which coincides with the rhythm of REM-episodes (Kleitman 1961). The hypothesis implies that an ongoing pattern of stronger cognitive and physical performance interrupted by sleep phases is conserved during adulthood, in a 90 minutes cycle. A strong intuitive argument in favor of the Brac hypothesis was supplied by the pathological state of narcolepsis, a continuation of explicit sleep episodes during daytime in an about ninety minutes pattern (Notili et al. 1996). A considerable list of motor performances, sensory acuity's, visceral and urinal functions with a period of about ninety minutes has been put forward as a support for the generality of the Brac hypothesis (Schulz and Lavie 1985). In line with the origin of the Brac hypothesis, ontogenetic studies in mice have been done, comparing REM-cycle length and periodicity in behavior. The results obtained by measuring avoiding tasks were interpreted as being in favor of the Brac concept (D'Olimpio and Renzi 1998). At the same time, doubts about the general validity of the Brac idea were raised. The likelihood of completely divergent mechanisms, for instance in the pontine REM-nonREM oscillator and gastric and renal regulation, was thought to be incompatible with a mono-causal concept like Brac (Aschoff and Gerkema 1985). Anyway, in humans a large range of periodicities around ninety minutes has been observed in "Brac" phenomena. In addition, some studies explicitly aimed at the establishment of Brac regulation for specific functions have failed to show those (eg. Okudaira et al. 1984; Neubauer and Freudenthaler 1995).

One could of course argue that because of the different frequencies among the ultradian patterns within a species, at least the coexistence of several Brac mechanisms has to be postulated. By doing so, each ongoing ultradian phenomenon should better be called a rest-activity-cycle. Why should it retain the epithet basal, if one gives up the connection of the REM cycle and this ultradian activity pattern? With all evidence available, how should one judge the existence of Brac? It is difficult to falsify such a Brac hypothesis in general. First, Brac could be true for even a single separate ultradian phenomenon but not for others. Second, the Brac concept suggests a general, but not defined, mechanism, and doesn't distinguish explicitly between the many ways in which a quite scattered range of ultradian periods can be achieved. Similarity in underlying mechanisms should have to be excluded for each combination of a specific ultradian rhythm and the REM cycle.

Allometry

In many ultradian rhythms, the period length (τ) varies with body mass (M): within birds and mammals (Lindstedt and Clader 1981; Daan and Aschoff 1982) and fish (Gerkema 1992) this can be approximated by the allometric equation $\tau \sim M^{0.25}$. The heartbeat of a house mouse is twenty times faster than that of an elephant (Figure 1). Likewise, the maximal life span of a house mouse is twenty times shorter, and from an internal, physiological point of view a 24-h day takes twenty times longer for the mouse than for the elephant. Internal circa rhythms (e.g. circadian, circannual, circatidal and circalunar) have similar periods for all organisms and reflect 'objective', environmental time. In contrast, internal non-circa rhythms reflect subjective, physiological time (Brody 1945). Because of the universality of the scaling exponent around 1/4, the non-circa rhythms within organisms tend to maintain rather fixed mutual relations, and the number of heartbeats and gutbeats per lifetime is nearly a constant for all mammalian species. The classic allometric rule for the relation between basic metabolic rate (BMR) and body mass (M): (BMR $\sim M^{3/4}$; Kleiber 1932) can also be expressed as the time needed to convert energy per gram body mass, and that again scales with an exponent of about 1/4. Thus, energy expenditure expressed per gram body mass is similar for an elephant and a vole, per fast muscle twitch, per heartbeat, and per lifetime.

Many physiological interpretations have been given to these allometric comparisons. Straightforward physiological theories started with rules derived from surface: volume ratios and were extended to rather complex dimensionality models (Heusner 1987). All this is based on the assumption that body mass has to be treated as an independent variable. Body mass itself, however, may be dynamically optimized (Daan and Tinbergen 1999). Instead of assuming a physiological mechanism, which determines the number of joules that can be used in a lifetime, Koslowski and Weiner (cited in Daan and

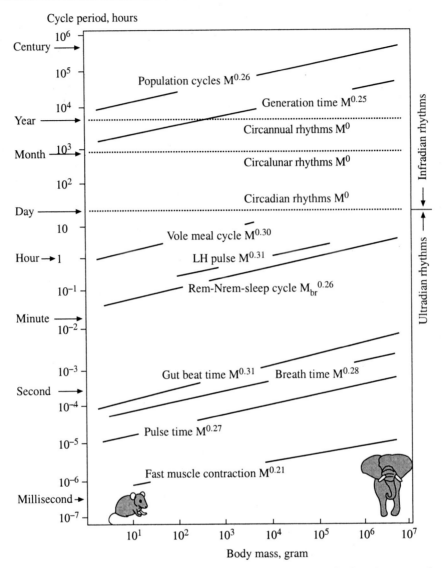

Fig. 1   **Allometry of cycle length duration for various non circa rhythms in mammals. Modified after Gerkema (1992)**

Tinbergen 1999) suggest that much of these allometric relationships can be explained as the outcome of evolutionary theory. Optimization processes in the allocation of energy to growth or reproduction result in a variety of optimal body masses. It remains to be seen whether an unambiguous explanation for the evolution of the allometric exponent 1/4 in non-circa rhythm does exist at all. The mass specificity in the ratio of time and energy is certainly a unifying principle (Peters 1983; Calder 1984), which allows very different rhythms in physiology and behavior to be tuned to each other.

Functions

Apart from its relationships with other non-circa rhythms, an ultradian rhythm may facilitate energetic optimization and the internal coordination of physiological processes (Aschoff and Gerkema 1985). These functions are reflected in several principles (table 1). Rhythms allow a temporal compartmentalization of mutually incompatible components of a process that have to take place in the same space, e.g. the conflicting phases of breathing, or several reaction steps in glycolysis and gluconeogenesis (Boiteux et al. 1980). Related to this is the economic principle not to spend energy continuously but to avoid overload and dissipation of energy by alternating rest and energy expenditure (Aschoff and Wever 1962). This principle applies not only to behavioral rhythms, but also to crucial cellular processes. Richter and Ross (1981) thus explain the extraordinary high yield of respiration of glucose in glycolysis. Some processes are needed on call, in response to randomly occurring stimuli. A rhythmic structure can help to avoid problems of atrophy (the wasting away of a system through lack of use; Aschoff and Wever 1962) and initial friction (the extra costs connected with starting up a process; von Holst 1949).

Table 1   Functional aspects of ultradian rhythms

Temporal organization
　　　　　　　　　　−compartmentalization of incompatible processes
Energetic efficiency
　　　　　　　　　　−avoidance of overload and dissipation
　　　　　　　　　　−avoidance of atrophy
　　　　　　　　　　−avoidance of initial friction
Information transfer
　　　　　　　　　　−stabilization of parameters (futile cycles)
　　　　　　　　　　−increase in predictability of parameters (idem)
　　　　　　　　　　−robustness of frequency modulation
　　　　　　　　　　−resistance to noise
　　　　　　　　　　−avoidance of down-regulation en habituation
Coordination
　　　　　　　　　　−internal coupling
　　　　　　　　　　−external (social) synchronization

Rhythmic regulation stabilizes body and cell parameters like pH and temperature (e.g. Meiske et al. 1978), replacing rapid and chaotic fluctuations around average values. Such energy consuming oscillations have been denounced in the past as 'futile cycles', due to a lack of understanding of its function. Later on, the stabilizing effect of such rhythmicity, which increases the predictability of parameters, has been appreciated (Ricard and Saolie 1982). Lisman and Goldring (1988) claim a crucial function of such 'futile cycles' in the consolidation of long term memory. Rhythmicity forms a reliable medium in the transfer of information by frequency modulation (Aschoff and Wever 1962).

Apart from the potential of such a modulation, rhythmic signals are more resistant to noise than tonic signals (Rapp et al. 1981) and avoid habituation, desensitization and down regulation of the receiver (Kupferman 1985).

A major advantage of rhythmic organization of behavior and physiology concerns the potential of coupling and synchronization of processes, a potential resulting from the 'physiological time' structure. Thus cells in a tissue can produce synchronized and effective output, in a heartbeat, in an unambiguous pulse of LH that controls the reproductive system. The hierarchy of frequencies (Aschoff and Wever 1962) enables the phase coupling of different functions, like that of locomotion and respiration and that of heartbeat and respiration (Bramble and Carrier 1983).

Ultradian rhythmicity has no counterpart in the abiotic environment. Yet, ultradian synchrony occurs with external factors, in the form of conspecifics with similar rhythmicity. This phenomenon of social synchrony is in fact wide spread in nature, varying from synchronizing amoebas (Alcantara and Monk 1974), the barnacle gees (Prop and Loonen 1986), voles (Gerkema and Daan 1985), cattle (Hughes and Reid 1951), and rhesus monkeys (Delgado-Garcia et al. 1976). Functionally, such synchrony has been related with diminishing predation risks by the principle of 'safety in numbers' (Daan and Aschoff 1981). Such a strategy is very successful in the case of the periodic cicada, reproducing synchronously each 13th or 17th year (Daan 1981). However, in less extreme situations predators can themselves respond with an increase in numbers, and thus overcome such a 'swamping' by vulnerable prey (Raptorgroup 1982). Whereas in the common vole synchronous activity in the population is observed (Daan and Aschoff 1981), the synchrony at the family level seems to be far more important in this species.

Rhythms in voles

Among the manifold examples of ultradian patterns in behavior, the activity rhythms of herbivore mammals are relatively well documented. This is true especially for the many vole species, in which the temporal organization of behavior is characterized by ultradian patterns of feeding activity. In species like *Microtus arvalis* and *Microtus agrestis* the balance between day and night activity changes slightly with season, but circadian aspects of behavior seem of secondary importance. This is also suggested by a high incidence of loss of circadian rhythmicity in constant light conditions. This loss of the circadian organization of behavior is accompanied by changes in neurochemical output of the CNS, central nervous system based central nervous system circadian pacemaker, *in vivo* and *in vitro* (Gerkema et al. 1994; Jansen et al. 1999).

Although the origin of ultradian rhythms in the herbivore voles probably will be related with digestive processing, food intake itself has been shown not to be a causative factor. Food quality and energetic demand do have long term effects on period length of activity rhythm in *Microtus arvalis*. (Gerkema and van der Leest 1991). Classical zeitgebers of behavior rhythms in the circadian range have no effect on the ultradian period of voles (Gerkema et al. 1993). So far, only direct body contact induced rapid changes of the ultradian rhythmicity, resulting in synchronization (Figure 2).

The social context has been shown to be of major importance in *Microtus arvalis,* with respect to synchronization of ultradian patterns of behavior. This is true in the rest phase of the ultradian rhythm, during which huddling allows a substantial reduction of energy expenditure. During the active phase of the ultradian rhythm, a lowering of predation risk has been indicated by behavioral experiments in which passive warning for aerial predators was shown. Thus, selection pressure on synchronization of ultradian rhythms of behavior may well be responsible for the persistence of these temporal patterns (Gerkema and Verhulst 1990).

Meanwhile, it has become clear that the causation of the ultradian patterns in the common vole does not depend on homeostatic types of regulation. On the contrary, behavioral evidence suggested clock regulation (Gerkema and van der Leest 1991). Results of lesion studies indicated specific medio basal hypothalamic areas, but not those responsible for circadian rhythmicity, to be involved in the generation of ultradian rhythms (Gerkema et al. 1990). Recently, *in vitro* studies confirmed these results by showing ultradian patterns of multi-unit activity in acute hypothalamic slices, obtained from common voles. It is unclear whether ultradian neurochemical output patterns, sometimes obtained in vole organotypic hypothalamic cultures (Gerkema et al. 1999), are related to the ultradian organization behavior in voles.

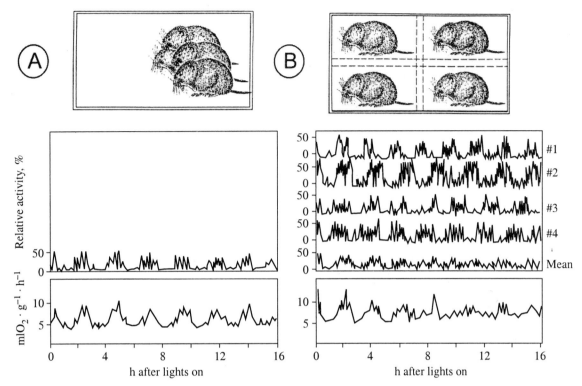

Fig. 2 Patterns of activity (passive infrared detection) and oxygen consumption during daytime in the common vole (*Microtus arvalis*), in a light:dark cycle of 16:8 hours, at 10°C. Panels A: measurements of four voles kept together. Panels B: measurements of the same four voles, separated by wire mesh. Activity is expressed as percentage of time with activity, per 2 minutes. In the separated condition, individual voles kept their individual rhythms, but note the decay in synchrony in the course of the day.

References

Alcantara, F., Monk, M. (1974) Signal propagation during aggregation in the slime mould *Dictyostelium discoideum*. J. Gen. Microbiol. **85**: 321–324.

Aschoff, J., Wever, R. (1962) Biologische Rhythmen und Regelung. In: Delius, L., Koepchen, H.P., Witzleb, E. (eds.) Probleme der zentralnervösen Regulation. Springer, Berlin, pp. 1–15.

Aschoff, J., Gerkema, M.P. (1985) On diversity and uniformity of ultradian rhythms. In: Schulz, H., Lavie, P. (eds.) Ultradian Rhythms in Physiology and Behavior. Springer, Berlin, pp. 321–334.

Aserinsky, E., Kleitman, N. (1953) Regularly occurring periods of eye motility and concomitant phenomena during sleep. Science **118**: 273.

Boiteux, A., Hess, B., Sel'kov, E.E. (1980) Creative function of instability and oscillations in metabolic systems. Current topics in cellular regulation **17**: 1171–2031.

Bornemann, H., Mohr, E., Ploetz, J., Krause, G. (1998) The tide as zeitgeber for Weddal seals. Polar Biol. **20**: 396–403.

Bramble, D.M., Carrier, D.R. (1983) Running and breathing in mammals. Science **219**: 251–256.

Brody, S. (1945) Bioenergetics and growth. Hafner, New York.

Calder, W.A. (1984) Size, Function and Life History, Harvard University Press, Cambridge, Massachusetts.

Cowcroft, P. (1954) The daily cycle in British shrews. Proc. Zool. Soc. London **123**: 715–729.

Daan, S. (1981) Adaptive daily strategies in behavior. In Aschoff, J. (ed.) Handbook of Behavioral Neurobiology, Plenum, New York, pp. 275–298.

Daan, S. (1987). Clocks and hour glass timers in behavioural cycles. In Hiroshige, T. and Hinama, K. (eds.) Comparative aspects of circadian clocks. Hokkaido University Press, Sapporo.

Daan, S., Aschoff, J. (1981) Short-term rhythms in activity. In: Aschoff, J. (ed.) Handbook of Behavioral Neurobiology, Plenum, New York, pp. 491–498.

Daan, S., Aschoff, J. (1982) Circadian contributions to survival. In: Aschoff, J., Daan, S., Groos, G.A. (eds.) Vertebrate circadian systems: structure and physiology. Springer, Berlin, pp. 305–321.

Daan, S., Tinbergen, J.M. (1999) Adaptation of Life Histories. In: Krebs, J.R., Davies, N.B. (eds.) Behavioural Ecology, An Evolutionary Approach. Blackwell Science, Oxford, pp. 311–333.

Delgado-Garcia, J.M., Grau, C., DeFeudis, P., Belpozo, F., Jeminez, M., Delgado, J.M.R. (1976) Ultradian rhythms in mobility and behavior of rhesus monkeys. Exp. Brain Res. **25**: 79–91.

D'Olimpio, F., Renzi, P. (1998) Ultradian rhythms in young and adult mice: further support for the basic rest-activity cycle. Physiol. Behav. **64**: 697–701.

Gerkema, M.P. (1992) Biological rhythms: Mechanisms and adaptive values. In: Ali, M.A. (ed.) Rhythms in Fishes. Plenum Press, New York, 27–37.

Gerkema, M.P. and Daan, S. (1985) Ultradian rhythms in behavior: the case of the common vole (*Microtus arvalis*). In: Schulz, H., Lavie, P. (eds.) Ultradian rhythms in physiology and Behavior. Springer, Berlin, pp. 11–31.

Gerkema, M.P., Groos, G.A., Daan, S. (1990) Differential elimination of circadian and ultradian rhythmicity by hypothalamic lesions in the common vole *Microtus arvalis*. J. Biol. Rhythms **5**: 81–95.

Gerkema, M.P., Verhulst, S. (1990) Warning against unseen predators: an experimental study in the common vole *Microtus arvalis*. Anim. Behav. **40**: 1169–1178.

Gerkema, M.P., Leest, F. van der (1991) Ultradian rhythms in the common vole *Microtus arvalis* during short deprivations of food, water and rest. J. Comp. Physiol. A **168**: 591–597.

Gerkema, M.P., Daan, S., Wilbrink, M., Hop, M.W., Leest, F. van der (1993) Phase control of ultradian feeding rhythms in the common vole (*Microtus arvalis*): the roles of light and the circadian system. J. Biol. Rhythms **7**: 151–171.

Gerkema, M.P., Van der Zee, E.A., Feitsma, L.E. (1994) Expression of circadian rhythmicity correlates with the number of arginine-vasopressin-immunoreactive cells in the suprachiasmatic nucleus of the common vole, *Microtus arvalis*. Brain Res. **639**: 93–101.

Gerkema, M.P., Shinohara, K., Kimura, F. (1999) Lack of circadian patterns in vasoactive intestinal polypeptide release and variability in vasopressin release in vole suprachiasmatic nuclei *in vitro*. Neursci. Lett. **259**: 107–110.

Godfrey, G.K. (1955) A field study of the activity of the mole (*T. europaea* L.). Ecology **36**: 678–685.

Grau, C., Escera, C., Cilveti, D., Garcia, R., Mojon, M., Fernandez, A., Hermida, R.C. (1995). Ultradian rhythms in gross motor activity of adult humans. Physiol. Behav. **57**: 411–419.

Halberg, F., Engeli, M., Hamburger, C., Hillmann, V.D. (1965) Spectral resolution of low-frequency, small amplitude rhythms in excreted 17-ketoseroids: probably androgen-induced circaseptan desynchronization. Acta Endocrinologica **103**: 1–54.

Hatyashi, M., Sato, K., Hori, T. (1994) Ultradian rhythms in task performance, self evaluation, and EEG activity. Percept. Mot. Skills **79**: 791–800.

Heusner, A.A. (1987) What does the power function reveal about structure and function in animals of different size? Am. Rev. Physiol. **49**: 121–133.

Holst, E.V. (1949) Zur Funktion des Statolitischenapparates im Wirbeltierlabyrinth. Naturwissenschaften **36**: 127–128.

Hughes, G.P., Reid, D. (1951) Studies on the behaviour of cattle and sheep in relation to the utilization of grass. J. Agr. Sci. **41**: 360–366.

Jansen, K., Van der Zee, E.A., Gerkema, M.P. (1999) Organotypic suprachiasmatic nuclei cultures of adult voles reflect locomotor behavior: differences in number of vasopressin cells. Chronobiol. Int. **16**: 745–750.

Kleiber, M. (1932) Body size and metabolism. Hilgardia **6**: 315–353.

Kleitmann (1961) The nature of dreaming. In Wolstenholme, G.E.W., O'Connor, M. (eds.). The nature of sleep. Churchill, London, p. 349.

Kupferman, I. (1985) Hypothalamus and lymbic system I: peptidergic neurons, homeostasis, and emotional behavior. In: Kandel, E.C., Schwartz, J.H. (eds.) Principles of neural science. Elsevier, New York, pp. 611–625.

Lindstedt, S.L., Calder, W.A. (1981) Body size, physiological time, and longevity of homeothermic animals. Quart. Rev. Biol. **56**: 1–16.

Lisman, J.E., Goldring, M.A. (1988) Feasibility of long-term storage of graded information by the Ca^{2+}/calmoduli-dependent protein kinase molecules of the postsynaptic density. Proc. Nat. Acad. Sci. USA **85**: 5320–5324.

Lloyd, D., Stupfel, M. (1991) The occurrence and functions of ultradian rhythms. Biol. Rev. **66**: 275–299.

Meiske, W., Glende, M., Nurnberg, G., Reich, J.G. (1978) On the influence of rapid periodic parameter oscillations on the long-term behaviour of cell metabolism. J. Theor. Biol. **71**: 11–19.

Neubauer, A.C., Freudenthaler, H.H. (1995) Ultradian rhythms in cognitive performance: no evidence for a 1.5-h rhythm. Biol. Psychol. **40**: 281–298.

Notili, L., Ferrilo, F., Besset, A., Rosadini, G., Schiavi, G., Billiard, M. (1996) Ultradian aspects of sleep in narcolepsy. Neurophysiol. Clin. **26**: 51–59.

Okudaira, N., Kripke, D.F., Webster, J.B. (1984) No basic rest-activity cycle in head, wrist or ankle. Physiol. Behav. **32**: 843–845.

Peters, R.H. (1983) The ecological implications of body size. Cambridge University Press.

Poon, A.M.S., Wu, B.M., Poon, P.W.F., Cheung, E.P.W., Chan, F.H.Y. (1997) Effect of cage size on ultradian locomotor rhythms of laboratory mice. Physiol. Behav. **62**: 1253–1258.

Prop, J., Loonen, H.J.J.E. (1986) Goose flocks and food exploitation: the importance of being first, XIX Int. Ornith. Congr. Ottawa, pp. 1878–1887.

Rapp, P.E., Mees, A.I., Sparrow, C.T. (1981) Frequency encoded biochemical regulations is more accurate than amplitude dependent control. J. Theor. Biol. **90**: 531–544.

Raptorgroup RUG/RIJP (1982) Timing of vole hunting in aerial predator. Mammol. Rev. **12**: 169–181.

Refinetti, R. (1996) Ultradian rhythms of body temperature and locomotor activity in wild-type and tau mutant hamsters. Anim. Biol. **5**: 111–115.

Ricard, J., Saolie, J.M. (1982) Self organization and dynamics of an open futile cycle. J. Theor. Biol. **95**: 105–121.

Richter, P.H., Ross, J. (1981) Concentration oscillations and efficiency: glycolysis. Science **211**: 715–717.

Schulz, H., Lavie, P. (1985) Ultradian rhythms in Physiology and Behavior. Springer, Berlin.

Simon, C. (1998) Ultradian pulsatility of plasma glucose and insulin secretion rate: circadian and sleep modulation. Horm. Res. **49**: 185–190.

Sinha, M.K., Sturis, J., Ohannesian, J., Magosin, S., Stephens, T., Polonsky, K.S., Caro, J.F., (1996) Ultradian oscillations of leptin secretion in humans. Biochem. Biophys. Res. Commun. **228**: 733–738.

Stupfel, M., Gourlet, V., Peramon, A., Merat, P., Putet, G., Court, L. (1995) Comparison of ultradian and circadian oscillations of carbon dioxide production by various endotherms. J. Am. Physiol. **268**: R253–265.

Van Cauter, E., Honinckx, E. (1985) Pulsatility in pituitary hormones. In: Schulz, H., Lavie, P. (eds.) Ultradian Rhythms in Physiology and Behavior. Springer, Berlin, pp. 41–61.

Wollnik, F. (1989) Physiology and regulation of biological rhythms in laboratory animals: an overview. Lab. Animal **23**: 107–125.

Biological Rhythms
Edited by V. Kumar

18. Biological Rhythms in Arctic Animals

E. Reierth* and K.-A. Stokkan

Department of Arctic Biology and Institute of Medical Biology,
University of Tromsø, N-9037 Tromsø, Norway

With increasing latitude the daily light-dark cycle becomes progressively distorted during substantial parts of the year. At high-Arctic latitudes (77–81°N) there is continuous darkness (polar night) between November and February and continuous light (polar day) from April to September. Circadian mechanisms generally rely heavily on the synchronizing or entraining effect of the daily light/dark cycle, and it is therefore important to study animals living under conditions where this zeitgeber is absent. Migratory birds visiting the Arctic in summer to breed apparently perceive sufficient environmental rhythmic information to remain entrained. Humans and resident animals such as ptarmigan and reindeer do not, but whereas humans show persistent circadian freerunning sleep/wake rhythms, reindeer and ptarmigan become continuously active. This is also revealed by their secretion of melatonin, which is markedly reduced at those times of the year when the light/dark cycle is absent. Presumably, their endogenous biological clocks or circadian machinery is flexible and becomes dampened to such an extent as to allow these animals to exploit their environment maximally at those times of the year when there is no marked differences between day and night.

Introduction

Precise timing of seasonal events such as breeding and migration, deposition of energy and growth of insulation and camouflage are extremely important to Arctic animals, where ambient temperature rises above freezing for only a few weeks in summer. On the other hand, the continuous daylight that these animals experience in summer as well as the lack of daylight in winter may reduce the importance of circadian rhythms with respect to night-and daytime preparations. While such rhythms are of fundamental importance to organisms living at lower latitudes they may be less so when there is only marginal differences between day and night. The life histories of resident Arctic animals, and also of plants, therefore, exemplify in an extreme form the general phenomenon of biological timing.

Few biological rhythms are driven directly by environmental factors. Instead, most overt rhythms are caused by self-sustained endogenous oscillators, acting as biological clocks. These clocks are synchronised or entrained by external factors to ensure a proper phase relationship between particular body events and the appropriate changes in the environment. The most important factor serving as a time cue, or zeitgeber, is the daily light-dark cycle, which varies in an entirely predictable manner throughout the year.

However, with increasing latitude the daily light-dark cycle becomes progressively distorted during parts of the year, and may virtually disappear for long periods each winter and summer. In northern Scandinavia (70°N) the sun does not set for two months in summer. Likewise, it does not rise above

---

*E-mail: eirikr@fagmed.uit.no

the horizon for nearly two months in winter, giving only 3–4 hours of twilight around noon at the winter solstice. Further north, on the high-Arctic archipelago of Svalbard (77–81°N) there is a continuous "polar night" from November to February when the sun remains permanently lower than 6° below the horizon. From April to September, on the other hand, there is a continuous "polar day" when the sun never sets (Fig.1). Biological clocks of Arctic species presumably evolved at much lower latitudes where there is always a marked daily light/dark cycle. It is, therefore, of interest to investigate if these mechanisms have changed as these animals adapted to conditions where the principal zeitgeber is markedly obscured for weeks or even months at a time each year.

This short review will focus mainly on the daily and circadian rhythms of one high-Arctic resident; the Svalbard ptarmigan (*Lagopus mutus hyperboreus*). In these birds, which are the only avian residents on this high-Arctic archipelago, laboratory experiments have revealed physiological and behavioural mechanisms that may enhance their fitness to the extreme light and climatic conditions. The photoperiodic control of seasonal events such as breeding, moult and energy turnover including fat deposition, has also been extensively studied in these birds (Mortensen et al. 1983; Mortensen and Blix 1989; Stokkan et al. 1986a; Stokkan et al. 1986b; Lindgård and Stokkan 1989) as well as in their more southern relatives, the willow ptarmigan (*Lagopus lagopus lagopus*) (Stokkan 1992). These aspects of daylength-controlled biological rhythms will not be covered here.

Activity rhythms in the Arctic

Many field studies report on persistent diurnal activity rhythms among migratory birds during the Arctic summer (Marshall 1938; Karplus 1952; Cullen 1954; Brown 1963), while the resident willow ptarmigan has been described as arrhythmic (Pulliainen 1978). Fewer studies have actually recorded the daily locomotor rhythms in animals exposed to outdoor light conditions at very high latitudes. Stokkan et al. (1986a) recorded activity patterns in caged ptarmigan at Svalbard for one whole year, while Krüll (1976a) and Johnsson et al. (1979) studied snowbuntings and humans, respectively, but only for a few weeks around mid summer at these latitudes.

Svalbard ptarmigan showed diurnal rhythms of behaviour during spring and autumn, foraging only during the daylight hours. In contrast to terrestrial vertebrates in general, this 24-hour rhythmicity disappeared under the continuous light-conditions of both the polar night and polar day and the birds became continuously active (Stokkan et al. 1986a). When human volunteers were kept isolated from social influences and without clocks in mid-summer at Svalbard, they displayed marked circadian, freerunning sleep/wake and body temperature cycles (Johnsson et al. 1979). This occurred despite the fact that they reported full mental awareness of the passing of time, such as e.g. observing the sun circling the horizon. When captive snowbuntings (*Plectrophenax nivalis*) or greenfinches (*Carduelis chloris*) were studied outdoors under similar conditions, their locomotor activity rhythms appeared to be entrained by the solar cycle (Krüll 1976a). Their resting period occurred around midnight, even though the cages were located such that this was the only time of day that they received direct sunshine. Thus, humans maintained endogenous body rhythms but did not perceive sufficient rhythmic environmental information to become entrained while snowbuntings that breed on the archipelago, and greenfinches that never visit this latitude, perceived entrainable zeitgeber information, presumably from the solar cycle. The activity rhythm in ptarmigan, by contrast, appeared to become uncoupled from both endogenous and environmental influences during the Arctic summer, as well as in the winter.

Recently, studies of locomotor activity in free-ranging reindeer have been reported (Van Oort et al. 1999) both at Svalbard (78°N; *Rangifer tarandus platyrhynchus*) and in northern Norway (70°N; *Rangifer tarandus tarandus*). At Svalbard, reindeer showed ultradian bouts of activity throughout both the polar day and night without signs of circadian rhythmicity. On the Norwegian mainland, the daily activity rhythm appeared to be synchronised by the daily twilight cycle throughout winter, but was intermittently continuous throughout the summer, without signs of circadian rhythmicity. Thus, reindeer appear to behave similar to ptarmigan. Presumably, the daily light/dark cycle may also impose strong masking effects on their locomotor and foraging rhythmicity (Van Oort et al. 1999).

Although the polar day has continuous daylight, the solar cycle still provides obvious 24-h rhythms, such as e.g. the changing azimuth position of the sun circling the horizon. Except at the pole, there are also marked changes in both intensity and spectral composition of light throughout the day (Krüll 1976a) since the sun is closer to the horizon at night than during the day. Experiments have shown that all these changes are possible zeitgebers for passerine birds that migrate to the Arctic in summer to breed (Demmelmayer and Haarhaus 1972; Krüll 1976b, c; Pohl 1999).

In an effort to investigate how subtle photoperiodic changes may affect the activity rhythm, Reierth and Stokkan (1998b) compared data from Svalbard ptarmigan caged outdoors at two different latitudes; 79°N (previously published by Stokkan et al. 1986a) and 70°N, where these birds normally do not live. Activity was intermittently continuous throughout the 24-h day from mid-November until end of January and from early April until the beginning of September at 79°N, and from beginning of May until mid August at 70°N (Fig. 1). Whenever there was a diel light-dark cycle, all birds showed a diurnal activity pattern, but during winter at 70°N, the morning peak of activity became less pronounced and virtually disappeared around the solstice. At both latitudes, night-time activity generally began to appear in October and lasted until March. Activity was always more intense in summer than in winter. When rhythmicity reappeared in autumn (79°N), activity onset occurred at a higher light intensity than in spring. Thereafter, onset rapidly phase-advanced relative to sunrise, contributing to a significantly longer activity period in autumn than in spring. Such "hysteresis", giving the longest activity period in autumn (not seen at 70°N), is opposite that generally found in birds (Daan and Aschoff 1975; Binkley and Mosher 1992).

Svalbard ptarmigan deposit large amounts of fat in autumn (Mortensen et al. 1983) and the ability to extend their daily search for food at this time of year may be of adaptive significance. Presumably, this pattern of behaviour is aided by a significant entrainment by feeding itself (Reierth and Stokkan 1998a). Putative feeding entrainable oscillators may tend to keep the morning and evening peaks of feeding activity at about the same time every day, while the rapidly decreasing daylength tends to shift them together (Reierth and Stokkan 1998a, b).

The more southern willow ptarmigan prepare for long inactive winter nights in snow dens by intense feeding and crop filling in the evening (Irving et al. 1967; Höglund 1980). This evening feeding pattern was also revealed in captivity (West 1968) and is similar to what was observed in winter in Svalbard ptarmigan caged at 70°N (Fig. 1). At 79°N, feeding was intermittent and arrhythmic from November to February, which coincides with the midwinter decline in body mass and the annual low in food intake (Stokkan et al. 1986a). Thus, continuous feeding does not necessarily reflect high food intake. Field data are lacking for Svalbard ptarmigan but it is believed that the pattern seen in captivity reflects the behaviour of freeliving birds. This is based on a comparison between Svalbard and willow ptarmigan, both showing high daytime crop contents associated with diurnal rhythmicity. Freeliving Svalbard ptarmigan have low daytime crop content both in summer, when food intake is high and in winter when food intake is low (Mortensen et al. 1983). Willow ptarmigan, on the other hand, that feed continuously only during summer (West 1968), have low daytime crop content in summer and high in winter (Irving et al. 1967).

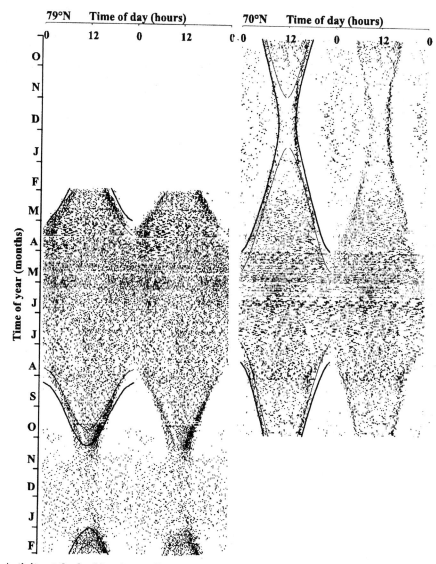

Fig. 1   **Activity at the food box in two Svalbard ptarmigan kept outdoors at two latitudes. Lines indicating start/end of civil twilight (thick lines), and sunrise/sunset (thin lines) are included in the left panel of each double-plotted actogram (From Reierth and Stokkan 1998b).**

Pineal secretion of melatonin

The pineal production of melatonin is a significant component of the circadian system and is believed to act as a physiological link between the organism and its photoperiodic environment, serving as a clock and a calendar (Reiter 1993). In birds melatonin often influences rhythms in e.g. locomotor-and feeding activity (Oshima et al. 1987; Chabot and Menaker 1992; Heigl and Gwinner 1995). One distinct feature of melatonin secretion is its rhythmicity, and with the exception of the nocturnal barn owl (*Tyto alba*; Van't Hof et al. 1998), regardless of whether the animal is diurnal or nocturnal, plasma level of melatonin is low during the day and high during the night.

Plasma melatonin variations in Svalbard ptarmigan reflect their concomitant photoperiodic conditions (70°N; Fig.2; Reierth et al. 1999), revealing arrhythmic and low secretion in summer and a low-amplitude rhythm in mid-winter. Plasma melatonin levels increased and remained elevated throughout the dark phase of the day in all months of the year that had a clear light-dark cycle. In contrast to findings reported in penguins (Cockrem 1991; Miche et al. 1991) and reindeer (Stokkan et al. 1994), plasma melatonin levels in ptarmigan were significantly reduced at mid-day at the winter solstice. This may reflect the birds' photoperiodic history or that the period of twilight at mid-day was sufficient to suppress melatonin production. Day-to-day weather conditions affect light conditions significantly at this time of year and may explain why Eloranta et al. (1992) observed a drop in melatonin in reindeer around midday at the winter solstice (69°N), while Stokkan et al. (1994) did not.

These findings support the notion that melatonin secretion depends on the relative changes in light intensity throughout the day rather than light intensity per se (Lynch et al. 1981; Reiter et al. 1983; Meyer and Millam 1991), and which may be an important feature of circadian rhythm control in Arctic animals.

At 70°N, Svalbard ptarmigan are mainly active during the brief, daily period of twilight around the winter solstice, but they also show increased activity at night, which does not occur in autumn and spring (Reierth and Stokkan 1998b). Although there was a clear plasma melatonin rhythm at the winter solstice, the amplitude as well as the daily mean production of melatonin was significantly reduced compared with autumn and spring (Fig. 2). At Svalbard (79°N), ptarmigan activity was intermittently continuous around the clock during the more than two months of continuous darkness in mid-winter (Stokkan et al. 1986a; Reierth and Stokkan 1998b). This was similar to the pattern seen during the continuous daylight of summer at this latitude, but the amount of activity in mid-winter was lower than in summer. Taken together, these findings support the notion of a causal relationship between melatonin and locomotor activity, in the sense that high activity may depend on low levels of plasma melatonin (Hendel and Turek 1978; Oshima et al. 1987; Yamada et al. 1988). However, it has not been shown any direct effects of melatonin on either seasonal or daily rhythms in Svalbard ptarmigan (Hanebrekke et al. 1998).

Adaptations to Arctic light conditions

Reduced melatonin amplitudes, as seen in ptarmigan and reindeer around summer and winter solstices, may be comparable to that observed in birds during migration. In migrants, this presumably reduces sensitivity to environmental zeitgebers, and, thereby, increases their ability to adjust to changing photoperiod as they cross longitudes (Gwinner et al. 1997). The reduced day-night amplitude and mean daily melatonin secretion shown in ptarmigan and reindeer in mid-winter, as well as their nearly undetectable melatonin secretion in summer, may reflect a similar adaptation to life in the Arctic. Reduced melatonin amplitudes and production could explain the absence of 24-h activity rhythms in these animals, due to damping of underlying oscillators. This hypothesis indicates that the intermittently continuous activity pattern seen in both ptarmigan and reindeer in mid-winter at Svalbard is causally related to the reduced and arrhythmic production of melatonin. Thus, both ptarmigan and reindeer may reflect the malleability of the vertebrate circadian system, as discussed by Menaker and Tosini (1996) with regard to its phylogenetic development and diversity. At those times of the year when the environment becomes arrhythmic and unpredictable, the circadian system is flexible and increase fitness by allowing these herbivorous animals to forage whenever physical conditions are favourable.

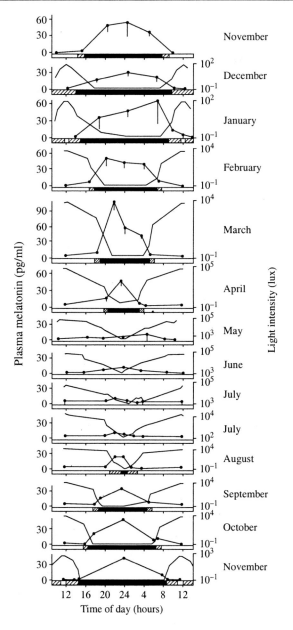

Fig. 2   **Daily variations in plasma melatonin concentrations (closed symbols; mean ± SEM) in Svalbard ptarmigan throughout one year. Black bars indicate the duration of night, shaded bars show civil twilight and white bars indicate sun above the horizon. Lines without symbols (not shown in top panel) show the actual light intensity in lux (from Reierth, Van't Hof and Stokkan 1999).**

References

Binkley, S., Mosher, K. (1992) Activity rhythms in house sparrows exposed to natural lighting for one year. J. Interdiscipl. Cycle Res. **23**: 17–33.

Brown, R.G.B. (1963) The behaviour of the willow warbler *Phylloscopus trochilus* in continuous daylight. Ibis **105**: 63–75.

Chabot, C.C., Menaker, M. (1992) Effects of physiological cycles of infused melatonin on circadian rhythmicity in pigeons. J. Comp. Physiol. A **170**: 615–622.

Cockrem, J.F. (1991) Plasma melatonin in the Adelie penguin (*Pygoscelis adeliae*) under continuous daylight in Antarctica. J. Pineal. Res. **10**: 2–8.

Cullen, J.M. (1954) The diurnal rhythm of birds in the Arctic summer. Ibis **96**: 31–46.

Daan, S., Aschoff, J. (1975) Circadian rhythms of locomotor activity in captive birds and mammals: Their variations with season and latitude. Oecologia (Berl.). **18**: 269–316.

Demmelmeyer, H., Haarhaus, D. (1972) Die Lichtqualität als Zeitgeber für Zebrafinken (*Taeniopygia guttata*). J. Comp. Physiol. **78**: 25–29.

Eloranta, E., Timisjärvi, J., Nieminen, M., Ojutkangas, V., Leppäluoto, J., Vakkuri, O. (1992) Seasonal and daily patterns in melatonin secretion in female reindeer and their calves. Endocrinol. **130**: 1645–1652.

Gwinner, E., Hau, M., Heigl, S. (1997) Melatonin: Generation and modulation of avian circadian rhythms. Brain. Res. **44**(4): 439–444.

Hanbrekke, T.L., Reierth, E., Sharp, P.J., Stokkan, K.A. (1998) Melatonin treatment does not affect long-day induced changes in high-Arctic ptarmigan (abstract). Sixth Meeting SRBR 49A.

Heigl, S., Gwinner, E. (1995) Synchronization of circadian rhythms of house sparrows by oral melatonin: Effects of changing period. J. Biol. Rhythms **10**(3): 225–233.

Hendel, R.C., Turek, F.W. (1978) Suppression of locomotor activity in sparrows by treatment with melatonin. Physiol. Behavior. **21**: 275–278.

Höglund, H.N. (1980) Studies on the winter ecology of the willow grouse *Lagopus lagopus lagopus* L. Swedish Sportsmen's Association, **11** (5): 248–270.

Irving, L., West, G.C., Peyton, L.J. (1967) Winter feeding program of Alaska willow ptarmigan shown by crop contents. Condor **69**: 69–77.

Johnsson, A., Englemann, W., Klemke, W., Ekse, A.T. (1979) Free-running human circadian rhythms in Svalbard. Z. Naturforsch. **34**: 470–473.

Karplus, M. (1952) Bird activity in the continuous daylight of Arctic summer. Ecology **33**: 129–134.

Krüll, F. (1976a) Zeitgebers for animals in the continuous daylight of high arctic summer. Oecologia (Berl.) **24**: 149–158.

Krüll, F. (1976b) The position of the sun is a possible zeitgber for Arctic animals. Oecologia (Berl.) **24**: 141–148.

Krüll, F. (1976c) The synchronizing effect of slight oscillations of light intensity on activity period of birds. Oecologia (Berl.) **25**: 301–308.

Lindgård K, Stokkan KA (1989) Daylength control of food intake and body weight in Svalbard ptarmigan. Ornis. Scand. **20**: 176–180.

Lynch, H.J., Rivest, R.W., Ronsheim, P.M., Wurtman, R.J. (1981) Light intensity and the control of melatonin secretion in rats. Neuroendocrinol. **33**:181–185.

Marshall, A.J. (1938) Bird and animal activity in the Arctic. J. Animal Ecol. **7**: 248–250.

Meyer, W.E., Millam, J.R. (1991) Plasma melatonin levels in Japanese quail exposed to dim light are determined by subjective interpretation of day and night, not light intensity. Gen. Comp. Endocrinol. **82**: 377–385.

Menaker, M., Tosini, G. (1996) The evolution of vertebrate circadian systems. In . Honma, K., Honma, S. (eds.). Circadian Organization and Oscillatory Coupling. Hokkaido University Press, Sapporo, pp. 39–52.

Michè, F., Vivien-Roels, B., Pevet, P., Spehner, C., Robin, J.P., LeMaho, Y. (1991) Daily pattern of melatonin secretion in an Antarctic bird, the emperor penguin, *Aptenodytes forsteri*: Seasonal variations, effect of constant illumination and of administration of isoproterenol or propranolol. Gen. Comp. Endocrinol. **84**: 249–263.

Mortensen, A., Blix, A.S. (1989) Seasonal changes in energy intake, energy expenditure, and digestibility in captive Svalbard rock ptarmigan and Norwegian willow ptarmigan. Ornis. Scand. **20**: 22–28.

Mortensen, A., Unander, S., Kolstad, M., Blix, A.S. (1983) Seasonal changes in body composition and crop content of Spitzbergen ptarmigan *Lagopus mutus hyperboreus*. Ornis. Scand. **14**: 144–148.

Oshima, I., Yamada, H., Sato, K., Ebihara, S. (1987) The phase relationship between circadian rhythms of locomotor activity and circulating melatonin in the pigeon (*Columba livia*). Gen. Comp. Endocrinol. **67**: 409–414.

Pohl, H. (1999) Spectral composition of light as a zeitgeber for birds living in the high Arctic summer. Physiol. Behav. **67**: 327–337.

Pulliainen, E. (1978) Behaviour of a willow grouse *Lagopus l. lagopus* at the nest. Ornis. Fennica. **55**: 141–148.

Reierth, E., Stokkan, K.A. (1998a) Dual entrainment by light and food in the Svalbard ptarmigan (*Lagopus mutus hyperboreus*). J. Biol. Rhythms **13** (**5**): 393–402.

Reierth, E., Stokkan, K.A. (1998b) Activity rhythm in high-arctic Svalbard ptarmigan (*Lagopus mutus hyperboreus*). Can. J. Zool. **76**: 2031–2039.

Reierth, E., Van't Hof, T.J., Stokkan, K.A. (1999) Seasonal and daily variations in plasma melatonin in the high-Arctic Svalbard ptarmigan (*Lagopus mutus hyperboreus*). J. Biol. Rhythm **14**(**4**): 314–319.

Reiter, R.J. (1993) The melatonin rhythm: both a clock and a calendar. Experientia **49**: 654–664.

Reiter, R.J., Steinlechner, S., Richardson, B.A., and King, T.S. (1983) Differential response of pineal melatonin levels to light at night in laboratory raised and wild-captured 13-lined ground squirrels (*Spermophilus tridecemlineatus*). Life Sci. **32**: 2625–2629.

Stokkan, K.A. (1992) Energetics and adaptations to cold in ptarmigan in winter. Ornis. Scand. **23**: 366–370.

Stokkan, K.A., Mortensen, A., Blix, A.S. (1986a) Food intake, feeding rhythm, and body mass regulation in Svalbard rock ptarmigan. Am. J. Physiol. **251**: R264–R267.

Stokkan, K.A., Sharp, P.J., Unander, S. (1986b) The annual breeding cycle of the high-Arctic Svalbard ptarmigan (*Lagopus mutus hyperboreus*). Gen. Comp. Endocrinol. **61**: 446–451.

Stokkan, K.A., Tyler, N.J.C., Reiter, J.R. (1994) The pineal gland signals autumn to reindeer (*Rangifer tarandus tarandus*) exposed to the continuous daylight of the Arctic summer. Can. J. Zool. **72**: 904–909.

Van Oort, B.E.H., Stokkan, K.A., Tyler, N.J.C. (1999) Long-term patterns of activity in relation to photoperiod in free-ranging reindeer and sheep (abstract). 10th. Arctic Ungulate Conference. Tromsø, Norway.

Van't Hof, T.J., Gwinner, E., Wagner, H. (1998) A highly rudimentary circadian melatonin profile in a nocturnal bird, the barn owl (*Tyto alba*). Naturwissenschaften **85**: 402–404.

West GC (1968) Bioenergetics of captive willow ptarmigan under natural conditions. Ecology **49**: 1035–1045.

Yamada, H., Oshima, I., Sato, K., Ebihara, S. (1988) Loss of the circadian rhythms of locomotor activity, food intake and plasma melatonin concentration induced by constant bright light in the pigeon (*Columba livia*). J. Comp. Physiol. A **163**: 459–463.

Biological Rhythms
Edited by V. Kumar
Copyright © 2002, Narosa Publishing House, New Delhi, India

19. Diversity in the Circadian Response to Melatonin in Mammals

S.M.W. Rajaratnam*[#] and J.R. Redman

[#]Centre for Chronobiology, School of Biomedical and Life Sciences, University of Surrey,
Guildford GU2 7XH, UK

Department of Psychology, Monash University, Victoria 3800, Australia

Early studies in rats showed that timed melatonin administration could entrain and phase advance free-running activity rhythms, and affect the rate and direction of reentrainment following light-dark cycle phase-shifts. Based on these studies, it was initially thought that the mammalian circadian system was responsive to melatonin only within a narrow window of sensitivity, near the late subjective day. More recent studies in mice, palm squirrels and humans suggest that there are other times of the day when the circadian system is sensitive to the hormone. There appears to be much diversity and complexity in the responses to melatonin across species. In view of this, it is surprising that melatonin phase response curves have been described for only four different mammals, and entrainment studies have been reported in only a handful of species. There are likely to be differences between diurnal and nocturnal species in response to melatonin, although this possibility has not been thoroughly investigated. Until such issues are resolved and the mechanisms of melatonin action are elucidated, application of the hormone in the clinical setting may be premature.

Introduction

Historically, it was thought that the light-dark (LD) cycle was crucial for normal entrainment processes in mammals, and that nonphotic changes in the animal's physiology and behaviour were unimportant. It is now well established that a number of centrally acting drugs and endogenous substances, as well as nonphotic changes to an animal's environment, may influence the circadian system. Over the past two decades, there has been increasing evidence to suggest that exogenous melatonin may act as a potent time cue for mammalian circadian systems. This paper will review major developments in the study of exogenous melatonin's phase-shifting properties in mammals.

Entrainment to melatonin

Redman and colleagues (1983) first demonstrated that wheel-running activity rhythms of male Long-Evans rats held under constant darkness (DD) are entrained to daily injections of melatonin (1 mg/ kg, s.c.). Entrainment to melatonin in rats was found to be dose-dependent (median effective dose 5.45 ± 1.33 μg/kg) and quantal (Cassone et al. 1986a).In addition to wheel running activity, entrainment of a number of other circadian rhythms to melatonin injections (s.c.) has been demonstrated, including

---

*E-mail: s.wilson-rajaratnam@surrey.ac.uk

drinking, locomotor activity and body temperature measured concurrently and pineal N-acetyltransferase (NAT) activity (see references cited in Redman 1993).

In order to entrain circadian rhythms, melatonin administration must be appropriately timed. Short pulses of melatonin delivered by s.c. injection (Redman et al. 1983), orally (drinking water, 2h/day) (Redman and Francis, unpublished) or by infusion (s.c., 1 h) (Slotten et al. 1999) will entrain rat activity rhythms when the onset of the active phase coincides with the time of administration. In contrast, when rats have continuous access to drinking water containing melatonin, no entrainment occurs (Armstrong 1989b). The effect of the duration of melatonin administration (1, 8 or 16 h infusions) on entrainment was recently investigated (Pitrosky et al. 1999). All rats entrained to 1 and 8 h infusions of melatonin, whereas only 50 percent of rats entrained to 16 h infusions. The phase angle of entrainment, measured using activity onset and onset of the infusion as the phase markers, was found to become increasingly negative as the duration of the infusion increased.

Daily injections of melatonin were able to synchronise or partially synchronise activity rhythms in 13/15 male rats that showed rhythm disruption under constant light (LL, 250–300 lux) after being exposed to an exotic lighting regime (Chesworth et al. 1987). However, 5 rats which showed intact rhythms under such conditions were not entrained to melatonin injections. These findings were taken as support for the proposition that melatonin may act on the coupling between oscillators which drive the activity rhythm, and suggest that melatonin may be more effective in rats with disrupted rhythms than those with intact rhythms under LL.

Female rats held in LL (20 lux) are not entrained to daily melatonin injections, whereas female rats held in DD are entrained (4/5 rats) (Thomas and Armstrong 1988). This difference was explained on the basis of free-running period (τ) Under LL, females showed a pre-injection τ ranging from 24.8 to 25.4 h, whereas in DD τ ranged from 24.1 to 24.4 h. In view of the relatively modest amplitude of the melatonin phase response curve (PRC) in rats (range = 15–52 mins, Armstrong 1989a), it is conceivable that τ exceeded the upper limit for entrainment under LL conditions (Thomas and Armstrong 1988).

In contrast to the earlier study with male Long-Evans rats, Marumoto et al. (1996) reported that melatonin injections did not synchronise disrupted activity rhythms of male Sprague-Dawley rats kept in bright LL (300 lux), but did entrain intact free-running rhythms of rats kept in dim LL (3 lux). The discrepancy between the findings of these two studies may be attributed to the different recording devices used for measuring activity (wheel-running vs gross locomotor activity), strain differences, or the different lighting regimes.

In the original study that reported melatonin entrainment in rats, two control rats showed susceptibility to the injection procedure (Redman et al. 1983). One of the rats entrained to saline injections, and the other showed evidence of synchronisation for a number of days. It has therefore been speculated that melatonin's effects may be mediated by changes in general arousal levels (Mrosovsky 1988). Redman and Roberts (1991) examined this hypothesis by preventing wheel-running in rats that were injected with melatonin. They found that entrainment persisted even under such conditions. Furthermore, when rats were immobilised for 3 h after melatonin injections, there was no effect on entrainment (Marumoto et al. 1996). Therefore, it is unlikely that the melatonin entrainment mechanism is mediated by alterations in activity levels.

In Syrian hamsters, melatonin injections entrain activity rhythms when administered via the suprachiasmatic nucleus (SCN)-lesioned mother to the foetus (Davis and Mannion 1988), or directly to the neonate until postnatal day 6 (Grosse et al. 1996). When a single dose of melatonin is administered to adult Syrian hamsters without handling the animals, melatonin has no effect (Hastings et al. 1992). Therefore, sensitivity to the circadian effects of melatonin in Syrian hamsters appears to depend (at least in part) on developmental age. This hypothesis was further supported by a study showing that daily melatonin injections entrain the restored activity rhythms of adult Syrian hamsters bearing foetal SCN grafts (Grosse and Davis 1998). Entrainment was observed when melatonin injections were given 1 week after transplantation, but not when given 6 weeks after transplantation.

The phase of melatonin entrainment may vary between different hamster strains. Six or 8 h daily (s.c.) infusions of melatonin entrain wheel running activity rhythms of pinealectomised (Px) adult Syrian hamsters and Px Djungarian hamsters held in dim LL (0.1–5 lux) (Kirsch et al. 1993). With Djungarian hamsters, as in previously examined nocturnal species, entrainment occurred when the beginning of the melatonin infusion coincided with activity onset (CT12). In contrast, in Syrian hamsters, entrainment occurred when the onset of activity preceded the beginning of the infusion period by 3–5 h. This difference was explained in terms of the species difference in endogenous melatonin. In Djungarian hamsters, as in rats, melatonin synthesis starts at the beginning of the dark phase, whereas in Syrian hamsters, melatonin synthesis occurs 3–5 h after dark onset (Kirsch et al. 1993). In both hamster strains, activity onset occurs at approximately dark onset. Therefore, from these data it would seem that circadian sensitivity to melatonin depends on the phase of the endogenous melatonin rhythm rather than that of the activity rhythm.

Few studies have examined the entraining properties of melatonin in diurnal (nonhuman) mammals. We have administered melatonin in food (1 mg/kg, 17 days), at either zeitgeber time (ZT) 0 (light onset) or ZT12 (dark onset), to diurnal Indian palm squirrels (*Funambulus pennanti*) maintained initially under LD cycles and then under LL (~14 lux) (Rajaratnam and Redman 1997). We found that squirrel activity rhythms synchronised to melatonin at both times of the day tested. One explanation for why melatonin is effective in squirrels at (at least) two circadian phases, in contrast to rats which only respond at activity onset, is based on the species τ. Palm squirrels may free-run with $\tau <$ or > 24 h, whereas Long Evans rats usually free-run with $\tau > 24$ h. If melatonin is to entrain activity rhythms of palm squirrels, either phase delays or phase advances must be induced depending on whether the animal's τ is less than or greater than 24 h. Therefore, the high degree of variability in τ for this species may explain why its circadian system responds to melatonin at more than one circadian phase.

Some species do not appear to entrain to exogenous melatonin. For example, only 1/10 Asian chipmunks (*Tamias asiaticus*) maintained under LL (200 lux) entrained to daily melatonin injections (1 mg/kg) despite the fact that their τs were relatively close to 24 h (23.6 ± 0.8 h) (Murakami et al. 1997). Similarly, we have found that daily melatonin injections (1 mg/kg) do not entrain free-running activity rhythms of Israeli sand rats (*Psammomys obesus*) maintained under dim LL (~0.1 lux) (Rajaratnam and Redman, unpublished). Although direct comparisons are difficult because of the variation in methodology, there do appear to be species differences in response to melatonin.

A number of human studies have reported that melatonin administration stabilises free-running sleep-wake rhythms of some blind and sighted subjects (see references cited in Arendt et al. 1997; Lockley et al. 2000). However, in such studies melatonin's effects on the sleep-wake rhythm were not accompanied by effects on endocrine rhythms, suggesting only partial entrainment of the circadian system, or that the apparent entrainment was actually a manifestation of melatonin's sleep-promoting effects. Lockley et al. (2000) recently demonstrated clear melatonin-induced entrainment of free-running cortisol rhythms in 3/7 blind subjects. Interestingly, there appeared to be a relationship between the circadian phase at which melatonin administration commenced and whether or not entrainment subsequently occurred; those who entrained all began their treatment during the phase advance portion of the melatonin PRC.

Melatonin administration under LD entrainment

Under LD cycles, Long-Evans rats often show a negative phase angle difference (PAD) of up to several hours. When such rats were injected with melatonin or its agonist S-20098 1 h before dark onset for either 8 or 9 days, activity onset was phase advanced toward dark onset (Armstrong 1989b; Armstrong et al. 1993). However, when the negative PAD was greater than 3 h, melatonin failed to advance the rhythm, probably because the injections fell outside the circadian window of sensitivity to melatonin.

Daily oral administration of melatonin (2 mg) to human subjects at 1700 h for 3–4 weeks was shown to phase advance the rhythm of endogenous melatonin by 1 to 3 h in 5/11 subjects (Arendt et al. 1985). Subjects self reported an increase in evening fatigue and in some cases evening sleep during melatonin treatment, suggesting that melatonin advanced the sleep-wake cycle. In patients suffering from delayed sleep phase syndrome (DSPS), daily melatonin administration at 1930h (Tzischinsky et al. 1993) or 2200h (Dahlitz et al. 1991) advanced the sleep-wake cycle. The data from healthy subjects and DSPS patients are therefore consistent with melatonin's phase advancing effects in rats.

Djungarian hamsters (*P. sungorus*) also show responses to timed injections of melatonin which are consistent with its phase advancing effects in rats. In adult Djungarian hamsters maintained under LD 16:8, daily melatonin injections given 4 h prior to dark onset phase advanced the onset of activity and extended the duration of activity (α), but only in animals that showed physiological responses to melatonin administration (e.g., gonadal regression, moult, body weight loss) (Puchalski and Lynch 1988). When Djungarian hamsters are maintained under LD 9:15, animals show large negative PADs. In such animals, activity onset was phase advanced by ~5 h when melatonin was injected each day 3 h after dark onset (Margraf and Lynch 1993). Finally, daily melatonin injections have been shown to phase advance the onset of the serum melatonin rhythm by up to 4.5 h in Djungarian hamsters injected with melatonin either 2 or 9 h before dark onset under a 16:8 LD regime (Yellon 1996).

The effects of exogenous melatonin on reentrainment following phase-shifts of the LD cycle have been investigated. An early study revealed that continuous administration of melatonin to rats via an implant near the SCN accelerated reentrainment of the adrenocortical rhythm after reversal of the LD cycle (Murakami et al. 1983). Since then, it has been shown that the reentrainment mechanism is differentially responsive to melatonin depending on the phase of administration and also the magnitude of the LD phase-shift. Both the direction and rate of reentrainment can be altered by appropriately timed melatonin administration. When melatonin was injected at the old dark onset time after a 8 h phase advance, all rats reentrained by phase advancing, whereas all control rats (whether vehicle-injected or unhandled) reentrained by phase delaying (Redman and Armstrong 1988). After a 5h phase advance, melatonin-injected rats (1 mg/kg, s.c.), reentrained faster than vehicle-injected controls when melatonin was administered at the time of maximum sensitivity for entrainment, that is, at the pre-shift dark onset (Redman and Armstrong 1988). Following an 8 h phase advance, pineal NAT rhythms of rats treated with melatonin at the new dark onset time reentrained faster than those of controls (Illnerová et al. 1989).

Melatonin's effects on rate of reentrainment have been confirmed in other rodents. Sharma et al. (1999b) showed that melatonin administration (1 mg/kg, s.c., 3 days) at either ZT 4 or ZT 22 (relative to the new LD cycle) can accelerate the rate of reentrainment of activity rhythms in the nocturnal field mouse *Mus booduga* after a 6 h LD phase advance or phase delay, respectively. Van Reeth et al. (1998) studied the effects of S-20098 (20 mg/kg, i.p.) on activity rhythms of the diurnal Nile grass rat *Arvicanthus mordax* subjected to 4, 6 or 8 h LD phase advances, and reported that S-20098 accelerated reentrainment in all cases.

In humans, several field studies and laboratory simulations have reported that melatonin administration reduces the subjective feelings associated with jet lag and accelerates reentrainment of circadian rhythms to real or simulated transmeridian travel (see references cited in Arendt et al. 1997). However, in a recent large-scale randomised, double blind trial, in which three alternative regimens of melatonin or placebo were given to 257 subjects who travelled from New York to Oslo (6 h eastward), symptoms of jet lag did not significantly differ between treatment groups (Spitzer et al. 1999). One limitation of this study, as the authors note, is that subjects had only arrived in New York (originally from Oslo) 4–5 days before their return flight to Oslo, which may have been an insufficient amount of time for the circadian system to fully adjust to New York time prior to the shift. Consequently, melatonin administration to these subjects may have been inappropriately timed (Arendt et al. 1997).

Melatonin phase response curves

To date, behavioural PRCs to melatonin have been described for four different mammals: Long Evans rats (Armstrong 1989a), humans (Lewy and Sack 1997; Zaidan et al. 1994), C3H/HeN mice (Benloucif and Dubocovich 1996) and field mice (Sharma et al. 1999a). In Long-Evans rats (50μg/kg, s.c.) and C3H/HeN mice (90μg, s.c. for 3 days), melatonin injections induce clear and consistent phase advances of activity rhythms in the late subjective day (~CT10). Phase advancing effects have also been reported in humans administered melatonin either orally (revised PRC by Lewy and Sack 1997: 0.5 mg, CT14 = endogenous dim light melatonin onset) or intravenously (Zaidan et al. 1994: 20 mg for 3 h). However, in humans, the phase advance portion is between approximately CTs 6 and 14.

In all species tested so far except rats, phase delays to melatonin have also been reported. In C3H/HeN mice, phase delays occur between CTs 0 and 2 and in humans between CTs 18 and 6.

The melatonin PRC in *M. booduga* appears to be somewhat different to other species (Sharma et al. 1999a). In *M. booduga*, a single dose of melatonin (1 mg/kg or 10 mg/kg) dose-dependently evoked phase advances in the locomotor activity rhythm when given in the late subjective night (maximal advances CT22) and phase delays when given in the early subjective day (maximal delays CT4). Therefore, the general shape and time course of the melatonin PRC do not appear to be consistent across all species.

Melatonin applied *in vitro* to a rat hypothalamic slice preparation containing the SCN phase-shifted the time of peak neuronal firing rate according to a PRC (McArthur et al. 1997). Phase advances of more than 3 h were observed when melatonin was applied at dusk (CTs 10–14) and also at dawn (CTs 23–0). This is in contrast to the behavioural PRC to melatonin in rats, according to which an injection of melatonin in the late subjective night did not consistently induce phase-shifts (Armstrong 1989a). However, Armstrong did report that 1/3 of rats phase advanced at CT22, and 1/5 of rats phase delayed at CT18. To date, our subsequent behavioural studies have been unable to confirm this second melatonin-sensitive phase in rats (Redman and Rajaratnam unpublished).

Site(s) of melatonin action

Several lines of evidence suggest that the SCN are the site of melatonin action in the mammalian circadian system. Early studies showed that in rats, whereas Px and enucleation have no observable effects on melatonin entrainment (Armstrong 1989a; Warren et al. 1993; Marumoto et al. 1996), entrainment does not occur following complete bilateral lesions to the SCN (Cassone et al. 1986b).

Second, studies using 2-[^{125}I]-iodomelatonin have localised high affinity melatonin binding sites to the rat SCN (Vanecek et al. 1987), and a daily variation in the density of melatonin SCN binding sites has been observed (Gauer et al. 1993). Molecular cloning studies have identified two mammalian melatonin receptor subtypes in the SCN (see Reppert 1997).

Third, injections of melatonin (s.c.) *in vivo* altered 2-deoxyglucose uptake in the rat SCN (Cassone et al. 1988). As mentioned previously, the rhythm in SCN neuronal activity is phase advanced when melatonin is applied *in vitro* to a hypothalamic brain slice preparation containing the SCN (McArthur et al. 1997). Finally, melatonin administered in the late subjective night (CT22) induces *c-fos* expression in the SCN of rats held in DD (Kilduff et al. 1992), and melatonin administered in the late subjective day immediately phase advances the SCN rhythm in photic induction of *c-fos* (Sumová and Illnerová 1996).

Recently the role of melatonin receptor subtypes MT_1 (Mel_{1a}) and MT_2 (Mel_{1b}), in the phase advancing effects of melatonin has been investigated using melatonin antagonists (Dubocovich et al. 1998; Weibel et al. 1999). While the MT_2 (Mel_{1b}) subtype seems a likely candidate for mediating melatonin's circadian effects (Lui et al. 1997; Dubocovich et al. 1998; Hunt et al. 2001), this proposition needs further confirmation.

Conclusions

It is well established that exogenous melatonin may serve as a zeitgeber for mammals. However, there appear to be significant interspecies differences in response to the hormone. One possibility is that the phase of sensitivity to exogenous melatonin may be linked to the endogenous melatonin rhythm of the species, rather than the phase of activity onset. Apart from species differences, a number of factors appear to influence melatonin's phase-shifting ability, including dose and route of administration, timing of administration and ambient lighting levels. At least in hamsters, the melatonin response also varies with strain and developmental age of the animal. Whether there are nocturnal-diurnal differences in response to melatonin is still not clear. Measurement techniques that avoid behavioural masking such as those used by Drijfhout and colleagues (1999) will aid the investigation of some of these issues. Further development of selective ligands of melatonin receptors is required, in order to determine the contribution of receptor subtypes to melatonin's circadian effects.

Acknowledgements

The authors are grateful to Dr. D-J Dijk for reviewing an earlier version of this article.

References

Arendt, J., Bojkowski, C., Folkard, S. (1985) Some effects of melatonin and the control of its secretion in humans. In: Evered, D., Clark, S. (eds.) Photoperiodism, Melatonin and the Pineal. Ciba Foundation Symposium. Pitman, London, pp. 266–283.

Arendt, J., Skene, D.J., Middleton, B., (1997) Efficacy of melatonin treatment in jet lag, shift work, and blindness. J. Biol. Rhythms **12(6)**: 604–617.

Armstrong, S.M. (1989a) Melatonin and circadian control in mammals. Experientia **45**: 932–938.

Armstrong, S.M. (1989b) Melatonin: The internal zeitgeber of mammals? Pineal Res. Rev. **7**: 157–202.

Armstrong, S.M., McNulty, O.M., Guardiola-Lemaitre, B. (1993) Successful use of S20098 and melatonin in an animal model of delayed sleep-phase syndrome (DSPS). Pharmacol. Biochem. Behav. **46**: 45–49.

Benloucif, S, Dubocovich, M.L. (1996) Melatonin and light induce phase shifts f circadian activity rhythms in the C3H/HeN mouse. J. Biol. Rhythms **11(2)**: 113–125.

Cassone, V.M., Chesworth, M.J., Armstrong, S.M. (1986a) Dose-dependent entrainment of rat circadian rhythms by daily injection of melatonin. J. Biol. Rhythms. **1(3)**: 219–229.

Cassone, V.M., Chesworth, M.J., Armstrong, S.M. (1986b) Entrainment of rat circadian rhythms by daily injection of melatonin depends upon the hypothalamic suprachiasmatic nuclei. Physiol. Behav. **36**: 1111–1121.

Cassone, V.M., Roberts, M.H., Moore, R.Y. (1988) Effects of melatonin on 2-deoxy-[1-14C] glucose uptake within rat suprachiasmatic nucleus. Am. J. Physiol. **255**: R332–R337.

Chesworth, M.J., Cassone, V.M., Armstrong,. S.M. (1987). Effects of daily melatonin injections on activity rhythms of rats in constant light. Am. J. Physiol. **253**: R101–R107.

Dahlitz, M.J., Alvarez, B., Vignau, J. (1991) Delayed sleep-phase syndrome: response to melatonin. Lancet **337**: 1121–1124.

Davis, F.C., Mannion, J. (1988) Entrainment of hamster pup circadian rhythms by prenatal melatonin injections to the mother. Am. J. Physiol. **255(3)**: R439–R448.

Drijfhout, W.J., de Vries, J.B., Homan, E.J. (1999) Novel non-indolic receptor agonists differentially entrain endogenous melatonin rhythm and increase its amplitude. Eur. J. Pharmacol. **382(3)**: 157–166.

Dubocovich, M.L., Yun, K., Al-Ghoul, W.M. (1998) Selective MT_2 melatonin receptor antagonists block melatonin-mediated phase advance of circadian rhythm FASEB J. **12**: 1211–1220.

Gauer, F., Masson-Pévet, M., Skene, D.J. (1993) Daily rhythms of melatonin binding sites in the rat pars tuberalis and suprachiasmatic nuclei; evidence for a regulation of melatonin receptors by melatonin itself. Neuroendocrinol. **57**: 120–127.

Grosse, J., Davis, F.C. (1998) Melatonin entrains the restored circadian activity rhythms of Syrian hamsters bearing fetal suprachiasmatic nucleus grafts. J. Neurosci. **18(19)**: 8032–8037.

Grosse, J., Velickovic, A., Davis, F.C. (1996) Entrainment of Syrian hamster circadian activity rhythms by neonatal melatonin injections. Am. J. Physiol. **270(3)**: R533–R540.

Hastings, M.H., Mead, S.M., Vindlacheruvu, R.R. (1992) Non-photic phase shifting of the circadian activity rhythm of Syrian hamsters: the relative potency of arousal and melatonin. Brain Res. **591**: 20–26.

Hunt, A.E., Al-Ghoul, W.M., Gillette, M.V., Dubocovich, M.L. (2001) Activation of MT (2) melatonin receptors in rat suprachiasmatic nucleus phase advances the circadian clock. Am J Cell Physiol, **280(1)**: C 110–C118.

Illnerová, H., Trentini, G.P., Maslova, L. (1989) Melatonin accelerates reentrainment of the circadian rhythm of its own production after an eight-hour advance of the light-dark cycle. J. Comp. Physiol. **166**: 97–102.

Kilduff, T.S., Landel, H.B., Nagy, G.S. (1992) Melatonin influences fos expression in the rat suprachiasmatic nucleus. Mol. Brain Res. **16**: 47–56.

Kirsch, R., Belgnaoui, S., Gourmelen, S. (1993) Daily melatonin infusion entrains free-running activity in Syrian and Siberian Hamsters. In: Wetterberg L (ed.) Light and Biological Rhythms in Man. Oxford/New York/Seoul/Tokyo, Pergamon Press, pp 107–120.

Lewy, A.J., Sack, R.L. (1997) Exogenous melatonin's phase-shifting effects on the endogenous melatonin profile in sighted humans: a brief review and critique of the literature. J. Biol. Rhythms **12**: 588–594.

Liu, C., Weaver, D.R., Jin, X. (1997) Molecular dissection of two distinct actions of melatonin on the suprachiasmatic circadian clock. Neuron **19(1)**: 91–102.

Lockley, S.W., Skene, D.J., James, K. (2000) Melatonin administration can entrain the free-running circadian system of blind subjects. J. Endocrinol. **164**: R1–R6.

Margraf, R.R., Lynch, G.R. (1993) Melatonin injections affect circadian behavior and SCN neurophysiology in Djungarian hamsters. Am. J. Physiol. **264(3)**: R615–R621.

Marumoto, N., Murakami, N., Katayama, T. (1996) Effects of daily injections of melatonin on locomotor activity rhythms in rats maintained under constant bright or dim light. Physiol. Behav. **60(3)**: 767–773.

McArthur, A.J., Hunt, A.E., Gillette, M.U. (1997) Melatonin and signal transduction in the rat suprachiasmatic circadian clock: activation of protein kinase C at dawn and dusk. Endocrinol. **138**: 627–634.

Mrosovsky, N. (1988) Phase response curves for social entrainment. J. Comp. Physiol. A **162**: 35–46.

Murakami, N., Hayafuji, C., Sasaki, Y. (1983) Melatonin accelerates the reentrainment of the circadian adrenocortical rhythm in inverted illumination cycle. Neuroendocrinol. **36**: 385–391.

Murakami, N., Marumoto, N., Nakahara, K. (1997) Daily injections of melatonin entrain the circadian activity rhythms of nocturnal rats but not diurnal chipmunks. Brain Res. **775(1–2)**: 240-243.

Pitrosky, B., Kirsch, R., Malan, A. (1999) Organization of rat circadian rhythms during daily infusion of melatonin or S20098, a melatonin agonist. Am. J. Physiol. **277(3)**: R812–R828.

Puchalski, W., Lynch, G.R. (1988) Daily melatonin injections affect the expression of circadian rhythmicity in Djungarian hamsters kept under a long-day photoperiod. Neuroendocrinol. **48**: 280–286.

Rajaratnam, S.M.W., Redman, J.R. (1997) Effects of daily melatonin administration on circadian activity rhythms in the diurnal Indian palm squirrel (*Funambulus pennanti*). J. Biol. Rhythms **12**(4): 339–347.

Redman, J., Armstrong, S., Ng, K.T. (1983) Free-running activity rhythms in the rat: entrainment by melatonin. Science **219**: 1089–1091.

Redman, J.R. (1993) Circadian effects of melatonin in rats: an update. In: Touitou, Y., Arendt, J., Pevet, P. (eds.) Melatonin and the pineal gland- From basic science to clinical application. Elsevier Science, pp. 143–150.

Redman, J.R., Armstrong, S.M. (1988) Reentrainment of rat circadian activity rhythms: Effects of melatonin. J. Pineal Res. **5**: 203–215.

Redman, J.R., Roberts, C.M. (1991) Entrainment of rat activity rhythms by melatonin does not depend on wheel-running activity. Soc. Neurosci. Abstracts **17**: 673.

Reppert, S.M. (1997) Melatonin receptors: Molecular biology of a new family of G protein-coupled receptors. J. Biol. Rhythms **12**(6): 528–531.

Sharma, V.K., Singaravel, M., Subbaraj, R. (1999a) Locomotor activity rhythm in the field mouse *Mus booduga* phase-shifts to melatonin injections in a dose-dependent manner. Biol. Rhythm Res. **30**(3): 313–320.

Sharma, V.K., Singaravel, M., Subbaraj, R. (1999b) Timely administration of melatonin accelerates reentrainment to phase-shifted light-dark cycles in the field mouse *Mus booduga*. Chronobiol. Int. **16**(2): 163–170.

Slotten, H.A., Pitrovsky, B., Pévet, P. (1999) Influence of the mode of daily melatonin administration on entrainment of rat circadian rhythms. J. Biol. Rhythms **14**(5): 347–353.

Spitzer, R.L., Terman, M., Williams, J.B. (1999) Jet lag: Clinical features, validation of a new syndrome-specific scale, and lack of response to melatonin in a randomized, double-blind trial. Am. J. Psychiat. **156**(9): 1392–1396.

Sumová, A., Illnerová, H. (1996) Melatonin instantaneously resets intrinsic circadian rhythmicity in the rat suprachiasmatic nucleus. Neurosci. Lett. **218**(3): 181–184.

Thomas, E.M.V., Armstrong, S.M. (1988) Melatonin administration entrains female rat activity rhythms in constant darkness but not in constant light. Am. J. Physiol. **255**: R237–R242.

Tzischinsky, O., Dagan, Y., Lavie, P. (1993) The effects of melatonin on the timing of sleep in patients with delayed sleep phase syndrome. In: Touitou, Y., Arendt, J., Pevet, P. (eds.) Melatonin and the pineal gland: From basic science to clinical application. Amsterdam, Elsevier Science Publishers, pp. 351–354.

Van Reeth, O., Olivares, E., Turek, F.W. (1998) Resynchronisation of a diurnal rodent circadian clock accelerated by a melatonin agonist. Neuroreport **9**(8): 1901–1905.

Vanacek, J., Pavlik, A., Illnerová, H. (1987) Hypothalamic melatonin receptor sites revealed by autoradiography. Brain. Res. **435**: 359–362.

Warren, W.S., Hodges, D.B., Cassone, V.M. (1993) Pinealectomized rats entrain and phase shift to melatonin injections in a dose-dependent manner. J. Biol. Rhythms **8**(3): 233–245.

Weibel, L., Rettori, M.C., Lesieur, D. (1999) A single oral dose of S22153, a melatonin antagonist, blocks the phase advancing effects of melatonin in C3H mice. Brain Res. **829**: 160–166.

Yellon, S.M. (1996) Daily melatonin treatments regulate the circadian melatonin rhythm in the adult Djungarian hamster. J. Biol. Rhythms **11**(1): 4–13.

Zaidan, R., Geoffriau, M., Brun, J. (1994) Melatonin is able to influence its secretion in humans: description of a phase-response curve. Neuroendocrinol. **60**(1): 105–112.

20. Light Sensitivity of the Biological Clock

S. Rani, S. Singh and V. Kumar*

Department of Zoology, University of Lucknow, Lucknow-226 007, India

Light is a ubiquitous input from the environment used by most species in one way or the other in regulation of their short and/ or long term activities. A response to light, the photoperiodic response, is the result of the interpretation of light input by the neuroendocrine machinery, collectively called the photoperiodic response system (PRS). Apart from the duration, gradual shifts in the intensity and wavelength of daily light are critical in regulation of the light (photic) sensitivity of the PRS. There is a direct relationship between the rate of initiation of a photoperiodic response and the intensity of light until the threshold is reached. A light wavelength to which PRS is maximally sensitive, or to which it has greater access, will induce a maximal response. There can also be differential effects of wavelength and intensity of light on circadian process(es) involved in the entrainment and induction of the photoperiodic clock, which may have adaptive implications. Synchronization to daily light-dark (LD) cycle may be achieved at dawn or dusk, depending whether the animal is day- or night-active, when there is relatively low intensity of light. By contrast, photoperiodic induction in many species occurs during long days of spring and summer when plenty of daylight at higher intensity is available later in the day.

Introduction

In organisms, adaptive advantage is key of synchronizing short- and or long-term activities such that these occur at the most favourable time of the day and of the year. On-going process of natural selection ensures that an organism selects the most reliable and highly predictable environmental cue(s) which helps it to anticipate the "appropriate season". At given latitude, the annual variation in light-dark (LD) is extremely predictable, and used by most species in one way or the other. The three important characteristics of daily light which provide information about the time-of-day and the time-of-year to an organism are the duration (the length of light period, hence of dark period), the quantity (light intensity) and the quality (colour, spectral composition). The duration is the hours of light available to an organism at and above the level of perception. Generally, it is the period between sunrise and sunset plus twilight periods at both ends. The intensity refers to the amount of photons that is available from a given light source. The colour is the sensation experienced as a result of the activation of certain class(es) of photoreceptor by a light wavelength from visible light spectrum (380 to 760 nm), commonly expressed as VIBGYOR. In VIBGYOR, violet, indigo and blue correspond to short wavelengths, green and yellow to mid wavelengths, and orange to red to long wavelengths. In nature, light is available in sinusoidal form. There is a gradual alteration in the intensity and the wavelength of light, which can easily be seen during morning and evening twilight hours when there are large changes in irradiance and very precise changes in light spectrum. This twilight period is the

*E-mail: drvkumar@sancharnet.in

critical component of daily light that provides temporal information from the environment to an animal.

A response to light, the photoperiodic response, is the result of the interpretation of light input by the neuroendocrine machinery, collectively called the photoperiodic response system (PRS). The PRS is conceptualized comprising three components: an input pathway (photoreceptor), a central processor (oscillatory, functions as a clock), and an output pathway (expressed in neuroendocrine events, e.g. reproduction, body fattening, fur colour, wool growth in sheep etc.). In mammals, all three components of the PRS are located in distinct tissues. Eyes are only photoreceptive organs that transmit light information exclusively through retinohypothalamic tract (RHT) to the suprachiasmatic nuclei (SCN, the central oscillatory unit) of the anterior hypothalamus. This may not be the case in non-mammalian vertebrates; for example, birds have more than one photic input pathway. They can perceive light environment by the retina of lateral eyes, by the pineal gland, and by the photoreceptors in hypothalamus. Circadian oscillators are present at all the three levels.

Light can act at any level in the PRS, but its importance is recognized mostly at two levels: the input pathway, the clock, or both. Earlier chapters in this book have focussed mainly either on input pathways (photoreception and photoentrainment, chapter 10), or on how the duration/ timing of light can effect daily and seasonal phenomena, from flowering in plants (chapter 15) to reproduction in vertebrates (chapter 16). In this chapter, therefore, we propose to briefly summarise and review the progress of research that has been made in studying the effects of wavelength and intensity of light on the biological clock, since this has not been adequately dealt elsewhere. We have used examples from higher vertebrates (birds and mammals) to illustrate the spectrum of wavelength- and intensity-effects of light on the biological clock with special reference to the photoperiodic time measurement (the photoperiodic clock), but similar and, in some cases, further advances have been made using a variety of species from other groups. At the end, we also give an overview of the light effects on other circadian (non-reproductive) and non-circadian functions.

The photoperiodic clock

The precision with which daily and seasonal activities are temporally organized clearly suggests the involvement of a clock mechanism that is fine-tuned by inputs from the environment. Experimental evidence conclude that endogenous *circa*-rhythms, synchronized to time-of-day (circadian rhythm; *circa*—about, *dies*—day) help measure the length of the day. Clearly, these rhythms are light-sensitive. Hence, the photoperiodic entrainment and induction of these light-sensitive rhythms, called circadian rhythm of photoperiodic photosensitivity (CRPP), by photoperiod are key to the timekeeping process. Under natural lighting conditions (NDL), photoperiodic effects occur either due to coincidence between phases of the photoperiodic oscillator and the light:dark (LD) cycle—the external coincidence, or due to interaction between two or more independent oscillators as a result of exposure to LD cycle-the internal coincidence (Pittendrigh 1972). It is, therefore, logical that light intensity and wavelength, the two main characteristics of 24 h daily cycle of illumination, will have effects on the photoperiodic entrainment and induction of the CRPP.

A large number of studies have investigated the intensity effects of light on photoperiodic induction of gonadal growth and development, a function believed to be mediated by the CRPP (Kumar and Follett 1993). It was more than six decades ago, when Bissonnette (1931) demonstrated the role of light intensity in the manipulation of European starling's reproductive response to day length. Since then, the effects of light intensity on photosexual response has been studied in some birds, house sparrow (Bartholomew 1949; Menaker et al. 1970; Underwood and Menaker 1970; Menaker and Eskin 1967), bobwhite quail (Kirkpatrick 1955), white-crowned sparrow (Farner 1959), domestic duck (Benoit 1964), domestic turkey (Nester and Brown 1972), Japanese quail (Oishi and Lauber 1973; Siopes and Wilson 1980; Follett and Millette 1982), and blackheaded and redheaded buntings (Kumar and Rani 1996; Rani and Kumar 2000). Of these, studies on house sparrow are the most comprehensive.

A photoperiodic response is maximally induced by light wavelengths at which photoreceptor(s) mediating such effect is most sensitive or by wavelengths which have greater access to the photoreceptors. When blackheaded buntings (*Emberiza melanocephala*) were subjected to a stimulatory day length of 13L:11D in white, green (528 nm) and red (654 nm) colours at 100 lux intensity, photostimulation (weight gain and testis growth) occurred in all groups but the responses were significantly greater in birds that were exposed to red light. Similar responses occur also in other bird species (for details see Kumar and Rani 1999). In mammals too, wavelength- and intensity-dependent photoperiodic induction has been reported. In the Syrian hamster, 1 h pulse of green, blue or near ultraviolet light, but not red or yellow light, presented to animals during dark period blocked the collapse of the reproductive system. Also, at equal irradiances, different bandwidths of light have different effects on short photoperiod induced collapse of the hamster reproductive system (Brainard et al. 1983; Brainard et al. 1984). The effects of light spectrum have been reported in lower vertebrates, as well. Joshi and Udaykumar (1998) found that the effects on ovarian follicular kinetics in the frog, *Rana cyanophlyctis*, were maximum in red light, followed by yellow and green light.

Photoreceptors seem to have precise spectral sensitivity, determined by exposing subjects to light of different wavelengths at equal irradiance. This is called an action spectrum study, and helps in the characterization of the pigment(s) mediating a photoperiodic response. However, action spectra do not necessarily tell how many photoreceptors are involved in mediating the physiological response. Brain photoreceptors were discovered more than five decades ago by Benoit and his colleagues but the action spectra for the photoperiodic responses have been studied only for a few species as yet (see especially Homma and Sakakibara 1971; Homma et al. 1977; Oliver and Bayle 1982).

The sensitivity of the photoperiodic clock to light intensity and wavelength can also be examined by measuring melatonin, which is produced rhythmically and, at least in mammals, plays a significant role in mediating the effects of light on reproduction. Irrespective whether an animal is diurnal, nocturnal or crepuscular, plasma melatonin levels are low during day and high at night, in both LD and constant conditions (continuous darkness [DD] or continuous dim light [dimLL]) *in-vivo* and *in-vitro*. At equal irradiance, the effects of light on melatonin secretion could be different depending on the wavelength of light used. Blue-green had maximum inhibitory effects on melatonin secretion in albino rats (Cardinalli et al. 1972) and hamsters (Brainard et al. 1984). In a recent study on European starlings (*Sturnus vulgaris*), daytime light intensity had some effects on the circadian pattern of plasma melatonin (for details and references see Kumar et al. 2000a). There was a difference in the phasing of the melatonin rhythm: peak melatonin levels tended to occur earlier in dimLD (1: 0.2 lux) than in brightLD, (500 : 0.2 lux). The relationship between the light intensity at day and/or night and the pattern of plasma melatonin and/ or enzymes involved in the melatonin synthesis have been studied also in a few mammalian species (Minnemann et al. 1974; Vanecek and Illnerova 1982; Trinder et al. 1996; Griffith and Minton 1992; Lynch et al. 1981).

It may be noted, however, that there can also be differential effects of light depending how it is applied. Light in wide range of wavelengths is stimulatory when applied directly near to or on to photoreceptors, but not when applied away from the photoreceptors. For example, ducks respond to both blue and red lights when light is introduced through a quartz rod directly into the brain (Benoit 1964). Similarly, Japanese quail (*Coturnix coturnix japonica*) respond to both blue (455 nm) and orange-yellow (575 nm) light when radioluminescent paint was implanted directly in specific brain areas (Homma and Sakakibara 1971). Considered together, all observations from spectral studies on birds are consistent with the idea that the difference in photoperiodic effects at a given intensity by different light wavelengths is due to the difference in number of photons received by the photoreceptors (Vriend and Lauber 1973). At equal energy levels, the number of photons emitted is larger in red light than in green or blue light and the penetration to brain tissues, thereby access to the photoreceptors (capture of large number of photons by the photopigments), of red light (long wavelengths) is far more faster and deeper than of blue or green light (short wavelengths). (Foster and Follett 1985; Oishi and Lauber 1973; Benoit 1964; Oliver and Bayle 1982; Bissonnette 1931).

CRPP: phasic effects

A photoperiodic clock under free-running condition is expressed as a circadian rhythm (Kumar and Follett 1993; Kumar et al. 1996). It consists of two distinct phases-an entraining phase (subjective day) and an inductive phase (subjective night), best illustrated by interrupted-night LD cycles where a brief light pulse of 1 min to 2 h is introduced at night of subjects maintained on short days (e.g. 8L:16D). The first long light pulse entrains the CRPP and the second short light pulse stimulates a photoperiodic response. Obviously, light exerts a phase-control over the circadian oscillation to enable its dual actions, of phasing the oscillation (entrainment) and of effecting the photoperiodic induction by extending (or not extending) into the photoinducible phase (ϕi), located early in the subjective night (external coincidence). There is also phase-control by light in the situation where exposure to light leads interaction between independent oscillators (internal coincidence).

Ironically, so much experimental effort has gone into testing the external coincidence (originally put forward by Bünning 1936) or internal coincidence hypotheses without making any significant effort in understanding how light exerts phase control over the CRPP. As yet, far less is known concerning the intensity and spectral sensitivity of the phase-dependent circadian effects of light? Indirect evidence from a few species nonetheless show that light at a threshold intensity is required for the entrainment and induction of the biological clock mediating photoperiod-induced gonadal growth and development. In the migratory bunting (*Emberiza melanocephala* and *Emberiza bruniceps*) held in a skeleton photoschedule, the rate of gonadal growth is very slow at 100 lux light intensity (Kumar and Rani 1996; Rani and Kumar 2000). Also, in Japanese quail, 1 h night-interruption at an intensity of 850 lux stimulates significantly larger ($p<0.02$) testes than those exposed to at 250 lux (Follett and Millette 1982). There is strong evidence of the role of light intensity as synchronizer of clock-regulated long-term seasonal responses. At equator where annual variation in day length is so small to serve as the reliable photoperiodic index of the changing season, animals seem to rely on seasonal changes in light intensity to regulate their physiological activities. This has been experimentally shown in an equatorial species of stonechats (*Saxicola torquata axillaris*) using simulation of high and low intensity of light cycle that occurs in nature (Gwinner and Scheuerlein 1998). It has been shown that stonechats show a synchronized annual gonadal cycle under constant LD cycle with changes in light intensity.

Recently, we began assaying the phasic effects on CRPP of light intensity and light spectrum using highly photosensitive bird models, the black- and red-headed buntings. We employed a skeleton paradigm with 6-and 1 h long and short light pulses, respectively (6L:6D:1L:11D). In a series of experiments, we have shown that the photoperiodic entrainment and induction were intensity-and spectrum-dependent (Kumar and Rani 1996; Rani and Kumar 1999, 2000). At given intensity, red light is more effective than white light and white light is more effective than green light. In other words, short light wavelengths are less effective than long light wavelengths in stimulation of avian photoneuroendocrine (PNE)-system (Kumar and Rani 1996; Rani and Kumar 2000).

Induction vs. entrainment

Differential responses to LD cycles that contain either identical entraining and varying inductive, or varying entraining and identical inductive, pulse indicate a wavelength-and intensity-dependent effects on the photoperiodic clock. There could be differences in light intensity thresholds between entraining and inducing light pulses (Kumar and Rani 1996), which may be taken to suggest that the entrainment and induction of the CRPP are two separate photoperiodic phenomena. The threshold light intensity is lower for the entrainment than for the induction. In the migratory redheaded bunting, for example, a red light at 5 lux may be effective as entraining but not as inducing agent; so is the 20 lux white light (Rani and Kumar 2000). In any case, it may be much easier to entrain the circadian system than to convert the photoinductive effects into neuroendocrine events, viz. lipogenesis (body fattening) and gametogenesis (testicular growth). Whether the difference is at the level of the photoperiodic clock, or at the level of the endocrine system, or at the both levels, is unclear. Also, whether the photoperiodic entrainment and induction are mediated by different classes of photoreceptors is unknown. A speculation though could be made. Very recently, Saldanha et al. (2001) have shown a direct innervation of GnRH neurons by deep brain photoreceptors (DBPs) in ring doves (*Streptopelia roesogrisea*). This might mean that DBPs, which are necessary and sufficient for the detection of changes in day length that regulates avian reproduction, are not linked to the reproductive axis via the circadian system. In other words, in birds "the circadian system is not a necessary intermediary between the sensory and reproductive components" (Saldanha et al. 2001). Since photoperiodic birds can make fine discrimination of even small changes in the characteristics of a photoperiod, it can be reasoned that DBPs also act as the photoperiodic clocks.

Temporal photosensitivity

Intensity and spectrum of light are not the only factors that induce seasonal responses in birds. If that were the case, regardless of the day length (short or long day length), light illumination at given spectrum and intensity should evoke similar physiological effects. But it does not: light pulses given at different times in the photosensitive phase of CRPP induce different response. Rani and Kumar (1999) have shown a duration-and time-dependent effects of the light pulse, introduced in the ϕi, on the rate and magnitude of the testicular response in the redheaded bunting (*Emberiza bruniceps*). In buntings, illumination of a larger portion of ϕi results in higher rates of gonadal growth but there is a duration limit above which there will be no further increase of testicular response. All this means that an "appropriate" photoperiod is a far more important factor, of course after threshold for light intensity is attained. In other words, a higher light intensity or a longer light wavelength cannot act as a substitute for a long photoperiod. The study of Burger (1939) clearly shows that regardless of the increase or decrease in light intensity, a photoperiod of 10.5L:13.5D (a non-stimulatory photoperiod *per se*) did not stimulate spermatogenesis in male European starlings. However, recently, Bentley et al. (1998) have shown that the decrease in light intensity alters the perception of day length in starlings. When starlings were kept on 18L:6D at 3–, 13–, 45– and 108-lux light intensities, the photoperiodic responses observed in these groups were comparable to those exposed to 11L:13D, 13L:11D, 16L:8D and 18L:6D photoperiods, respectively. Thus, there can be a fascinating possibility that light intensity and photoperiod may have synergistic effects. Very recently, we have shown that the action on the PRS of light wavelengths can be compensatory (Kumar et al. 2000b). A 12 h (weakly inductive) photoperiod at long light wavelength produced gonadal response similar to, or somewhat larger than, that induced by a longer 14 h (strongly inductive) photoperiod at short wavelengths. This difference in the photoperiodic response was achieved when photoperiods at short light wavelengths were longer by 2 h in duration and higher by 4 to 6-fold in light intensity than photoperiods at long light wavelengths (Kumar et al. 2000b).

There can also be differential effects of light intensity, depending upon the nature of LD cycle used. In a recent significant finding, Kumar and Rani (1999) have shown that action of a single long light pulse in the complete photoperiod, extending simultaneously to different phases, is different from that of two short light pulses in the skeleton photoperiod, falling discretely at two different phases of the photoperiodic oscillator. Blackheaded buntings were subjected to 12L:12D and 6L:5D:1L:12D daily or alternately interposed with constant darkness (12L:12D/DD, 6L:5D:1L:12D/DD) for a period of 21 weeks. There was clear differential response to two photoperiods in the rate and magnitude of photoperiod-induced body fattening and gonadal growth.

Non-reproductive circadian functions

Light effects on circadian functions, other than reproduction, are known extensively. Light intensity is an important factor in entrainment of the clock, as shown in majority of cases by its effects on the circadian activity rhythms. Even under constant conditions of light (LL), the intensity can affect the expressed period (represented by the Greek letter, tau τ) of the freerunning rhythm (freerun is a chronobiological jargon that denotes the expression of a rhythm with its period under constant conditions). J. Aschoff first summarized these effects more than three decades ago in what became known as "Aschoff's Rule" (for detail see Aschoff 1981), which states that in LL as light intensity increases, τ lengthens for nocturnal species and shortens for diurnal species. With other effects of light intensity on the circadian system, namely the effects on activity (represented by Greek letter alpha, α) and rest (represented by Greek letter rho, ρ) of the daily activity rhythm, added to Aschoff's Rule, it is known as "Circadian Rule". This then states that in LL as light intensity increases, tau shortens, activity level increases, and the ratio of active phase to rest phase ($\alpha : \rho$) increases for diurnal animals. The opposite effects are true of nocturnal animals. This rule holds pretty well for most groups of vertebrates that have been studied: fish, reptiles, amphibians, birds and nocturnal mammals. The rule is less successful with most invertebrates, curiously seems not to apply to diurnal mammals, many of which respond as if they were, in fact, nocturnal.

Fresh water alga *Chlamydomonas* (Kondo et al. 1991) and the marine alga *Gonyaulax* (Roenneberg and Deng 1997) are good examples of how is spectral information used in photoperiodic timing even by organisms with simple organization. Among vertebrates, a number of findings suggest spectral effects of light on the circadian system. For example, in phase advancing the activity rhythm of evening-active wild rabbits, blue light increments are more effective than blue decrements and yellow decrements are more effective than yellow increments (Nouber et al. 1983). Hamster's circadian system is maximally sensitive to light wavelengths near 500 nm (Takahashi et al. 1984). Based on the response of bat's circadian system to different wavelengths of light, Joshi and Chandrashekaran (1984) conclude that delay and advance phase shifts to light are mediated respectively by two different classes of photoreceptors. Whereas the photoreceptors that mediate delay phase shifts are maximally sensitive to light at the wavelength of 430 nm, the photoreceptors that mediate advance phase shift are maximally sensitive to light at the wavelength of 520 nm. In Djungarian hamsters, single weak red light pulse given 2 h before regular "lights on" had acute and long term effects persisting for several days, indicating high sensitivity of the circadian system in this species to red light pulses during later part of the night (Klante and Steinlechner 1995).

In human also, light at night could have found profound influences on rhythms of core temperature (Dijk et al. 1991). Further, melatonin secretion is inhibited by 1 h monochromatic light at 509 nm at various intensities (Brainard et al. 1985). Systematic studies concerning spectral effects of light on circadian rhythms of core temperature and melatonin secretion in humans are lacking. However, a series of experiments using fluorescent light (Morita et al. 1995; Morita and Tokura 1996; Morita et al. 1997) show that even at the same light intensity, long light wavelengths (e.g. low color temperature of 3000K and red light) had little influence on human biological rhythms. On the other hand, mid-short light wavelengths (e.g high color temperature of 6500K and green and blue light) had a greater influence. Green or blue light inhibited the nighttime decrease and promoted the morning increase of core temperature. This is because these light colours inhibited the increase at night, and promoted the decrease in the morning, of melatonin secretion from the pineal body (Morita et al. 1997).

Despite several convincing findings, spectral sensitivity of the vertebrate circadian system still needs to be more thoroughly studied. For instance, we do know that the circadian system of mouse is sensitive to both green light and near-UV irradiance (Provencio and Foster 1995), but we do not know how these spectral signals are utilized by the mammalian circadian system.

Non-circadian functions

Apart from the behavioural and reproductive rhythms, the effects of light intensity and wavelength have been recorded on many physiological funcitons. The spectrum of light intensity effects in birds, for example, includes the effects on food and feeding, immune system to general health (Hollwich 1979; Scott and Siopes 1994). Both low and high light intensities can cause convulsions and death in the African weaver finch (Rollo and Domm 1943). Wiltschko and his colleagues (1993, 1995) have investigated the effects of light wavelengths on magnetoreception in avian migrants (e.g. Tasmanian silver eyes, *Zosterops l. lateralis*). Their findings show that these migrants were well oriented in their appropriate migratory direction under 'white' (full spectrum), blue and green light but were disoriented under 633 nm light falling in red region of the light spectrum (Wiltschko et al. 1993; Wiltschko and Wiltschko 1995; Munro et al. 1997). Light-dependent compass orientation has also been reported in other vertebrates, such as fishes (Quinn 1980), amphibians (Lohmann 1991; Phillips and Borland 1992), reptiles (Phillips and Borland 1994) and mammals (Marhold et al. 1991), although wavelength-dependence of them seems to differ from that of birds.

In primitive chordate like river lamprey, *Lampetra japonica*, neuronal activity seems to be influenced by the light colour. It is inhibited by the light of short wavelengths and excited by middle to long wavelengths; the maximum sensitivities of the inhibitory and excitatory responses occur at about 380 and 540 nm, respectively. Since environmental light contains both inhibitory and excitatory components, the neurons would keep both sensitivities during daytime and could measure the variation in the spectral composition. Tosini and Avery (1996) have reported that the spectral composition of light might be an important variable in mediating the thermoregulatory processes in lizards. Light illuminance and wavelength also affect growth of birds (Barrett and Pringle 1951; Cherry and Barwick 1962; Wabeck and Skoglund 1973; Osol et al. 1980). When large white turkey hens at the age of 30 weeks were exposed to blue, green and red or incandescent light equalized at a photon output, a color-dependent effect of light was found on the cellular and humoral immune responses, but not on stress status (Scott and Siopes 1994). The light colour has also been found to affect body temperature, and the functions of kidney adrenal function and hypophysis in birds (Hollwich 1979).

Remarks

As yet, it has been difficult to produce a PRC for the photoperiodic oscillator because it cannot be monitored directly. Formal analyses of the photoperiodic clock are done employing various LD cycle paradigms (Kumar and Follett 1993) and examining their effects on the physiological and/ or behavioural markers of the clock system, assuming that these markers faithfully predict the photosensitive phase of the circadian photoperiodic clock (Hamner and Enright 1967; Elliott et al. 1972). However, studies on few species including Japanese quail, which show dissociation of photoperiodic responses (LH rise) from other circadian locomotor activity rhythm, question if the properties of photoperiodic oscillator can be studied using any other marker of the circadian system (Pickard and Turek 1983; Hakim et al. 1991; Juss et al. 1995).

It is intriguing that the CRPP responsible for photoperiodic induction of reproductive responses is stimulated at such a high light intensity when other circadian functions (e.g. activity rhythms) are affected by very low light intensities. A further conundrum in the photoperiodic literature is that while red light is the most effective agent for stimulation of the PNE-system in birds, it is considered as the "safe light" (i.e. non-inductive) for stimulation of the PNE-system in mammals. Is it because of the differences in the sensitivity of photoreceptors or because of the differences in the degree of penetration of red light through brain tissues overlying photoreceptors in two vertebrate classes? It is more puzzling when basic characteristics of the photoreceptors among birds and mammals have great similarity with each other. For example, although the anatomy of their eyes differs, the cones of human and fowl are both trichromatic (Cornsweet 1970) and the relative spectral luminosities of chicken, pigeon and man are the same (Bowmaker and Knowles 1977; Blough 1957).

Conclusions and implications

It is important to examine the significance of light in the environment in terms of its effects on biology of animals, including health, comfort and biological rhythms in humans. For example, laboratory experiments (a few of them cited above) clearly suggest differential effects of the light intensity and light wavelength on circadian processes, mediating photoperiod-induced physiological responses in birds, which may have adaptive implications. For example, in a long day species, photoperiodic entrainment at low intensity is achieved early in the day, and the photoperiodic induction occurs at relatively high intensity later in the day during spring and summer when days are sufficiently long to extend into the photosensitive (= photoinducible) phase of the CRPP. Furthermore, an avian photoperiodic clock may not be maximally sensitive to red light (Foster and Follett 1985), but shows maximal response to it. This may be in view of the fact that red light has greater access to the avian PRS because of being dominant in the light environment and having greater penetrance through the brain tissues. Similarly, light intensity and spectrum regulate many other physiological processes in animals. There could be wide species variation in response to light wavelengths and light intensities. Species-specificity of spectral- and intensity-sensitivity of the PRS could have tremendous ecological implications. Species sharing same environment may adopt different temporal strategies for their survival and optimal performance.

In humans, the range of oscillations in core temperature and/ or melatonin is strongly coupled with sleep sensation. A light pulse during night could influence core temperature and melatonin and, hence, could affect sleep physiology. Experimental data suggest that light with a low color temperature should be used for low-level lighting at night, and light with a high color temperature should be used for morning when higher levels of illumination are required. There could always be physical and mental disorder due to discrepancy between phase of biological rhythm and living life-time, regulated by light. This is clearly reflected in situations where is the insufficient synchronization to light-dark cycle due to lack of light or difference in light phase. Examples are: jet lag syndrome caused by rapid movement across time zones, insomnia in night workers, and seasonal affective disorder that occurs mostly in winters.

An important consideration in making best use of light is the phasing of daily light with respect to biological oscillations. In other words, it is important to consider how light from morning to midday and also during night should be treated, because fluctuations in daily light may be directly related with the strengthening of range of oscillations in biological rhythms that mediate different physiological functions. In a living environment, therefore, it will be necessary to examine further the influence on biological rhythms of repeated and daily irradiation of light with a particular spectral distribution and of any effects of chromatic adaptation.

Acknowledgement

Financial assistance from a SERC research grant from the Department of Science and Technology, New Delhi to VK were utilized in preparation of this review paper.

References

Aschoff, J., ed., Biological rhythms. Handbook of behavioral neurobiology, Vol 4 (New York: Plenum) 1981

Barott, H.G., Pringle, E.M. (1951) The effect of environment on growth and feed and water consumption of chicks. IV. The effect of light on early growth. J. Nutr. **45**: 265–274.

Bartholomew, G.A. Jr. (1949) The effect of light intensity and daylength on reproduction on reproduction in the English sparrow. Bull. Mus. Comp. Zool. **101**: 431476.

Benoit, J. (1964) The role of the eye and of the hypothalamus in the photostimulation of gonads in the duck. Annals of the New York Academy of Science **117**: 204–216.

Bentley, G.E., Goldsmith, A.R., Dawson, A., Briggs, C., Pemberton, M. (1998) Decreased light intensity alters the perception of day length by male European starlings (*Sturnus vulgaris*). J. Biol. Rhythms **13**: 148–158.

Bissonnette, T.H. (1931) Sexual periodicity. Quart. Rev. Biol. **11**: 371–376.

Blough, D.S. (1957) Spectral sensitivity in the pigeon. J. Optical Soc. Amer. **47**: 827–833.

Bowmaker, J.K., Knowles, A. (1977) The visual pigments and oil droplets of the chicken retina. Vision Res. **17**: 755–764.

Brainard, G.C., Richardson, B.A., King, T.S., Matthews, S.A., Reiter, R.J. (1983) The suppression of pineal melatonin content and N-acetyltransferase activity by different light irradiance in the Syrian hamster: A dose response relationship. Endocrinol. **113**: 293–296.

Brainard, G.C., Richardson, B.A., King, T.S., Reiter, R.J. (1984) The influence of different light spectra on the suppression of pineal melatonin content in the Syrian hamster. Brain Res. **294**: 333–339.

Brainard, G.C., Richardson, B.A., Menaker, M., Fredrikson, R.H., Miller, L.S., Weleber, R.G., Cassone, V., Hudson, D. (1985) Effect of light wavelength on the suppression of nocturnal plasma melatonin in normal volunteers. Ann. N.Y. Acad. Sci. **453**: 376–378.

Bünning, E. (1936) Die endogene Tagesrhythmik als Grundlage der Photoperiodische Reaktion. Ber. Deut. Bot. Ges. **54**: 590–607.

Burger, J.W. (1939) Some aspects of the roles of light intensity and the daily length of exposure to light in the sexual photoperiodic activation of the male starling. J. Exp. Zool. **81**: 333–341.

Cardinali, D.P., Larin, F., Wurtman, R.J. (1972) Action spectra for effects of light on hydroxyindole-o-methyltransferases in rat pineal, retina and harderian gland. Endocrinol. **91**: 877–886.

Cherry, P., Barwick, M.W. (1962) The effect of light on broiler growth. 1. Light intensity and colour. British Poult. Sci. **3**: 31–39.

Cornsweet, T.N. (1970) Visual Perception. Academic Press, London.

Dijk, D., Cajochen, C., Borbely, A.A. (1991) Effect of a single 3-hour exposure to bright light on core body temperature and sleep in humans. Neuronsci. Lett. **121**: 59–62.

Elliot, J.A., Stetson, M.H., Menaker, M. (1972) Regulation of testis function in golden hamsters: A circadian clock measures photoperiodic time. Science **178**: 771–773.

Farner, D.S. (1959) Photoperiodic and related control of annual gonadal cycles. In: Withrow, R.B. (ed.)

Photoperiodism and Related Phenomena in Plants and Animals. Am. Assoc. Advance Sci., Washington, D.C. pp. 716–750.

Follett, B.K., Millette, J.J. (1982) Photoperiodism in quail: testicular growth and maintenance under skeletal photoperiod. J. Endocrinol. **93**: 83–90.

Foster, R.G., Follett, B.K. (1985) The involvement of a rhodopsin-like photopigment in the photoperiodic response of Japanese quail. J. Comp. Physiol. A **157**: 519–528.

Griffith, M.K., Minton, J.E. (1992) Effect of light intensity on circadian profiles of Melatonin, Prolactin, ACTH and Cortisol in pigs. J. Anim. Sci. **70**: 492–498.

Gwinner, E., Scheuerlein, A. (1998) Seasonal changes in day-light intensity as a potential zeitgeber of circannual rhythms in equatorial stonechats. J. Ornithol. **139**: 407–412.

Hakim, H., DeBernardo, A.P., Silver, R. (1991) Circadian locomotor rhythms, but not photoperiodic responses, survive surgical isolation of the SCN in hamsters. J. Biol. Rhythms **6**: 97–113.

Hamner, W.M., Enright, J.T. (1967) Relationship between photoperiodism and circadian rhythms of activity in the house finch. J. Exp. Biol. **46**: 43–61.

Hollwich, F. (1979) The influence of ocular light perception on metabolism in man and animal, Springer, New York.

Farner, D.S. (1959) Photoperiodic and related control of annual gonadal cycles. In: Withrow, R.B. (ed.) Photoperiodism and Related Phenomena in Plants and Animals. Am. Assoc. Advance Sci., Washington, D.C. pp. 716–750.

Homma, K., Sakakibara, Y. (1971) Encephalic photoreceptors and their significance in photoperiodic control of sexual activity in Japanese quail. In: Menaker, M. (ed.) Biochronometry. Natl. Acad. Sci., Washington, D.C. pp. 333–341.

Homma, K., Ohta, M., Sakakibara, Y. (1977) in First int symp avian endocrinol, Follett, B.K. (ed.) (University College of North Wales, UK), pp. 25.

Joshi, D., Chandrashekaran, M.K. (1984) Bright light flashes of 0.5 milliseconds reset the circadian clock of a microchiropteran bat. J. Exp. Zool. **230**: 325–328.

Joshi, B.N., Udaykumar, K. (1998) Changes in ovarian follicular kinetics in intact and blinded and parietal shielded frogs exposed to different spectra of light. Gen. Comp. Endocrinol. **109**: 310–314.

Juss, T.S., Wing, V.M., Kumar, V., Follett, B.K. (1995) Does an unusual entrainment of the circadian system under T36h photocycles reduce the critical daylength for photoperiodic induction in the Japanese quail. J. Biol. Rhythms **10**: 17–32.

Kirkpatrick, C.M. (1955) Factors in photoperiodism of Bobwhite quail. Physiol. Zool. **28**: 255–264.

Klante, G., Steinlechner, S. (1995) A short red light pulse during dark phase of LD-cycle perturbs the hamster's circadian clock. J. Comp. Physiol. A **177**: 775–780.

Kondo, T., Johnson, C.H., Hastings, J.W. (1991) Action spectrum for resetting the circadian phototaxis rhythm in the CW15 Strain of *Chlamydomonas*. I: Cells in darkness. Plant Physiol. **95**: 197–205.

Kumar, V., Follett, B.K. (1993) The nature of photoperiodic clock in vertebrates. Proc. Zool. Soc. Calcutta; J.B.S. Haldane Commemoration Vol. pp. 217–227.

Kumar, Rani, S. (1996) Effects of wavelength and intensity of light in initiation of body fattening and gonadal growth in a migratory bunting under complete and skeleton photoperiods. Physiol. Behav. **60**: 625–631.

Kumar, V., Rani, S. (1999) Light sensitivity of the photoperiodic response system in higher vertebrates: Wavelength and intensity effects. Indian J. Exp. Biol. **37**: 1053–1064.

Kumar, V., Jain, N., Follett, B.K. (1996) The photoperiodic clock in blackheaded buntings (*Emberiza melanocephala*) is mediated by self-sustaining circadian system. J. Comp. Physiol. A 179: 59–64.

Kumar, V., Gwinner, E., Van't Hof, T.J. (2000a) Circadian rhythms of melatonin in the European starling exposed to different lighting conditions: Relationship with locomotor and feeding rhythms. J. Comp. Physiol. A **186**: 205–215.

Kumar, V., Rani, S., Malik, S. (2000b) Wavelength of light mimics the effects of the duration and intensity of a long photoperiod in stimulation of gonadal responses in the male blackheaded bunting (Emberiza melanocephala). Curr. Sci **79**: 508–510.

Lohmann, K.J. (1991) Magnetic orientation by hatchling loggerhead sea turtles (*Caretta caretta*). J. Exp. Biol. **155**: 37–49.

Lynch, H.J., Rivest, R.W., Ronsheim, P.M., Wurtman, R.J. (1981) Light intensity and the control of melatonin secretion in rats. Neuroendocrinol. **33**: 181–185.

Marhold, S., Burda, Wiltschko, W. (1991) Magnetkompassorientierung und Richtungspraferenzen bei subterranen Graumullen, *Cryptomys hottentotus* (Rodentia). Verhandlungen der Deutschen Zoologischen Gesellschaft, **84**: 354.

Menaker, M., Eskin, A. (1967) Circadian clock in photoperiodic time measurement: a test of the Bünning hypothesis. Science **157**: 1182–1185.

Menaker, M., Roberts, R., Elliot, J., Underwood, H. (1970) Extraretinal light perception in sparrow. III: The eyes do not participate in photoperiodic photoreception. Proc. Natl. Acad. Sci. USA **67**: 320–325.

Minnemann, K.P., Lynch, H.J., Wurtman, R.J. (1974) Relationship between environmental light intensity and retina-mediated suppression of rat pineal serotonin N-acetyltransferase. Life Sci. **15**: 1791–1796.

Morita, T., Tokura, H. (1996) Effects of light of different color temperature on the nocturnal changes in core temperature and melatonin in humans. Appl. Human Sci. **15(5)**: 243–246.

Morita, T., Teramoto, Y., Tokura, H. (1995) Inhibitory effect of light of different wavelengths on fall of core temperature during the nighttime. Jpn. J. Physiol. **45**: 667–671.

Morita, T., Tokura, H., Wakamura, T., Park, S.J., Teramoto, Y. (1997) Effects of the morning irradiation of light with different wavelengths on the behavior of core temperature and melatonin in humans. Appl. Human Sci. **16(3)**: 103–105.

Munro, U., Munro, J.A., Phillips, J.B., Wiltschko, W. (1997) Effect of wavelength of light pulse magnetisation on different magnetoreception systems in a migratory bird. Australian J. Zool. **45**: 189–198.

Nester, K.E., Brown, K.I. (1972) Light intensity and reproduction of Turkey hens. Poultry Sci. **51**: 117–121.

Nouber, J.F.W., van Nuys, W.M., Steenbergen, J.C.V. (1983) Colour changes in a light regimen as synchronizers of circadian activity. J. Comp. Physiol. **151**: 359–366.

Oishi, T., Lauber, J.K. (1973) Photoreception in the photosexual response of quail: I. Site of photorecpetor. Amer. J. Physiol. **225**: 155–158.

Oliver, J., Bayle, J.Q. (1982) Brain photoreceptors for the photo-induced testicular responses in birds. Experientia **38**: 1021–1029.

Osol, J.G., Foss, D.C., Carew, L.B. (1980) Effect of light environment and pinealectomy on growth and thyroid function in the broiler cockerel. Poult. Sci. **59**: 647–653.

Phillips, J.B., Borland, S.C. (1992) Behavioural evidence for the use of a light-dependent magnetoreception mechanism by a vertebrate. Nature **359**: 142–144.

Phillips, J.B., Borland, S.C. (1994) Use of a specialized magnetoreception system for homing by the eastern red-spotted newt, *Notophthalmus viridescens*. J. Exp. Biol. **188**: 275–291.

Pickard, G.E., Turek, F.W. (1983) The suprachiasmatic nuclei: The circadian clocks. Brain Res. **268**: 201–210.

Pittendrigh, C.S. (1972) Circadian surfaces and the diversity of possible roles of circadian organization in photoperiodic induction. Proc. Natl. Acad. Sci. (Wash.) **69**: 2734–2737.

Provencio, I., Foster, R. (1995) Circadian rhythms in mice can be regulated photoreceptors with cone-like characteristics. Brain Res. **694**: 183–190.

Quinn, T.P. (1980) Evidence for celestial and magnetic compass orientation in lake migrating sockeye salmon. J. Comp. Physiol. A **137**: 243–248.

Rani, S., Kumar, V. (1999) Time course of senstivity of the photoinducible phase to light in the redheaded bunting. Biol. Rhythm Res. **30**: 555–562

Rani, S., Kumar, V. (2000) Phasic response of photoperiodic clock to wavelength and intensity of light in the redheaded bunting, *Emberiza bruniceps*. Physiol. Behav. **69**: 277–283.

Roenneberg, T., Deng, T.S. (1997) Photobiology of the *Gonyaulax* circadian system: I. Different phase response curves for red and blue light. Planta **202**: 484–501.

Rollo, M., Domm, L.V. (1943) Light requirements of weaver finch. I. Light period and intensity. Auk **60**: 357–367.

Saldanha, C.J., Silverman, A-J, Silver, R. (2001) Direct innervation of GnRH neurons by encephalic photoreceptors in birds. J. Biol. Rhythms **16**: 39–49.

Scott, R.P., Siopes, T.D. (1994) Light color: effect on blood cells, immune function and stress status in turkey hens. Comp. Biochem. Physiol. A **108**: 161–168.

Siopes, T.D., Wilson, F.E. (1980) Participation of the eyes in the photosexual response of Japanese quail (*Coturnix coturnix japonica*). Biol. Reprod. **23**: 342–357.

Takahashi, T.S., Decoursey, P.J., Baumen, L, Menaker, M. (1984) Spectral sensitivity of a novel photosensitive system mediating entrainment of mammalian circadian rhythms. Nature **308**: 186–188.

Tosini, G., Avery, R. (1996) Spectral composition of light influences thermoregulatory behaviour in a Lacertid lizard 9*Podarcis muralis*). J. Therm. Biol. **21**: 191–195.

Trinder, J., Armstrong, S.M., O'Brien, C., Luke, D., Martin, M.J. (1996) Inhibition of melatonin secretion onset by low levels of illumination. J. Sleep Res. **5**: 77–82.

Underwood, H., Menaker, M. (1970) Extraretinal light perception: entrainment of the biological clock controlling lizard locomotor activity. Photochem. Photobiol. **24**: 227–243.

Vanecek, J., Illnerova, H. (1982) Night pineal N-acetyltransferase activity in rats exposed to white or red light pulses of various intensity and duration. Experientia **38**: 1318–1320.

Vriend, J., Lauber, J.K. (1973) Light intensity, wavelength and quantum effects on gonads and spleen of the deer mouse. Nature **244**: 37–38.

Wabeck, C.J., Skoglund, W.C. (1973) Influence of radiant energy from fluorescent light sources on growth, mortality and feed conversion of broilers. Poult. Sci. **53**: 2055–2059.

Wiltschko, R., Wiltschko, W. (1995) Magnetic Orientation in Animals. Springer-Verlag: Berlin, Heidelberg, New York.

Wiltschko, W., Munro, U., Ford, Wiltschko R (1993) Red light disrupts magnetic orientation of migratory birds. Nature **364**: 525–527.

Subject Index